"十三五"高等职业教育计算机类专业规划教材

Android 应用程序设计基础

李华忠　陈 勖　但唐仁　主　编

刘业涛　刘立明　副主编

中国铁道出版社
CHINA RAILWAY PUBLISHING HOUSE

内 容 简 介

全书共分 6 章，前 3 章主要包括 Android 应用（Application）与活动（Activity）、Android 用户界面布局管理器（Layout Manager）、Android 常用控件和高级控件使用方法等 Android 应用编程基础理论知识和应用技巧；后 3 章为基础项目实训，主要包括 Android 应用编程基本项目实训、基于 Android 未来超市系统项目实训和博物馆智能导览系统项目综合实训，综合应用了前面 3 章所介绍的基础理论核心知识和关键技术。

本书符合教学规律和课堂要求，很好地反映了嵌入式和移动互联等行业出现的 Android 方面的新知识、新技术、新方法和新应用。每个基础知识点的讲解均配备相应的应用范例，能解决高校 Android 课程教学面临的迫切问题。

本书适合作为高职院校 Android 应用程序设计基础的教材，也可作为移动开发爱好者的自学参考书。

图书在版编目（CIP）数据

Android 应用程序设计基础 / 李华忠，陈勍，但唐仁主编. — 北京：中国铁道出版社，2016.2（2019.1 重印）
"十三五"高等职业教育计算机类专业规划教材
ISBN 978-7-113-21264-3

Ⅰ. ①A… Ⅱ. ①李… ②陈… ③但… Ⅲ. ①移动终端－应用程序－程序设计－高等职业教育－教材 Ⅳ. ①TN929.53

中国版本图书馆 CIP 数据核字(2016)第 026103 号

书　　名：Android 应用程序设计基础
作　　者：李华忠　陈勍　但唐仁　主编

策　　划：王春霞　　　　　　　　　读者热线：(010)63550836
责任编辑：王春霞　王　惠
封面设计：付　巍
封面制作：白　雪
责任校对：汤淑梅
责任印制：郭向伟

出版发行：中国铁道出版社（100054，北京市西城区右安门西街 8 号）
网　　址：http://www.tdpress.com/51eds/
印　　刷：三河市燕山印刷有限公司
版　　次：2016 年 2 月第 1 版　　2019 年 1 月第 2 次印刷
开　　本：787mm×1092mm　1/16　印张：19.75　字数：480 千
书　　号：ISBN 978-7-113-21264-3
定　　价：47.00 元（附赠光盘）

版权所有　侵权必究

凡购买铁道版图书，如有印制质量问题，请与本社教材图书营销部联系调换。电话：(010) 63550836
打击盗版举报电话：(010) 51873659

前言

Android 是由谷歌和开放手机联盟开发与倡导的，以 Linux 为基础的完整、开放、免费的手机平台。它由应用程序、应用程序框架、系统库、Android 运行时及 Linux 内核 5 部分组成，已经成为当前主流移动终端操作系统之一。

目前，我国很多院校的计算机软件、移动互联、嵌入式和物联网等相关专业，都将"Android 应用程序设计基础"作为一门专业核心课程。为了帮助院校老师能够比较全面、系统地讲授这门课程，使学生能够熟练地使用 Android 应用技术进行移动互联软件开发，我们几位长期在院校从事 Android 教学的教师和企业工程师，共同编写了《Android 应用程序设计基础》这本教材。

我们对本书的体系结构做了精心的设计，按照 Android 平台的技术体系结构和项目内容，如应用(Application)与活动(Activity)、布局管理器(Layout Manager)、Android 常用控件（TextView、EditText、Button、ImageButton、ToggleButton、RadioButton、RadioGroup、CheckBox、Switch、ImageView、AnalogClock、DigitalClock、Chronometer、TimePicker、DatePicker 和 RatingBar）和高级控件（AutoCompleteTextView、MultiAutoCompleteTextView、Spinner、ScrollView、ListView、GridView、ProgressBar、SeekBar、TabHost 和 Gallery）等基础理论知识，设计多个典型项目。每个典型项目又结合知识体系和实践技能细化为若干个知识点和实践案例，适应当前微课题教学需要，由浅入深，实用性强。最后，结合移动互联应用实际情况，安排相应的基础实训项目（Android 应用编程基本项目实训、基于 Android 未来超市系统项目实训和博物馆智能导览系统项目综合实训），在提高学生应用技能的同时，强化项目驱动，实施"工学结合"，提高基础理论知识教学和实践教学质量，充分满足高职院校对教学和学生自学的需求。在内容编写方面，难点分散、循序渐进；在文字叙述方面，言简意赅、重点突出；在实例选取方面，实用性强、针对性强。

本书每章都附有一定数量的强化练习题，可以帮助学生进一步巩固基础知识；配备实践性较强的实训项目，可以供学生上机操作时使用。本书配备了源代码、习题答案、PPT 课件、教学大纲、课程设计等丰富的教学资源，其中源代码和习题答案请见光盘，其他资源可到中国铁道出版社网站（http://www.51eds.com/）免费下载使用。本书的教学参考总学时为 64 学时，其中实践环节为 32 学时，各章的参考学时参见下面的学时分配表。

章 节	课程内容	学时分配	
		讲 授	实 训
第 1 章	Android 应用(Application)与活动(Activity)	4	4
第 2 章	Android 用户界面布局管理器(Layout Manager)	4	4
第 3 章	Android 常用控件和高级控件使用方法	6	6
第 4 章	Android 应用编程基本项目实训	6	6
第 5 章	基于 Android 未来超市系统项目实训	6	6
第 6 章	博物馆智能导览系统项目综合实训	6	6

本书由李华忠、陈勋、但唐仁担任主编，深圳市大雅新科技有限公司总经理刘业涛和深圳尚信物联有限公司研发总经理刘立明任副主编，深圳市亿道电子技术有限公司总监钟景洲参与本书的部分编写、策划和审核工作。另外，也得到很多同事的大力支持和帮助，在此表示衷心的感谢！尤其感谢深圳市大雅新科技有限公司、深圳尚信物联有限公司和深圳市亿道电子技术有限公司提供本书所需的企业级项目源代码。

由于移动开发技术发展日新月异，加之编者水平有限，书中难免存在疏漏和不妥之处，敬请广大读者批评指正。

编　者

2015 年 12 月

目录

第1章 Android 应用（Application）与活动（Activity） 1
1.1 学习导入 1
- 1.1.1 什么是 Android 1
- 1.1.2 什么是 Android 系统架构 1
- 1.1.3 什么是 Android 应用 3
- 1.1.4 什么是 Android 上下文 4
- 1.1.5 什么是活动 4
- 1.1.6 什么是意图 5

1.2 技术准备 5
- 1.2.1 快速构建 Android 开发环境 5
- 1.2.2 活动生命周期 6
- 1.2.3 Android 调试工具及其方法 9

1.3 案例实施 11
- 1.3.1 使用 ADT 创建第一个 Android 应用项目 11
- 1.3.2 利用 DDMS 观察 Android 活动的生命周期 22

1.4 知识扩展 24
- 1.4.1 活动的属性 25
- 1.4.2 应用的属性 26
- 1.4.3 手动创建活动的方法 28
- 1.4.4 启动和关闭活动的方法 29
- 1.4.5 扩展项目实训 30

本章小结 39
强化练习 39

第2章 Android 用户界面布局管理器（Layout Manager） 43
2.1 学习导入 43
- 2.1.1 什么是 Android 布局管理器（Layout Manager） 43
- 2.1.2 什么是 Android 界面组件（Component）结构层次 44

2.2 技术准备 45
- 2.2.1 线性布局管理器 45
- 2.2.2 相对布局管理器 53
- 2.2.3 帧布局管理器 58
- 2.2.4 表格布局管理器 65
- 2.2.5 网格布局管理器 71
- 2.2.6 绝对布局管理器 75

2.3 案例实施 77
- 2.3.1 利用线性布局显示贺知章《回乡偶书》 77

2.3.2　利用相对布局实现密码验证界面 81
　　　2.3.3　利用表格布局实现菜单 82
　　　2.3.4　利用网格视图布局浏览图片 85
　2.4　知识拓展 88
　　　2.4.1　Android 布局管理器的嵌套 88
　　　2.4.2　Android 抽象布局标签 90
本章小结 94
强化练习 94

第 3 章　Android 常用控件和高级控件使用方法 100

　3.1　学习导入 101
　　　3.1.1　什么是 Android 视图类 101
　　　3.1.2　什么是 Android 视图组类 102
　3.2　技术准备 103
　　　3.2.1　文本视图控件 103
　　　3.2.2　编辑框控件 111
　　　3.2.3　按钮和图片按钮 115
　　　3.2.4　状态开关按钮和开关控件 120
　　　3.2.5　单选按钮和复选框 123
　　　3.2.6　图像视图控件 128
　　　3.2.7　日期与时间控件 131
　　　3.2.8　模拟时钟和数字时钟 135
　　　3.2.9　计时器控件 138
　　　3.2.10　自动完成文本框 142
　　　3.2.11　下拉列表控件 145
　3.3　案例实施 152
　　　3.3.1　利用 DatePickerDialog 和 TimePicker 设置日期和时间 152
　　　3.3.2　创建基于多种控件的表单应用案例 155
　　　3.3.3　利用下拉列表（Spinner）控件实现歌曲选择功能 159
　3.4　知识拓展 161
　　　3.4.1　滚动视图控件 161
　　　3.4.2　列表视图控件 167
　　　3.4.3　网格视图控件 177
　　　3.4.4　进度条 183
　　　3.4.5　滑块控件 187
　　　3.4.6　评分条控件 192
　　　3.4.7　选项卡控件 195
　　　3.4.8　画廊控件 199
本章小结 202
强化练习 202

第 4 章　Android 应用编程基本项目实训 207

　4.1　利用动画资源文件实现程序过渡动画 207
　　　4.1.1　项目实训目标 207

	4.1.2 知识准备	207
	4.1.3 项目实践	211
4.2	利用Drawable资源文件美化程序UI界面	217
	4.2.1 项目实训目标	217
	4.2.2 知识准备	217
	4.2.3 项目实践	219
	4.2.4 Android数据存储Shared Preferences	226
4.3	利用列表控件实现九宫格菜单	228
	4.3.1 项目实训目标	228
	4.3.2 知识准备	228
	4.3.3 项目实践	229
4.4	利用列表控件实现Metro UI菜单	231
	4.4.1 项目实训目标	231
	4.4.2 知识准备	231
	4.4.3 项目实践	231
4.5	获取URL地址上的资源	237
	4.5.1 项目实训目标	237
	4.5.2 知识准备	237
	4.5.3 项目实践	242
4.6	利用天气预报接口编写天气查询软件	244
	4.6.1 项目实训目标	244
	4.6.2 知识准备	244
	4.6.3 项目实践	248
本章小结		251
强化练习		251

第5章 基于Android未来超市系统项目实训 ... 255

5.1	项目概述	255
5.2	项目设计	256
	5.2.1 项目总体功能需求	256
	5.2.2 项目总体设计	256
5.3	项目实施	257
	5.3.1 登录页面	257
	5.3.2 主页面	258
	5.3.3 环境监控页面	261
	5.3.4 物品入库页面	265
	5.3.5 具体设备页面	269
	5.3.6 物品出库页面	272
本章小结		275
强化练习		275

第6章 博物馆智能导览系统项目综合实训 ... 278

| 6.1 | 项目概述 | 278 |

  6.1.1 项目简要介绍 278
  6.1.2 项目背景 278
 6.2 目标分析与运行环境 279
  6.2.1 目标分析 279
  6.2.2 运行环境 279
 6.3 需求分析 280
  6.3.1 功能需求 280
  6.3.2 性能需求 281
 6.4 总体设计 281
  6.4.1 总体结构 281
  6.4.2 处理流程设计 282
 6.5 详细设计 284
  6.5.1 主要数据结构设计 284
  6.5.2 关键或难点技术的实现 285
  6.5.3 软件主要功能的使用说明 286
 6.6 项目实施 288
  6.6.1 IntelligentBrowsing 活动 288
  6.6.2 数据帧读取模块 SerialPort 292
  6.6.3 数据解码模块 ZigbeePackage 292
  6.6.4 数据转换模块 ConvertHelper 293
  6.6.5 滚动字幕模块 MarqueeText 295
  6.6.6 图片处理模块 GalleryFlow 296
  6.6.7 图文显示模块 Allshow 297
  6.6.8 视频显示模块 Allmovie 301
 本章小结 303
 强化练习 304

参考文献 307

第1章 Android 应用（Application）与活动（Activity）

【知识目标】
- 了解 Android（安卓）的基本概念和系统构架知识；
- 了解安卓活动（Activity）和意图（Intent）的基本概念和作用；
- 了解安卓应用（Application）和上下文（Context）的基本概念和作用；
- 理解 Android 应用程序的框架模型；
- 理解安卓活动生命周期管理框架模型；
- 理解 Android 活动栈管理模型。

【能力目标】
- 掌握搭建 Android 开发环境的基本方法；
- 掌握构建 Android 应用程序的基本方法；
- 掌握创建 Android 活动的基本方法；
- 掌握 Android 应用程序的启动入口定制方法；
- 掌握 Android 应用程序运行和调试方法。

【重点、难点】
- Android 项目的创建和运行方法；
- Android 活动生命周期及其创建方法；
- Android 应用程序的启动入口定制方法。

1.1 学习导入

1.1.1 什么是 Android

所谓 Android，中文俗称"安卓"，是一种以 Linux 为基础的开放源代码移动设备操作系统（Operating System, OS），主要用于智能手机和平板电脑，由谷歌（Google）成立的开放手机设备联盟（Open Handset Alliance, OHA）持续领导与开发。Android 一词的本义指"机器人"，早期由 Andy Rubin（Android 之父）创办，现已发布的最新版本为 Android 6.0（Android M）。

Android 系统具有开放性、挣脱运营商的束缚、丰富的硬件选择、不受任何限制的开发商和无缝结合的 Google 应用等五大优势特色，成为最近几年叱咤移动互联网最火的系统之一。

1.1.2 什么是 Android 系统架构

Android 采用了软件堆层（Software Stack，又名软件叠层）的架构，其系统架构主要由

Linux 内核（Kernel）、库（Libraries）和 Android 运行时（Runtime）、应用框架（Application Framework）和应用程序（Applications）等四个层次，如图 1.1 所示。

图 1.1　Android 系统架构图

1. Linux Kernel 层

Android 的 Linux Kernel 基于 Linux 内核，但并不包括全部的 Linux，主要负责提供系统核心服务，如进程管理、内存管理、电源管理、网络协议栈、驱动模型与安全性等。Linux Kernel 是位于硬件和软件堆之间的抽象层，它隐藏具体硬件细节而为上层提供统一的服务，具有高内聚、低耦合特征，主要作用为：

（1）提供核心服务：安全机制、内存管理、进程管理、网络、硬件驱动；

（2）扮演的是硬件层和系统其他层次之间的一个抽象层的概念；

（3）Android 操作系统的初始化和编程接口与标准 Linux 系统有所不同。

2. 系统库层

Android 系统库包含一个 C/C++库的集合，供 Android 系统的各个组件使用。这些功能通过 Android 的应用程序框架（Application Framework）暴露给开发者。系统库由一系列二进制动态库共同构成，通常使用 C、C++开发。与框架层的系统服务相比，系统库不能独立运行于线程中，需要被系统服务加载到其进程空间里，通过类库提供的 JNI 接口（Java Native Interface，Java 本地调用接口）进行调用。系统库分为函数库和 Android 运行时两部分。

Android 主要核心库包括：系统 C 库、多媒体框架（Media Framework）、桌面管理库（SurfaceManager）、嵌入式浏览器引擎（WebKit）、轻量级嵌入式关系数据库（SQLite）、FreeType 库等。

Android 包含一个核心库的集合，提供大部分在 Java 编程语言核心类库中可用的功能。每一个 Android 应用程序都是 Dalvik 虚拟机（Virtual Machine，VM）中的实例，运行在它们自己的进程中。Dalvik 虚拟机采用基于寄存器的架构，被设计成在一个设备中可高效运行多个虚拟机，其可执行文件格式是.dex。.dex 格式是专为 Dalvik 设计的一种压缩格式，适合内存和处理器速度有限的系统。一个.dex 文件通常会有多个.class 文件。由于.dex 文件有时必须进行最佳化，会使文件大小增加 1～4 倍，以 ODEX 结尾。dx 是一套工具，可将 Java .class 转换成.dex 格式。Dalvik 虚拟机依赖于 Linux 内核提供基本功能，如线程和内存管理。

3. 应用框架层

应用程序的体系结构旨在简化组件的重用，它是用户进行Android 开发的基础，是谷歌发布核心应用时所使用的 API 框架。任何应用程序都能发布它的功能，且任何其他应用程序可以使用这些功能（需要服从框架执行的安全限制）。通过提供开放的开发平台，Android 使开发者能够编制极其丰富和新颖的应用程序。该层开放给用户的主要组件包括：视图系统（Views System）、内容提供者（Content Providers）、资源管理器（Resource Manager）、通知管理器（Notification Manager）、活动管理器（Activity Manager）等。

4. 应用层

Android 装配一个核心应用程序集合，包括电子邮件（E-mail）客户端、短消息（SMS）程序、日历、地图、浏览器、联系人和其他设置。所有应用程序都是用 Java 编程语言编写的。

从上述内容可知，Android 的软件叠层架构，层次清晰，分工明确。Android 系统主要功能包的简要功能描述如表 1.1 所示。

表 1.1 Android 系统主要功能包的简要描述

序号	功能包名	功能简要描述
1	android.app	提供高层的程序模型和基本的运行环境
2	android.content	提供对各种设备上的数据的访问和发布接口
3	android.database	通过内容提供者浏览和操作数据库
4	android.graphics	封装底层的图形库，包括画布、颜色过滤、点、矩阵
5	android.location	封装定位和相关服务的类
6	android.media	提供管理多种音频、视频的媒体接口
7	android.net	提供帮助网络访问的类，超过通常的 Java.net.*接口
8	android.os	提供系统服务、消息传输和进程间通信（Inter-Process Communication，IPC）机制
9	android.opengl	提供 OpenGL 的工具
10	android.provider	提供访问 Android 内容提供者的类
11	android.telephony	提供与拨打电话相关的 API 交互
12	android.view	提供基础的用户界面接口（User Interface，UI）框架
13	android.util	涉及工具性的方法，如时间/日期的操作
14	android.webkit	默认浏览器操作接口
15	android.widget	包含各种 UI 元素（大部分是可见的）在应用程序屏幕布局中的使用

1.1.3 什么是 Android 应用

Android 应用（Application）是 Android 框架的一个系统组件，当 Android 程序启动时，Android 系统会自动创建一个 Application 对象，用来存储系统的一些信息。通常用户不需要指定一个 Application，这时系统会自动帮用户创建，如果用户需要创建自己的 Application，也很简单，只需创建一个子类继承 Application，并在 AndroidManifest.xml 的 application 标签中进行注册，即给 application 标签增加 name 属性，并将其值设置为用户创建的 Application 的名称。Application 是单例（singleton）模式的一个类，Android 系统会为每个程序运行时创建一个 Application 类的对象且仅创建一个，且 application 对象的生命周期是整个程序中最长的，它的生命周期就等于这个程序的生命周期。因为它是全局单例，因而在不同的活动（Activity）、服务（Service）等组件中获得的对象都是同一个对象，从而可通过 Application 来进行数据传递、数据共享、数据缓存等操作。Application 的生命周期包括进程在 Android 系统中从启动到终止的所有阶段，即 Android 程序启动到停止的全过程，完全由 Android 系统所调度和控制。

所谓 XML 是 Extensible Markup Language（可扩展标记语言）的缩写，是用户可用来创建自己标记的语言。

1.1.4 什么是 Android 上下文

Android 上下文（Context）描述了应用程序的环境信息，通过上下文可获取应用程序的资源和调用一些操作，如简易的消息提示（toast）、发送意图等。在 Android 中，上下文只有活动、应用和服务，拥有一个上下文就可访问该实例对应的资源，并有权调用其方法。

在 Android 应用框架中，根据作用域的不同，可以把上下文分为两种，一种是活动上下文（Activity Context），另一种是应用上下文（Application Context）。

活动上下文在活动启动时被创建，主要用于保存当前对活动资源的引用。它在活动控制器类中被使用，当用户需要加载或者访问活动相关资源时，会需要用到该活动的上下文对象。它的生命周期跟活动是同步的，即当活动被销毁时，其对应的上下文也被销毁了，同时，该上下文相关的控件对象也将被销毁并回收。因此，活动上下文主要用于串联 Android 应用之中的对象和组件。

应用上下文在整个应用开始时被创建，用于保存对整个应用资源的引用，在程序中可以通过 Activity Context 的 getApplicationContext() 方法或者 getApplication() 方法来获取。在实际应用时，用户通常会把应用上下文当作全局对象的引用来使用。

1.1.5 什么是活动

活动（Activity）是 Android 应用程序提供交互界面的一个重要组件，也是 Android 重要组件之一，是一个用来与用户交互的系统模块，几乎所有的活动都是和用户进行交互的。Android 的应用开发遵循经典的 MVC（Model View Controller，模型-视图-控制器）设计模式，其中，Model 表示应用程序核心，是应用程序中用于处理应用程序数据逻辑的部分；View 用来显示数据（数据库记录），是应用程序中处理数据显示的部分；Controller 主要负责处理输入（写入数据库记录），是应用程序中处理用户交互的部分。在 Android 中，活动主要用来做控制，它既可选择要显示的视图，也可从视图中获取数据，然后把数据传给 Android 程序的模型层进行处理，最后显示出处理结果。因此，活动是 Android 程序的表现层，显示可视化的用户界面，负责接收与用户交互所产生的界面事件。一个 Android 应用可包含一个或多个活动，一般在程序启动后会呈现一个活动，用于提示用户程序已经正常启动。活动在界面上的表现形式为：全屏窗体、非全屏悬浮窗体或对话框等。

1.1.6 什么是意图

Android 意图（Intent）是一个动作的完整描述，包含产生组件、接收组件和传递数据信息，它利用消息实现应用程序之间的交互机制，这种消息描述了应用中一次操作的动作、数据及附加数据，系统通过该意图的描述找到对应的组件，并将意图传递给调用的组件，完成组件的调用。在一个 Android 应用中，意图是对执行某个操作的一个抽象描述，意图负责提供组件之间相互调用的相关信息传递，实现调用者和被调用者之间的解耦。如果把活动比作积木，那么意图就是胶水，即把不同的积木粘连起来；或者说，如果活动是指不同的手机屏幕，则意图就是把不同的手机屏幕粘连起来的胶水。

因此，意图主要负责程序跳转和传递数据，它是一种在不同组件之间传递的请求信息，是应用程序发出的请求和意图。作为一个完整的消息传递机制，它不仅需要发送端，还需要接收端。

1.2 技术准备

1.2.1 快速构建 Android 开发环境

工欲善其事，必先利其器。本节介绍快速构建 Android 开发环境的方法。

步骤 1：安装 Java 开发工具（JDK）和配置 Java 开发环境。

安装 Java 开发工具（Java Development Kit，JDK）是学习编写 Android 应用程序设计的基础。安装和配置 Java 开发环境的步骤如下：

（1）从 http://www.oracle.com/technetwork/java/javase/downloads/index.html 下载最新版 JDK。

（2）安装 JDK。下载的 JDK 安装包中包含 JDK 和 JRE 两部分，建议将它们安装在同一个逻辑分区中。双击安装程序，选择安装目录，单击"下一步"按钮，等待安装程序自动完成安装即可。

（3）设置 Java 环境变量。右击【我的电脑】图标，在弹出的快捷菜单中选择【属性】命令，在弹出的对话框中选择【高级】选项卡，单击【环境变量】按钮，在弹出的【环境变量】对话框的【系统变量】列表中查找 Path 变量。如果找到，双击该变量打开设置对话框；如果未找到该变量，则新建一个 Path 变量，并在变量设置对话框中设置 JDK 中 bin 目录的路径 %JAVA_HOME%\bin；按照同样的方法设置 JAVA_HOME 变量，例如 C:\Program Files\Java\jdk 1.7.0_03。类似地将 CLASSPATH 变量设置为 %JAVA_HOME%\lib\tools.jar。

（4）安装配置完成后，启动 Windows 控制台。输入 java -version 命令，如果输出 JDK 的版本信息，则说明已经成功安装和配置了 Java 开发环境。

步骤 2：从学校的教师机器上复制已装好的 Eclipse 和 android-sdk-windows 文件夹（或从网上下载已制作好的安装包）到想安装的任意目录，例如 C:\Android-Develop-Tools，双击 eclipse.exe 文件启动 Eclipse。

步骤 3：配置 Android 软件开发工具包（Software Development Kit，SDK）路径。选择【Window】→【Preferences】→【Android】→【SDK Location】，输入：..\android-sdk-windows（见图 1.2），即完成了 Android 开发环境的快速安装。

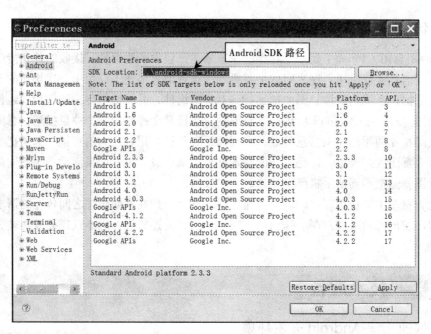

图 1.2　配置 Android SDK 路径

1.2.2　活动生命周期

正确理解 Android 活动（Activity）的生命周期是非常重要的，只有这样，才能确保应用程序提供一个符合逻辑的用户体验并有效管理其资源。Android 中的应用程序并不能管理自身的生命周期，而是由系统统一管理的，当然活动的生命周期也一样由系统负责统一管理。

1. 活动的四种状态

活动是 Android 程序的表现层，显示可视化的用户界面，并接收与用户交互所产生的界面事件活动。活动可表现为以下四种状态：

（1）运行（Running）状态：活动在用户界面中处于最上层，能完全被用户看到，能与用户进行交互；活动正处于运行状态。

（2）暂停（Paused）状态：活动在界面上被部分遮挡，该活动不再处于用户界面的最上层，且不能够与用户进行交互；活动正在执行中，在系统内存不够时被回收。

（3）停止（Stopped）状态：活动在界面上完全不能被用户看到，即该活动被其他活动全部遮挡；活动正在执行中，但已退出界面。

（4）销毁（Destroyed）状态：不在以上三种状态中的活动，如活动已被回收或未启动。

2. 活动交互机制

因活动直接涉及与用户交互界面的处理，而任意时刻与用户交互的界面只有一个，所以 Android 针对活动的管理采用了具有层次感的栈数据结构，其工作原理示意如图 1.3 所示。活动栈保存了已经启动并且没有终止的活动，并遵循"先进后出"的原则。例如，在某个时刻只有一个活动处于栈顶，当这个活动被销毁后，下面的活动才可能处于栈顶，或者是有一个新的活动被创建出来，则上一个活动就被压栈下去。

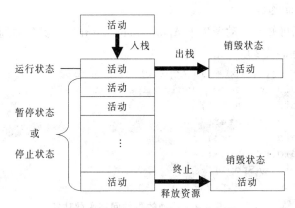

图 1.3　活动栈工作原理示意图

3. 活动的生命周期

所谓活动的生命周期也就是它所在进程的生命周期，图 1.4 描述了活动生命周期过程及状态改变，图中椭圆是活动可经历的主要状态，矩形框代表当活动在状态间发生改变时，可重写的回调方法。这些方法都是生命周期钩子，可通过重写它们完成适当工作。注意：onSaveInstanceState()和 onRestoreInstanceState()并不是生命周期方法，意味着它们并不一定会被触发。

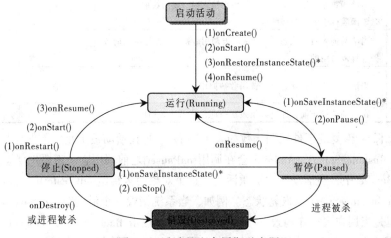

图 1.4　活动的生命周期示意图

活动的生命周期可分成三类：前台生命周期、可视生命周期和完整生命周期。

（1）完整生命周期：从 onCreate()开始直到 onDestroy()结束。在 onCreate()中初始化活动的状态，在 onDestroy()中释放所有资源。

（2）可视生命周期：从 onStart()开始到 onStop()结束。在这段时间里，用户可在屏幕上看见活动，但并不一定在前台和用户进行交互。可通过反复调用这两个方法来实现界面的显示和隐藏，例如，可在 onStart()中注册一些 BroadcastReceiver 来监听用户的操作，在 onStop()中注销这些监听事件，因用户已经无法看见操作界面。

（3）前台生命周期：从 onResume()开始到 onPause()结束。这段时间，活动处在最前端和用户进行交互。在这两个方法中，处理都应该尽量轻量级，避免耗时很多的操作，因为它们可能会被无数次的调用，如手机进入睡眠模式就会触发 onPause()。

4. 活动的事件回调函数

从 Android 源代码可知，活动派生于 ApplicationContext 类，其生命周期的事件回调方法如下（其主要功能描述表 1.2）：

```
1. public class Activity extends ApplicationContext {
2.     protected void onCreate(Bundle savedInstanceState);
3.     protected void onStart();
4.     protected void onRestart();
5.     protected void onResume();
6.     protected void onPause();
7.     protected void onStop();
8.     protected void onDestroy(); }
```

表 1.2　活动的事件回调函数功能

回调方法	描　　述	是否可终止	下一个操作
onCreate()	活动启动后第一个被调用的函数，常用来初始化活动，如创建视图、绑定数据或恢复信息等	否	onStart()
onStart()	当活动显示在屏幕上时，该函数被调用	否	onResume()
onRestart()	当活动从停止状态进入运行状态前，调用该函数	否	onResume()或onStop()
onResume()	当该函数能与用户交互，接收用户输入时，该函数被调用。此时该函数位于该函数栈的栈顶	否	onPause()
onPause()	当活动进入暂停状态时，该函数被调用。常用来保存持久的数据或释放占用的资源	是	onResume() 或 onStop()
onStop()	当活动进入停止状态时，该函数被调用	是	onRestart()或onDestroy()
onDestroy()	在活动被终止前，即进入销毁状态前，该函数被调用	是	无

活动状态保存/恢复的事件回调函数如表 1.3 所示。想在活动被中止之前保存状态，必须实现方法 onSaveInstanceState()，Android 会在调用 onPause()之前调用它。需要把状态保存在一个键值对对象 Bundle 中，当活动再次被激活时，这个对象会被传给 onCreate()以及稍后的另一个方法 onRestoreInstanceState()以恢复状态。例如，当系统销毁一个活动时，onSaveInstanceState()会被调用，但是当用户去操作销毁一个活动时（如用户点击 Back 按钮），onSaveInstanceState()就不会被调用。在这种场合，用户的行为决定了状态不需要被保存。因 onSaveInstanceState()不会经常被调用，所以只应该用它来保存一些临时的状态，而在 onPause()里做数据的持久化操作。

表 1.3　活动状态保存/恢复的事件回调函数

回调方法	描　　述	是否可终止
onSaveInstanceState()	Android 系统因资源不足中止活动前调用该函数，用于保存活动的状态信息，供 onRestoreInstanceState()或 onCreate()恢复之用	否
onRestoreInstanceState()	恢复 onSaveInstanceState()保存的活动状态信息，在 onStart()和 onResume()之间被调用	否

5. 活动的四种加载模式

在多活动开发中，有可能是自己应用之间的活动跳转，或者嵌入其他应用的可复用活动，或者可能希望跳转到先前某个活动实例，而不产生大量重复的活动。这就需要为活动配置特定的加载模式，而不是使用默认的加载模式。Activity 有四种加载模式，分别是 standard、singleTop、singleTask 和 singleInstance。

（1）standard：标准模式，一旦调用 startActivity()方法就会产生一个新实例。这种模式下的活动可被实例化多次。

（2）singleTop：拒绝堆叠模式，若已有一个实例位于活动栈顶部，就不产生新实例，只是调用活动中的 newInstance()方法。若不位于栈顶，则会产生一个新实例。这种模式下的活动也可被实例化多次。

（3）singleTask：独立门户模式，会在一个新任务中产生这个实例，以后每次调用都会使用该实例，而不会产生新实例。

（4）singleInstance：孤独寂寞模式，会在一个新任务中产生这个实例，以后每次调用都会使用该实例，而不会产生新的实例。该模式与 singleTask 基本相同，唯一的区别在于：在该模式下的活动实例所处的任务中，只能有这个活动实例，不能有其他实例。

活动启动模式需通过 AndroidManifest.xml 文件中 activity 元素的 android:launchMode 属性进行配置，其示例代码如下：

```
<activity android:name=".MainActivity" android:launchMode="singleTask">
</activity>
```

在 Eclipse ADT 图形界面中配置活动启动模式的操作方法如图 1.5 所示。

活动加载模式的 Intent 标识主要有：
- FLAG_ACTIVIYT_NEW_TASK（同 singleTask 效果）；
- FLAG_ACTIVITY_SINGLE_TOP（同 singleTop 效果）；
- FLAG_ACTIVITY_RESET_TASK_IF_NEEDED；
- FLAG_ACTIVITY_CLEAR_TOP。

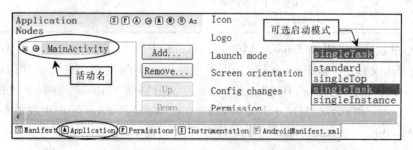

图 1.5　Eclipse ADT 图形界面中配置活动启动模式的操作方法

1.2.3　Android 调试工具及其方法

在 Android 应用程序开发中，最常用的调试工具是 Logcat。既可用 Logcat 命令来查看系统日志缓冲区的内容，也可通过 Logcat 打印程序的启动、初始化过程信息、状态信息和错误信息等。图形化 DDMS（Dalvik Debug Monitor Service，Dalvik 虚拟机调试监控服务）Logcat 最有效。当应用程序在模拟器中加载并启动时，Eclipse 会自动切换到 Debug 布局，可以通过【Window】→【Show View】→【Other】→【Android】→【Logcat】命令查看日志信息，如图 1.6 所示。

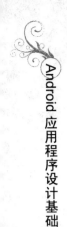

图 1.6 DDMS Logcat 界面

在 DDMS Logcat 视图的工具栏中可以看到标记为 V、D、I、W、E 和 F 的几个按钮，它们的作用是对消息进行过滤，如表 1.4 所示。

表 1.4　DDMS Logcat 视图的工具栏过滤标记的含义

序 号	标 记	描 述
1	V（Verbose）	显示所有类型的消息
2	D（Debug）	显示 Debug、Information、Warning 和 Error 消息
3	I（Information）	只显示 Information、Warning 和 Error 消息
4	W（Warning）	只显示 Warning 和 Error 消息
5	E（Error）	只显示 Error 消息
6	F（Fatal）	只显示致命错误信息

DDMS Logcat 视图中包含的列表信息如表 1.5 所示。

表 1.5　DDMS Logcat 视图工具栏列表信息

序 号	列 表 项	描 述
1	Time	用于显示消息产生的时间
2	Priority	消息的级别（取值为 D、I、W 或 E，分别代表 Debug、Information、Warning 和 Error），其详细说明见表 1.4
3	PID	产生消息的进程 ID
4	Tag	消息产生来源的简短描述
5	Message	消息的详细内容

在 Android 程序中采用 Android 工具包（android.util）中的日志记录类 Log 将消息写入 AVD 模拟器上的日志文件中，android.util.Log 可设置输出日志的级别，有五个常用的方法：Log.v()、Log.d()、Log.i()、Log.w() 和 Log.e()，其含义分别是 Verbose、Debug、Info、Warn 和 Error。Verbose 和 Debug 只用于开发过程，Info、Warn 和 Error 可出现在发布版本中。设 string TAG=this.getPackageName()，Log 用法示例如表 1.6 所示。

表 1.6　android.util.Log 用法示例

方 法	示 例 代 码	备 注
Log.v(TAG, strings)	Log.v(TAG, "详细消息")	调试颜色为黑色，任何消息都会输出
Log.i(TAG, strings)	Log.i(TAG, "提示信息")	输出颜色为绿色，一般提示性消息
Log.d(TAG, strings)	Log.d(TAG, "调试信息")	输出颜色为蓝色，仅输出 debug 调试消息
Log.w(TAG, strings)	Log.w(TAG, "警告信息")	输出颜色为橙色，仅输出 warning 警告信息
Log.e(TAG, strings)	Log.e(TAG, "错误信息")	输出颜色为红色，仅显示红色 error 错误信息

Android 应用程序 Debug 调试操作方法为：双击代码编辑器左侧设置断点，单击菜单栏中的【Run】菜单或按【F11】键便可开始程序调试。程序运行到断点处时会弹出一个对话框，单

击【yes】转到 Debug 视图，找到正在调试的类。主要调试操作命令如表 1.7 所示。

表 1.7 Android 应用程序 Debug 调试操作命令

序 号	菜 单 命 令	描 述	快 捷 键
1	Run→Step Into	逐行调试每条语句	F5
2	Run→Step Over	逐个调试每个过程（略过方法）	F6
3	Run→Step Return	单步返回或逐语句进入方法后跳出	F7
4	Run→Run To Line	运行到光标处	Ctrl+R
5	Run→Resume	断续运行到结束	F8

注意：支持调试需在<application>标签中增加调试属性：android:debuggable="true"。

1.3 案例实施

1.3.1 使用 ADT 创建第一个 Android 应用项目

本项目主要任务目标在于：
- 如何利用 ADT（Android Development Tools，安卓开发工具）新建一个 Android Application Project；
- 如何在 AVD（Android Virtual Device，Android 虚拟设备）上运行 Android 应用程序；
- 了解 Android Application Project 的结构，并掌握资源字符串和风格的定制方法。

1. 使用 ADT 自动向导生成 Android 应用程序

使用 ADT 自动向导生成 Android 应用程序的操作步骤如下：

（1）打开配置好 Android 开发环境的 Eclipse。

（2）选择【File】→【New】→【Android Application Project】命令，弹出【New Android Application】对话框。

（3）在【Application Name】文本框中输入应用名：Ch01_ex01_HelloWorld（应用名称第一个字母必须大写），在【Package Name】文本框中输入包名：sziit.lihz.ch01，在【Target SDK】下拉列表框中选择 Android SDK（软件开发工具）目标版本：API 17:Android 4.2(Jelly Bean)，单击【Next】按钮，如图 1.7 所示。

图 1.7 新建 Android 应用程序对话框

（4）直到出现【Activity Name】，根据个人习惯自己命名，这里命名为 MainActivity，单击【Finish】按钮。

2. 创建 AVD

Android 虚拟设备为运行和调试 Android 应用程序提供了有效、快捷的途径，其创建步骤如下（见图 1.8）：

（1）创建新的 AVD：选择【Window】→【Android Virtual Device Manager】命令（或单击工具栏中的 AVD 按钮），弹出【Android Virtual Device Manager】对话框，单击【New】按钮，在【AVD Name】文本框中输入 AVD422（根据个人习惯命名），在【Target】下拉列表框中选择 Android 4.2.2 –API Level 17（与用户选择的 Target SDK 保持一致）。

（2）在【Android Virtual Device Manager】对话框中选择已创建好的 AVD 名"AVD422"，单击【Start】按钮，启动 Android 模拟器。

图 1.8 创建新 AVD 对话框

3. 在 AVD 上运行 Android 程序

模拟器启动后，在 Project Explorer 中，将鼠标移动到【Ch01_ex01_HelloWorld】项目上，右击，在弹出的快捷菜单中选择【Run As】→【Android Application】命令，第一个 Android 程序就成功发布到模拟器上了，如图 1.9 所示。Ch01_ex01_HelloWorld 项目完整代码请参见配套光盘\代码\CH01\Ch01_ex01_HelloWorld.rar。

图 1.9 Ch01_ex01_HelloWorld 程序在 AVD 上的运行结果

Tips1.1：Eclipse 编译 Android 项目时，提示错误：Error generating final archive: Debug certificate expired on xxxxxx（日期）。解决办法：找到【Window】→【Preferences】→【Android】→【Build】中【Default debug keystore】显示的地址：C:\Documents and Settings\Administrator\.android\debug.keystore，删除此路径下的 debug.keystore 及 ddms.cfg（注：不同的用户安装

路径，此处的路径会不同）。

Tips1.2：更改模拟器语言为中文。解决办法：在模拟器的菜单中找到 Setting，然后向下滚动找到 Language&input 并单击，将语言设置为【中文（简体）】。

Tips1.3：由于 AVD 每次启动都比较慢，可以在打开 AVD 之后不要关闭，这样每次运行 Android 应用，ADT 就会自动使用这个模拟器，这样就不用每次都启动了。将 minSdkVersion 和 targetSdkVersion 均设置为 17（SDK 4.2.2），避免 ADT 运行 Android 程序时每次自动重新创建不同的 AVD。

4. 解析 Android 项目结构

在搭建 Android 开发环境及简单地建立一个 Ch01_ex01_HelloWorld 项目后，对利用 ADT 自动向导生成的项目结构做详细解析。如图 1.10 所示，Ch01_ex01_HelloWorld 项目的目录结构主要涉及如下项目：

- src 文件夹；
- gen 文件夹；
- Android 4.2.2 文件夹；
- assets 文件夹；
- res 文件夹；
- AndroidManifest.xml 文件；
- default.properties 文件。

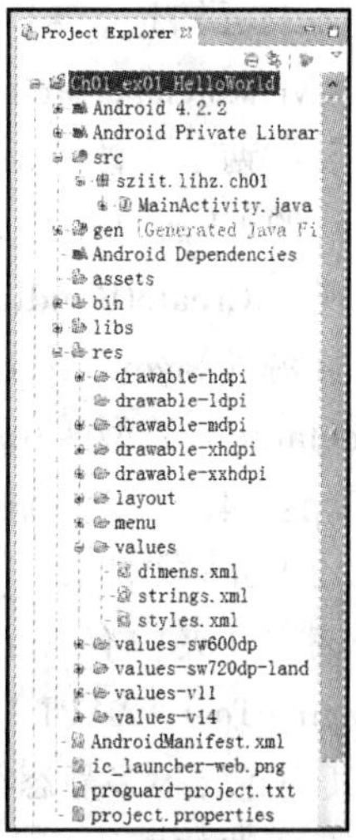

下面将分别介绍各级目录结构。

（1）src 文件夹

顾名思义，src 即 source code，该文件夹用于放置项目的源代码。打开 MainActivity.java 文件会看到如源代码清单 1.1 所示。

图 1.10 Ch01_ex01_HelloWorld 项目目录结构

源代码清单 1.1　\Ch01_ex01_HelloWorld\src\sziit\lihz\ch01_1\MainActivity.java

```
1.  /**
2.   * 第一个 Android 应用项目: Ch01_ex01_HelloWorld。
3.   * 作者: 李华忠(Li Huazhong)
4.   * 功能: 演示如何在 Eclipse 集成开发环境下，利用 ADT 自动向导快速构建
5.   *       Android 应用程序，并利用 AVD 运行该程序。
6.   * 说明: 该程序全部由 ADT 向导自动生成，代码中的注释由作者增加以帮助初学者
7.   *       阅读和理解源代码。
8.   */
9.  package sziit.lihz.ch01_1;          //用户自定义包名
10. import android.os.Bundle;           //导入安卓系统绑定包
11. import android.app.Activity;        //导入安卓应用活动包
12. import android.view.Menu;           //导入安卓视图菜单包
13. //从基类或父类 Activity(活动)公有(public)派生子类 MainActivity
14. public class MainActivity extends Activity {
15.     @Override
16.     protected void onCreate(Bundle savedInstanceState) {
17.                                     //子类重写基类创建活动方法
18.         super.onCreate(savedInstanceState);  //调用基类或父类 onCreate 方法
19.         /**通过 R.layout.activity_main 引用在 res\layout 中定义的布局文件
```

```
20.            *activity_main.xml*/
21.            setContentView(R.layout.activity_main);
22.                        //传递参数R.layout.activity_main,设置活动界面布局
23.        }
24.        @Override
25.        public boolean onCreateOptionsMenu(Menu menu) {//子类重写基类创建菜单方法
26.            //通过R.menu.main引用在res\menu目录中定义的菜单资源文件main.xml
27.            getMenuInflater().inflate(R.menu.main, menu);
28.                        //通过菜单资源文件R.menu.main,创建程序菜单
29.        return true;    }}
```

从中可知道，ADT 自动导入了三个类：android.app.Activity、android.os.Bundle 和 android.view.Menu，MainActivity 类公有继承自 Activity，且重写了 onCreateOptionsMenu()和 onCreate()两个方法。android.app.Activity 类负责创建窗口，用方法 setContentView(View) 将自己的 UI 放到里面，以全屏或浮动窗口方式展示给用户。几乎所有的 Activity 子类都会实现 onCreate(Bundle)方法，用来初始化活动。setContentView(int)方法调用用户定义的布局资源（layout resource）。android.os.Bundle 类负责从字符串值映射各种可打包的（Parcelable）类型（Bundle 是捆绑的意思）。

Tips1.4： 在方法前面加上 @Override，系统可帮助检查方法的正确性。

（2）gen 文件夹

该文件夹中有一个 R.java 文件，是在建立项目时由 Android 资源打包工具(Android Asset Packaging Tool，AAPT）工具自动生成，为只读模式，不能更改。R.java 文件中定义了一个 R 类，其中包含很多静态类，且静态类的名字都与 res 文件夹中的一一对应，即 R 类定义该项目所有资源的索引。其部分示意代码如清单 1.2 所示。

源代码清单 1.2　R.java 类中定义的常量与 res 资源一一对应

```
1.  /* AUTO-GENERATED FILE.  DO NOT MODIFY.... */
2.  package sziit.lihz.ch01_1;//定义包名
3.  public final class R {//public 说明属性公开的，final 表示最终的，不能修改
4.      ......
5.      public static final class drawable {//static 表明此类常量是静态的
6.          public static final int ic_launcher=0x7f020000;     图标资源索引
7.      }
8.      public static final class layout {
9.          public static final int activity_main=0x7f030000;   布局资源索引
10.     }
11.     public static final class menu {
12.         public static final int main=0x7f070000;
13.     }                                                       菜单资源索引
14.     public static final class string {                      字符串资源索引
15.         public static final int action_settings=0x7f050001;
16.         public static final int app_name=0x7f050000;
17.         public static final int hello_world=0x7f050002;
18.     }
19.     public static final class style {                       风格资源索引
20.         public static final int AppTheme=0x7f060001;  }}
```

一方面，通过 R.java 用户可很快地查找所需要的资源；另一方面，编译器也会检查 R.java 列表中的资源是否被用到，没有被使用的资源不会被编译进目标代码中，从而可减少应用占用的空间。该文件被删除后还会自动生成。

（3）Android 4.2.2 文件夹

该文件夹下包含一个 Java 归档文件 android.jar，其中包含构建应用程序所需的所有 Android SDK 库（如 Views、Controls）和 API。通过该文件将应用程序绑定到 Android SDK 和 Android Emulator，从而允许用户使用所有 Android 库和包，且使应用程序在适当的环境中调试，如源代码清单 1.1 MainActivity.java 源代码中的第 10～12 行就是从 android.jar 导入包。

（4）assets 文件夹

包含应用系统需要使用到的诸如 MP3、视频类的文件。

（5）res 文件夹

资源目录，包含 Android 项目中的资源文件并编译进应用程序。向此目录添加资源时，会被 R.java 自动记录。新建一个项目，res 目录下会有三个子目录：drawable、layout、values。资源名命名规则：只支持数字（0～9）、小写字母（a～z）、下画线（_）且必须以字母开头。

- drawable-?dpi：包含应用程序可用的图标文件(*.png、*.jpg)。
- layout：包含界面布局文件 activity_main.xml，其代码如源代码清单 1.3 所示。

源代码清单 1.3 \Ch01_ex01_HelloWorld\res\layout\activity_main.xml

- values：软件所需要显示的各种文字。可存放多个 *.xml 文件（如 strings.xml、dimens.xml、styles.xml），存放不同类型的数据，例如，strings.xml 源代码清单 1.4。

源代码清单 1.4 \Ch01_ex01_HelloWorld\res\values\strings.xml 源代码

字符串的定义及其引用方式：字符串资源文件采用键值对定义字符串，第 3 行定义了一个键值对，name 为 app_name，对应的值为"Ch01_ex01_HelloWorld"，在 AndroidManifest.xml 文件中，@string/app_name 引用的就是该字符串；第 5 行定义了一个键值对，name 为 hello_world，对应的值为"Hello world!"，在 activity_main.xml 布局文件中，@string/hello_world 引用的就是该字符串。

（6）AndroidManifest.xml 文件

Android 项目的全局配置文件，用于向 Android 系统提供关于应用程序的重要信息，负责注册应用中所使用的各种组件，并指定其用到的服务（如电话、互联网、短信、GPS 等服务）。AndroidManifest.xml 主要设置包括：application permissions、Activities、intent filters 等，其示意代码如源代码清单 1.5 所示。

源代码清单 1.5 Ch01_ex01_HelloWorld\AndroidManifest.xml

```
1.  <?xml version="1.0" encoding="utf-8"?>
2.  <manifest xmlns:android="http://schemas.android.com/apk/res/android"
3.      package="sziit.lihz.ch01_1"
4.      android:versionCode="1"          ← 包名
5.      android:versionName="1.0" >
6.      <uses-sdk   ← uses-sdk 标签
7.          android:minSdkVersion="17"   ← 最小 SDK 版本
8.          android:targetSdkVersion="17" />   ← 目标 SDK 版本
9.      <application   ← application 标签
10.         android:allowBackup="true"   ← 设置应用程序图标
11.         android:icon="@drawable/ic_launcher"
12.         android:label="@string/app_name"   ← 设置应用程序标题
13.         android:theme="@style/AppTheme" >   ← 设置应用程序主题风格
14.         <activity   ← activity 标签，注册活动
15.             android:name="sziit.lihz.ch01_1.MainActivity"   ← Activity 名
16.             android:label="@string/app_name" >   ← Activity 标题
17.             <intent-filter>   ← 意图过滤器
18. 设置行动名→ <action android:name="android.intent.action.MAIN" />   ← 主行动
19.                 <category android:name="android.intent.category.LAUNCHER"/>
20.             </intent-filter>   设置类名        ← 可加载类型
21.         </activity>
22.     </application>
23. </manifest>
```

<action android:name="android.intent.action.MAIN"/> 设置应用程序会首先加载 MainActivity 活动，因此在没有显式指定 application 的派生实例时，其 onCreate()方法会首先调用，类似 Android 程序的默认调用入口。

AndroidManifest.xml 的主要功能包括：

① 指定应用程序的 Java 包名，包名作为该应用程序的一个独特标识。

② 描述应用程序组件：该应用程序由哪些活动（Activity）、服务（Service）、广播接收器（BroadcastReceiver）和内容提供者（ContentProvider）组件构成，如图 1.11 所示。

③ 指定实现每个组件的类并公开发布它们的能力。
④ 决定进程将容纳哪些应用程序组件。
⑤ 声明应用程序必须拥有哪些权限（permission），以便访问被保护的 API 部分，能与其他应用程序交互。
⑥ 声明其他应用程序在与该应用程序交互时所需持有的许可。
⑦ 声明应用程序所需的 Android API 的最小版本和目标版本。
⑧ 列出应用程序必须链接的库。

图 1.11　Android 项目典型结构及其 AndroidManifest.xml 文件

Tips 1.5：<category android:name="android.intent.category.LAUNCHER"/>的含义是什么？

答：该属性表示该活动为可加载类型。若某个应用的 Activity 都没有设置 LAUNCHER，则虽然该应用可安装到设备或 AVD 虚拟器上，但在桌面上看不到该应用的图标，因此，只有设置 LAUNCHER 后，才能在桌面上看到该应用程序的图标。

Tips 1.6：<action android:name="android.intent.action.MAIN" />的含义是什么？

答：该属性表示该活动的行动为主启项，决定该应用程序最先启动哪个活动。如果该程序的活动虽然设置了 LAUNCHER，但没有任何活动设置 MAIN，则该应用程序不知启动哪个 Activity，故也不会有图标出现。当然，可给多个活动设置 MAIN，但只设定 MAIN，没有任何活动设置 LAUNCHER，该程序仍无法启动任何活动。因此，只有同时设置 MAIN 和 LAUNCHER，才能正常启动应用程序。如果在一个应用程序中，有多个活动同时设置 MAIN 和 LAUNCHER，则在桌面上可同时看到多个图标，点击不同图标，则分别启动其对应的活动。

Tips1.7：应用程序 application 的 android:label="@string/app_name"表示全局标题。若某个活动也定义了 android:label="@string/activity_title"，则该活动启动时，会覆盖应用程序的标题；当应用程序执行到其他没有定义 android:label 的活动时，则会显示全局标题。

（7）default.properties 文件

记录项目中所需要的环境信息，如 Android 的版本等。Ch01_ex01_HelloWorld 的 default.properties 文件代码如源代码清单 1.6 所示，代码中的注释已将 default.properties 解释得很清楚了。

源代码清单 1.6　Ch01_ex01_HelloWorld\project.properties

```
1. # This file is automatically generated by Android Tools.
2. # Do not modify this file -- YOUR CHANGES WILL BE ERASED!
3. ……
4. # Project target.
5. target=android-17     ← 项目目标版本
```

5. Android 应用构成

Android 主要包括四大核心组件：活动、服务、广播接收器和内容提供者。Android 应用可包含四大组件的任意组合，如图 1.12 所示。大部分 Android 应用包含至少一个活动，如图 1.13 所示。

图 1.12　Android 通用应用构成图　　　图 1.13　常用 Android 应用构成图

- 活动：Android 应用程序的表现层，主要负责实现应用程序的用户界面。
- 服务：不直接与用户进行交互，没有用户界面，能够长期在后台运行，且比活动具有更高的优先级，在系统资源紧张时不会轻易被 Android 系统终止。它不仅可以实现后台服务的功能，也可以用于进程间的通信。
- 广播接收器：一种全局监听器，用于监听系统全局的广播消息，可以非常方便地实现系统不同组件之间的通信。
- 内容提供者：将应用程序数据封装起来，只暴露出希望提供给其他程序的数据，以提供应用程序数据对外共享的统一访问方式，不必针对不同数据类型采取不同的访问策略，内容提供者中数据更改可被监听，可支持在多个应用中存储和读取数据，这也是跨应用共享数据的唯一方式。

注：在 Android 应用中所用任何活动、服务、广播接收器和内容提供者都必须在系统控制文件 AndroidManifest.xml 中注册声明。

6. 举一反三

（1）如何修改 Android 应用程序的标题，布局中显示的内容、字体和颜色？

步骤 1：按照项目 Ch01_ex01_HelloWorld 实施步骤，利用 ADT 自动生成项目 Ch01_ex02 ModifyPropterties。

步骤 2：打开字符串资源文件\Ch01_ex02ModifyPropterties\res\values\strings.xml，可利用可视化资源管理器协助修改字符串 hello_world 和 app_name 的内容，如图 1.14 所示。如源代码清单 1.7 所示。

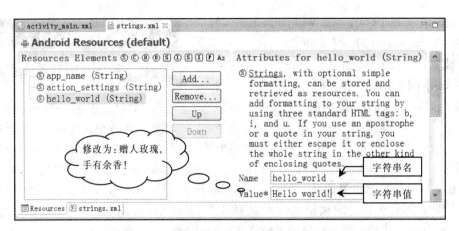

图 1.14 利用资源管理器修改字符串内容

源代码清单 1.7 \Ch01_ex02ModifyPropterties\res\values\strings.xml

```
1. <string name="app_name">Ch01_ex02 标题：谚语</string>
2. <string name="action_settings">设置</string>
3. <string name="hello_world">赠人玫瑰,手有余香!</string>  ← 修改后的内容
```

步骤3：打开布局资源文件\Ch01_ex02ModifyPropterties\res\layout\activity_main.xml，右击 TextView 控件选择【Show In】→【Propterties】命令，直接修改文本控件字体大小、背景颜色和显示颜色等属性，如图 1.15 所示，已修改源代码清单如 1.8 所示。

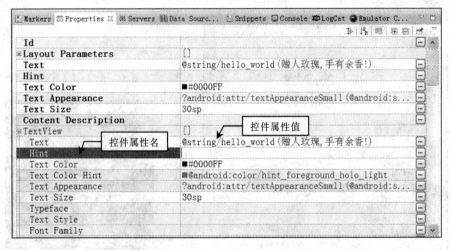

图 1.15 利用可视化方法设置文本控件的属性

源代码清单 1.8 \Ch01_ex02ModifyPropterties\res\layout\activity_main.xml

步骤4：选择【Ch01_ex02ModifyPropterties】并右击，选择【Run As】→【Android Application】命令（或直接单击工具栏中的运行按钮），程序运行结果如图1.16所示。该项目完整代码请参见配套光盘\代码\CH01\Ch01_ex02ModifyPropterties.rar。

图1.16　Ch01_ex02ModifyPropterties 结果

（2）如何修改隐藏应用程序标题栏最大化屏幕、定制显示风格和设置屏幕背景图片？

步骤1：按照项目 Ch01_ex01_HelloWorld 的步骤，利用 ADT 自动生成项目 Ch01_ex03Theme。

步骤2：双击 AndroidManifest.xml 文件→选择【Application】选项卡→单击【Theme】右边的【Browse】按钮，弹出对话框【Resource Chooser】→选择【System Resource】→输入【Theme.】→从系统预先定义的主题风格资源中选择【Theme.Light.NoTitleBar.Fullscreen】，从而在<application>标签中增加隐藏标题的主题风格（见图1.17），其自动生成如下源代码：

```
android:theme="@android:style/Theme.Light.NoTitleBar.Fullscreen"
```

步骤3：双击 activity_main.xml 文件→选择【Graphics_Layout】选项卡→在背景上右击→选择【Show In】→【Properties】命令→设置【Background】属性为@drawable/bg，其生成的代码为（事先要将准备好的背景图标 bg.jpg 复制到 drawable 目录）：

```
android:background="@drawable/bg"
```

步骤4：选择【Ch01_ex03Theme】并右击，选择【Run As】→【Android Application】命令（或直接单击工具栏中的运行按钮），程序运行结果如图1.18所示。该项目完整代码请参见配套光盘\代码\CH01\Ch01_ex03Theme.rar。

图1.17　设置隐藏标题主题风格

图1.18　隐藏标题栏风格和设置背景案例

以上案例演示了通过在 manifest 配置文件中设置主题风格来最大化屏幕，以及在布局文件中设置背景图片的 xml 方法，另外也可通过编写 Java 代码来设置，其示例源代码清单见1.9（详细项目源代码见配套光盘\代码\CH01\Ch01_ex04FULLSCREEN.rar）。

源代码清单1.9　实现隐藏标题、设置全屏和背景图片示例代码

```
1. public class MainActivity extends Activity {
2.                              //从基类 Activity 公有派生子类 MainActivity
```

```
3.    protected void onCreate(Bundle savedInstanceState) {//重写onCreate方法
4.        super.onCreate(savedInstanceState);      //调用基类或父类方法onCreate()
5.        //注意：这些方法必须放置在setContentView()方法前面，否则会出现异常现象
6.        //输入this.可智能感知可重写的方法，提高代码书写效率
7.        this.requestWindowFeature(Window.FEATURE_NO_TITLE);//设置无标题
8.        this.getWindow().setFlags(WindowManager.LayoutParams.FLAG_FULLSCREEN,
9.             WindowManager.LayoutParams.FLAG_FULLSCREEN);//设置全屏
10.       setContentView(R.layout.activity_main);
11.       this.setTitle("Android应用程序设计基础");//设置标题
12.       this.getWindow().setBackgroundDrawableResource(R.drawable.sziit);
13.                                                          //设置背景图片
14.   }
```

ADT项目自动向导也提供了实现隐藏标题栏、最大化屏幕的应用向导模板，例如，可按照项目Ch01_ex01_HelloWorld步骤自动生成Ch01_ex05FullScrActivity程序。不同之处在于：在【New Android Application】对话框的【Create Activity】中，选择【Fullscreen Activity】（见图1.19），然后修改其res/values/strings.xml文件，见源代码清单1.10（详细源代码见Ch01_ex05FullScrActivity.rar）。运行结果如图1.20所示。

源代码清单1.10 \Ch01_ex05FullScrActivity\res\values\strings.xml

```
1. <string name="app_name">Ch01_ex03 李白励志名言</string>
2. <string name="epigram_content">长风破浪会有时\n 直挂云帆济沧海</string>
```

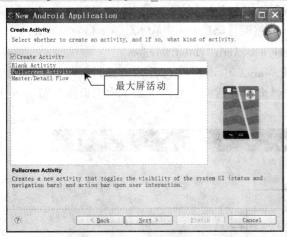

图1.19 选择创建Fullscreen Activity类型活动　　图1.20 ADT屏幕最大化模板结果

Tips 1.8 查看Android源代码的设置方法：将鼠标指针移动到想深入了解的Android类上，例如活动Activity，按住【Ctrl】键，【Activity】下面会出现链接下画线，单击此链接，如果已经配置源代码位置，则会直接跳转到相应的源代码处，否则会出现Class File Editor "Source not find"信息，点击【Attach Source】→【Source Attachment Configuration】，选择【External Location】→【External Folder】，选择源代码安装路径F:\Android\Android-Develop-Tools\android-sdk-windows\sources\android-17即可（注：具体路径取决于用户的安装路径）。

Tips 1.9 将源代码导入工作空间的方法：启动Eclipse,选择【File】→【Import】→【Android】→【Existing Android Code Into Workspace】，在弹出的【Import Projects】对话框中单击【Browse】按钮选择欲导入的工程文件，并选中【Copy projects into workspace】复选框，再单击【Finish】按钮，如图1.21所示。

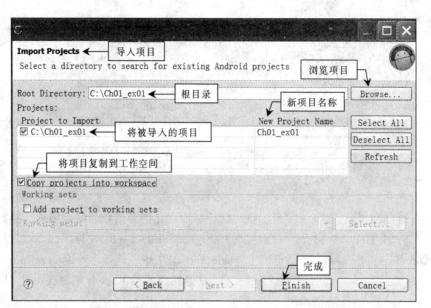

图 1.21 将已存在的项目源代码导入到工作空间

1.3.2 利用 DDMS 观察 Android 活动的生命周期

本项目重写活动方法，利用 DDMS 观察 Android 活动的运行情况，加深对 Android 活动的生命周期机制的理解，其实现步骤如下（完整源代码见 Ch01_ex06ActivityLifeCycle.rar）：

步骤 1：利用 ADT 向导生成项目 Ch01_ex06ActivityLifeCycle。

步骤 2：重写活动相关方法，先双击 MainActivity.java 打开该文件，选择【Source】→【Override/Implement Methods】命令，在弹出的对话框（见图 1.22）中选取要重写的方法，单击【OK】按钮即可。

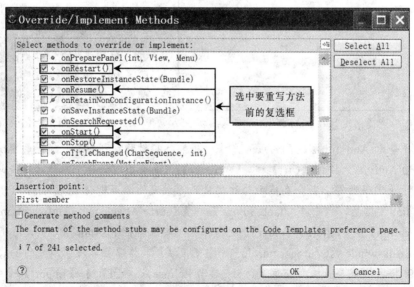

图 1.22 重写/实现方法对话框

步骤 3：重写 onCreate()、onStart()、onResume()、onPause()、onRestart()、onStop()和 onDestroy()

等方法,增加输出调试信息的Java代码,例如Log.v(TAG, "onCreate")等(见源代码清单1.11)。

源代码清单1.11   \Ch01_ex06ActivityLifeCycle\src\sziit\lihz\ch01_6\MainActivity.java

```java
1. public class MainActivity extends Activity {
2.                                //从基类Activity派生子类Main Activity
3.    private static final String TAG = "LifeCycle"; //定义日志TAG标记
4.    //@Override,表明下面的方法为从基类继承的方法,从而引导ADT检查其定义的合理性
5.    @Override
6.    protected void onCreate(Bundle savedInstanceState) {//子类重写onCreate方法
7.        super.onCreate(savedInstanceState);    //调用基类onCreate方法
8.        setContentView(R.layout.activity_main);//通过索引方法,设置屏幕布局资源
9.        Log.d(TAG, "onCreate");}    //输出日志信息
10.    protected void onStart() {//子类重写onStart方法,为Activity生命周期钩子方法
11.        super.onStart();        //调用基类onStart方法
12.        Log.d(TAG, "onStart");}//输出日志信息
13.    protected void onRestoreInstanceState(Bundle savedInstanceState) {
14.        //子类重写onRestoreInstanceState方法,非Activity生命周期钩子方法
15.        super.onRestoreInstanceState(savedInstanceState);
16.                                //调用基类onRestore InstanceState方法
17.        Log.d(TAG, "onRestoreInstanceState");}
18.    protected void onResume() {
19.                //子类重写onResume方法,为Activity生命周期钩子方法
20.        super.onResume();//调用基类onResume方法
21.        Log.d(TAG, "onResume");}    //输出日志信息
22.    protected void onPause() {
23.                //子类重写onPause方法,为Activity生命周期钩子方法
24.        super.onPause();        //调用基类onPause方法
25.        Log.d(TAG, "onPause");}    //输出日志信息
26.    protected void onRestart() {
27.                //子类重写onRestart方法,为Activity生命周期钩子方法
28.        super.onRestart();        //调用基类onRestart方法
29.        Log.d(TAG, "onRestart");}//输出日志信息
30.    protected void onSaveInstanceState(Bundle outState) {
31.            //子类重写onSave InstanceState方法,非Activity生命周期钩子方法
32.        super.onSaveInstanceState(outState);//调用基类onSaveInstanceState方法
33.        Log.d(TAG, "onSaveInstanceState");}//输出日志信息
34.    protected void onStop() {//子类重写onStop方法,为Activity生命周期钩子方法
35.        super.onStop();            //调用基类onStop方法
36.        Log.d(TAG, "onStop");}    //输出日志信息
37.    protected void onDestroy() {
38.                //子类重写onDestroy方法,为Activity生命周期钩子方法
39.        super.onDestroy();        //调用基类onDestroy方法
40.        Log.d(TAG, "onDestroy");}}//输出日志信息
```

步骤4:打开DDMS Logcat窗口,在AVD上运行此程序,可见到图1.23所示的调试信息。

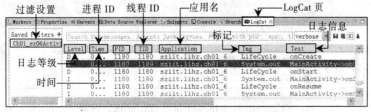

图1.23 DDMS Logcat查看活动生命周期日志

按【Ctrl+Shift+B】组合键设置断点，选择【Debug As】→【Android Application】命令，程序进入调试状态（见图1.24），按【F6】键可单步调试程序。

图 1.24　调试 Debug Android 程序

1.4　知识扩展

项目系统配置文件 AndroidManifest.xml 在 Android 应用项目中起到非常重要的作用，其中包含的各种组件的属性可以通过两种方式设置：①通过可视化法；②XML 编码法。

（1）通过可视化法：双击 Android 项目中的 AndroidManifest.xml 文件，分别选择【Manifest（清单）】【Application（应用）】【Permissions（许可）】和【Instrumentation】选项卡，设置相应组件的属性，例如设置 Application 的许可权限，如图 1.25 所示。

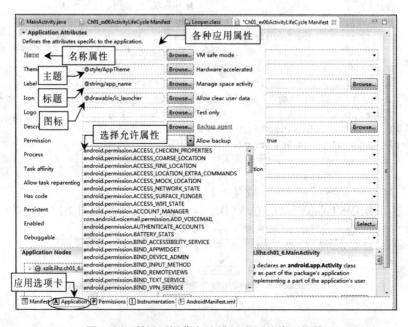

图 1.25　利用可视化方法设置应用的许可属性

（2）XML 编码法：双击 Android 项目中的 AndroidManifest.xml 文件，选择【AndroidManifest.xml】选项卡，利用 Ecplise 代码智能感知功能，在相应组件标签中输入相关属性，例如在<activity>标签中输入"android:"后，Ecplise 会自动出现可属性选列表，从中选择所需属性，再设置相关值即可，如图 1.26 所示。

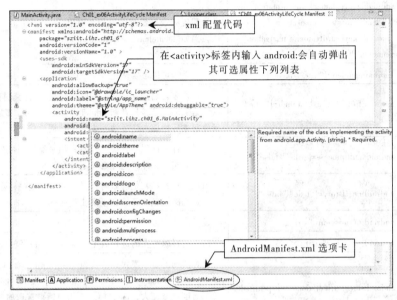

图 1.26　通过 XML 编码法设置活动的相关属性

1.4.1　活动的属性

在 Android 中，活动是作为一个对象存在的，有很多设置其 XML 属性的方法，如表 1.8 所示。

表 1.8　活动的属性定义及其功能描述

序号	XML 属性	描述
1	android:name="string"	设置当前活动类名
2	android:theme="resourceortheme"	设置活动主题样式，若没有设置，则活动的主题样式从属于应用程序，参见元素的 theme 属性
3	android:label="stringresource"	设置在桌面启动器（Launcher）中显示的名称
4	android:icon="drawableresource"	设置在桌面启动器中显示的图标
5	android:clearTaskOnLaunch=["true"|"false"]	根活动为 true 时，当用户离开任务并返回时，任务会清除直到根活动
6	android:excludeFromRecents=["true"|"false"]	是否被显示在最近打开的活动列表里，默认为 false
7	android:exported=["true"|"false"]	是否允许活动被其他程序调用
8	android:launchMode=["multiple"|"singleTop"|"singleTask"|"singleInstance"]	设置活动启动方式 "standard"、"singleTop"、"singleTask" 和 "singleInstance"，前两个为一组，后两个为一组
9	android:allowTaskReparenting=["true"|"false"]	是否允许 Activity 更换从属的任务，如从短信息任务切换到浏览器任务

续表

序 号	XML 属性	描 述
10	android:alwaysRetainTaskState=["true"|"false"]	是否保留状态不变，例如切换回 home，再重新打开，活动处于最后状态；若用户不希望丢失中间状态，此属性应设置为 true
11	android:configChanges=[oneormoreof:"mcc""mnc""locale""touchscreen""keyboard""keyboardHidden""navigation""orientation""fontScale"]	当配置列表被修改时，是否调用 onConfigurationChanged()方法，例如"locale|navigation|orientation"
12	android:enabled=["true"|"false"]	活动是否可以被实例化
13	android:finishOnTaskLaunch=["true"|"false"]	当用户重新启动这个任务的时候，是否关闭已打开的活动，默认是 false；若该属性和 allowTaskReparenting 都是 true，这个属性起主导作用，活动的 taskAffinity 属性将被忽略
14	android:multiprocess=["true"|"false"]	是否允许多进程，默认是 false；用来配置是否在多个进程中有不同的实例；配置进程之间是否为单例
15	android:noHistory=["true"|"false"]	当用户切换到其他屏幕时，是否需要从栈中移除这个活动。默认是 false，活动不会留下历史痕迹
16	android:permission="string"	设置活动许可名
17	android:process="string"	活动运行时所在的进程名（默认名应用程序包名）
18	android:screenOrientation=["unspecified"|"user"|"behind"|"landscape"|"portrait"|"sensor"|"nonsensor"]	设置活动的显示模式；默认为 unspecified：由系统自动判断显示方向；landscape 横屏模式，宽度比高度大；portrait 竖屏模式，高度比宽度大；user 模式，用户当前首选的方向；behind 模式：在活动堆栈中与该活动下面的那个活动方向一致；sensor 模式：由物理传感器来决定，若用户旋转设备则屏幕会横竖屏切换；nosensor 模式：忽略物理传感器
19	android:stateNotNeeded=["true"|"false"]	活动被销毁或者成功重启时是否保存状态
20	android:taskAffinity="string"	活动的亲属关系，默认同一个应用程序下的活动有相同关系
21	android:windowSoftInputMode=[oneormoreof:"stateUnspecified""stateUnchanged""stateHidden""stateAlwaysHidden""stateVisible""stateAlwaysVisible""adjustUnspecified""adjustResize""adjustPan"]	设置活动主窗口与软键盘的交互模式（软键盘弹出的模式），用来避免输入法面板遮挡或隐藏问题；stateUnspecified：未指定软键盘状态时，系统将选择合适状态或依赖设置的主题；stateUnchanged：当该活动出现时，软键盘一直保持在上一个活动里的状态，无论是隐藏还是显示；stateHidden：用户选择活动时，软键盘总被隐藏；stateAlwaysHidden：当该活动主窗口获取焦点时，软键盘总被隐藏；stateVisible：软键盘通常是可见的；stateAlwaysVisible：用户选择活动时，软键盘总是显示状态；adjustUnspecified：默认设置，通常由系统自行决定是隐藏还是显示；adjustResize：该活动总是调整屏幕大小以便留出软键盘空间；adjustPan：当前窗口内容将自动移动以便当前焦点从不被键盘覆盖和用户总能看到输入的内容

1.4.2 应用的属性

　　Android 很关键的核心特点在于：如果一个应用允许，其他应用能够使用该应用的元素，且其他应用既不需要包含该应用代码，也不要连接该应用代码，只是当其他应用需要该应用的代码时，就启动该应用的相应代码（不是让该应用全部启动）。为了达到此目的，当一个应用的任何部分被需要时，系统必须能启动这个应用进程，并且将这个部分实例化成 Java 对象。因此，和其他大多数操作系统不同的是，Android 应用程序没有一个单独的程序入口（例如，没有主函数 main()），相反，Android 应用有必要的组件以便需要时系统能实例化并运行它。Android 中有四种组件：活动、服务、广播接收器、内容提供者。无论何时，一个请求都应该由一个特定的组件来处理，Android 系统来确保包含这个组件的应用进程运行，如果需要就启动它，并为这个组件创建一个实例。一些应用的属性可以用来控制和修改这些行为，如表 1.9 所示。

表 1.9 应用的属性定义及其功能描述

序 号	XML 属性	描 述
1	android:allowClearUserData= ["true" | "false"]	用户是否能自行清除数据，默认为 true；程序管理器包含允许用户清除数据选项；当为 true 时，用户可自己清理数据
2	android:allowTaskReparenting= ["true" | "false"]	是否允许活动更换从属的任务，例如从彩信任务切换到浏览器任务
3	android:backupAgent="string"	设置该应用的备份，属性值应是一个完整的类名，且必须指定类名；此属性没有默认值
4	android:debuggable=["true"| "false"]	设置是否允许调试应用程序；当设置为 true 时，表明该应用在手机上可以被调试。默认为 false，在 false 的情况下调试该应用会报错
5	android:description="string resource"	用于描述获取该许可的程序可做哪些事情，以让用户知道如果同意程序获取该权限，该程序可做什么；通常用两句话来描述许可：第一句描述该许可，第二句警告用户如果批准该权限会可能有什么不好的事情发生
6	android:enabled=["true"| "false"]	Android 系统是否能实例化该应用程序的组件；如果为 true，每个组件的 enabled 属性决定哪个组件是否可被 enabled；如果为 false，它覆盖组件指定的值；所有组件都是 disabled
7	android:hasCode=["true"| "false"]	表示该应用是否包含任何代码，默认为 true；若为 false，则系统在运行组件时，不会去尝试加载任何应用代码
8	android:icon="drawable resource"	声明整个应用的图标，图片一般都放在 drawable 文件夹下
9	android:killAfterRestore= ["true" | "false"]	在整个系统恢复出厂设置期间，应用程序设置重建时是否终止应用程序。单一的应用程序恢复设置时通常不会引起程序的终止。整个系统恢复出厂设置通常只在第一次使用手机时才会发生。第三方应用程序通常不需要设置此属性
10	android:label="string resource"	该属性是为许可提供的，均为字符串资源，当用户去看许可列表（android:label）时，该字符串资源就可显示给用户；label 应尽量简短，只需告知用户该许可是在保护什么功能即可；用户可见的关于应用程序的总体描述
11	android:manageSpaceActivity= "string"	一个系统可启动让用户来管理该应用程序的内存使用的活动子类的完整类名。该活动必须在 <activity> 标签中声明
12	android:name="string"	为应用程序所实现的 Application 子类的全名；当应用程序进程开始时，该类在所有应用程序组件之前被实例化；若该类（如 mainApp 类）直接声明在 package 下，则可直接声明 android:name="mainApp"，若该类是在 package 下面的子包，则必须声明为全路径或 android:name="package 名称.子包名称.mainApp"
13	android:permission="string"	设置许可名，该属性若在 <application> 上定义，是一个给应用程序的所有组件设置许可的便捷方式，当然它是被各组件设置的许可名覆盖的

续表

序号	XML 属性	描述
14	android:persistent=["true"| "false"]	该应用程序是否应该在任何时候都保持运行状态，默认为 false；因为应用程序通常不应该设置本标识，持续模式仅仅应该设置给某些系统应用程序才是有意义的
15	android:process="string"	应用程序运行的进程名，默认值为 <manifest>元素中设置的包名，当然每个组件都可通过设置该属性来覆盖默认值；若希望两个应用程序共用一个进程，则可设置它们的 android:process 相同，但前提条件是它们共享一个用户 ID 且被赋予相同证书
16	android:restoreAnyVersion=["true" | "false"]	用来表明应用是否准备尝试恢复所有的备份，甚至该备份是比当前设备上更新的版本，默认是 false
17	android:taskAffinity="string"	拥有相同的 affinity 的活动理论上属于相同的任务，应用程序默认的 affinity 的名字是 <manifest>元素中设置的 package 名
18	android:theme="resource or theme"	设置资源风格，定义一个默认的主题风格给所有的活动，当然也可在自己的主题 theme 中设置，类似风格 style

1.4.3 手动创建活动的方法

在创建 Android 项目时，ADT 会自动创建一个默认的活动，但是，如何手动创建活动呢？下面详细说明手动创建活动的步骤。

步骤 1：在加载了 Android 项目的 Eclipse 中，选择【File】→【New】→【Class】命令，或者单击工具栏中的按钮，弹出图 1.27 所示的【New Java Class】对话框。

步骤 2：在该对话框中选择源文件夹、包，并输入活动名称，然后单击【Superclass】文本框后面的【Browse】按钮，在弹出的【Superclass Selection】对话框中找到 android.app.Activity 基类，如图 1.28 所示。

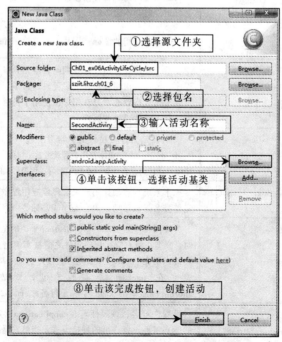

图 1.27 【New Java Class】对话框

图 1.28 【Superclass Selection】对话框

步骤 3：单击【Superclass Selection】对话框中的【OK】按钮，返回【New Java Class】对话框，单击【Finish】按钮，即可创建一个活动，已创建的活动及其源代码清单 1.12 如下，然后按图 1.22 所示活动重写/实现方法增加其类成员函数。

源代码清单 1.12　SecondActivity

```
1. package sziit.lihz.ch01_6;//生命包名
2. import android.app.Activity;//引入 Android 应用活动包
3. public class SecondActivity extends Activity {//从基类 Activity 公有派生子类
4. }
```

步骤 4：务必记住在 AndroidManifest.xml 中注册新建的 SecondActivity 活动，否则将会抛出异常信息：ActivityNotFoundException。具体代码如下：

`<activity android:name="sziit.lihz.ch01_6.SecondActivity"></activity>`

1.4.4　启动和关闭活动的方法

1. 启动活动的方法

从一个活动启动另一个活动主要包括两种方法：①显式启动；②隐式启动。启动活动的方法主要通过意图，分为显式和隐式意图两种。

1）显式启动活动

显式启动活动分成三种启动情形：①类启动；②包名.类名启动；③组件名（ComponentName）启动。

（1）类启动

使用情形：应用内部跳转，频率较高，示例代码如下：

```
1. Intent intent = new Intent(MainActivity.this, SecondActivity.class);
   //新建意图对象 intent
2. intent.putExtra("key","value"); //启动第二个活动时传的参数
3. startActivity(intent);//启动活动，完成从主活动到第二个活动的跳转
```

以上意图构造器包含两个参数：第一个参数是上下文（Context）类型，因活动类是 Context 类的子类，所以这里使用 MainActivity.this；第二个参数是 Class 类型，它指向一个应用程序

组件，系统会把 Intent 对象发送给这个应用程序组件。

（2）包名.类名启动

使用情形：内部跳转+外部跳转，示例代码如下：

```
1. Intent intent= new Intent(); //新建意图对象intent
2. intent.setFlags(Intent.FLAG_ACTIVITY_NEW_TASK);//设置活动启动模式
//设置另一个活动的包名和类名
3. intent.setClassName("sziit.lihz.ch01_1","sziit.lihz.ch01_1.SecondActivity");
4. intent.putExtra("key","value"); //向活动SecondActivity传递一些参数
5. startActivity(intent);
```

（3）组件名启动

使用情形：内部跳转+外部跳转，示例代码如下：

```
1. Intent intent = new Intent();//新建意图对象intent
2. intent.setComponent(new ComponentName("sziit.lihz.ch01_1", sziit.lihz.
   ch01_1.SecondActivity "));
3. intent.putExtra("key","value"); //向活动SecondActivity传递一些参数
4. startActivity(intent);//启动新活动
```

2）隐式启动活动

使用情形：内部跳转+外部跳转，方便定制该 action 跳转，示例代码如下：

```
//该参数为要启动的活动中定义的行动名action,该action已在AndroidManifest.xml中预先定义好
1. Intent intent = new Intent("sziit.lihz.ch01_3.MainActivity.action");
2. intent.putExtra("key","value"); //设置跳转时传递的参数
3. startActivity(intent); //启动新活动
```

注意事项：在 AndroidManifest.xml 中为要跳转的活动增加属性 android:exported="true"，允许外部跳转，此时能够跳转成功；否则会遇到权限问题，程序运动时会出现异常。

2. 关闭活动的方法

Android 开发中关闭活动的几种常用方法包括：

1）finish()方法

该方法是活动类的方法，其仅针对该活动。当调用 finish()方法时，其只是将该活动推向后台，并没有立即释放内存，该活动的资源并没有从活动栈中被清理。调用 finish()方法会执行 Activity.onDestroy()方法，以结束该活动生命周期。

2）Dalvik 虚拟机（VM）本地方法

（1）android.os.Process.killProcess(android.os.Process.myPid())：通过进程 API 获取该程序的进程 PID，然后将其杀掉，但结束其他进程有权限限制。

（2）System.exit(0)：该方法是 Java 程序的标准退出法，返回值为 0 代表正常退出，可在 onDestory()方法中调用，其示例代码如下：

```
1. protected void onDestroy() { //派生类重写销毁方法onDestroy
2.   super.onDestroy();           //调用基类或父类方法onDestroy
3.   System.exit(0);//调用Java方法直接结束程序或者调用以下killProcess方法杀掉进程
4.   //android.os.Process.killProcess(android.os.Process.myPid());
5. }
```

1.4.5 扩展项目实训

本小节通过两个扩展项目实训，强化本章学习目标和要掌握的重点及难点知识。

1. 两个活动相互切换

（1）项目实训目标

① 掌握利用 ADT 创建 Android 应用项目的操作方法；
② 掌握手动创建活动类的方法；
③ 掌握利用重写/实现对话框生成活动生命周期接口的方法；
④ 掌握利用 DDMS LogCat 输出日志观察活动生命周期的方法；
⑤ 掌握从一个活动启动另一个活动及关闭活动的方法

该项目完整源代码参见 Ch01_ex07FirstToSecondActivity.rar。

（2）项目实施步骤

步骤1：根据1.3.1节介绍的方法操作，利用 ADT 创建项目 Ch01_ex07FirstToSecondActivity，在【New Android Application】→【Blank Activity】自动向导页面，将活动名设为 FirstActivity，屏幕布局文件设置为 first_layout.xml，单击【Finish】按钮，如图1.29所示。FirstActivity.java 和 first_layout.xml 文件源代码清单分别如1.13和1.14所示。

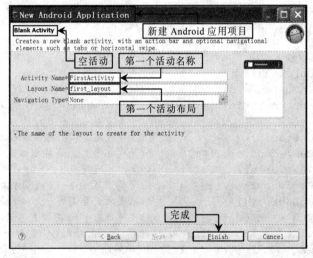

图 1.29　定制新建项目活动和布局

源代码清单 1.13　FirstActivity.java

```
1.  public class FirstActivity extends Activity {
                            //从基类Activity公有派生子类FirstActivity
2.    private static final String TAG="FirstActivity";//定义私有、静态和只读字符串
3.    /** 当活动创建时首先被调用 */
4.    public void onCreate(Bundle savedInstanceState) {//子类重写onCreate方法
5.      super.onCreate(savedInstanceState);     //基类或父类调用onCreate方法
6.      setContentView(R.layout.first_layout);  //设置第一个活动的屏幕布局
7.      //按住Ctrl键，移动鼠标，出现链接线，单击此链接可跳转到屏幕布局资源定义文件
8.      Log.d(TAG, "onCreate(1)");   //使用日志Log类输出调试信息
9.      System.out.println("FirstActivity->onCreate(1)");
                              //采用Java经典方法输出调试信息
10.     Button btnStart = (Button)findViewById(R.id.startBtn);
                              //通过按钮资源ID绑定按钮控件
11.     //按住Ctrl键，移动鼠标，出现链接线，单击此链接可跳转到按钮资源定义文件
12.     btnStart.setOnClickListener(new View.OnClickListener() {
```

```
                                              //设置按钮控件的单击监听接口
13.     public void onClick(View v) {//重写该按钮的单击回调事件处理方法
14.       //构造意图对象 intent,第一个参数为当前上下文 FirstActivity.this,
15.       //第二个参数为希望启动的活动类名 SecondActivity.class
16.       Intent intent = new Intent(FirstActivity.this,SecondActivity.
          class);
17.       startActivity(intent); //启动意图设置的活动 SecondActivity
18.     }});  }
19.   protected void onStart() {         //子类重写 onStart 方法
20.     super.onStart();                 //基类或父类调用 onStart 方法
21.     Log.d(TAG, "onStart(1)");        //使用日志 Log 类输出调试信息
22.     System.out.println("FirstActivity->onStart(1)");}
23.   protected void onRestart() {       //子类重写 onRestart 方法
24.     super.onRestart();               //基类或父类调用 onRestart 方法
25.     Log.d(TAG, "onRestart(1)");      //使用日志 Log 类输出调试信息
26.     System.out.println("FirstActivity->onRestart(1)");}
27.   protected void onResume() {        //子类重写 onResume 方法
28.     super.onResume();                //基类或父类调用 onResume 方法
29.     Log.d(TAG, "onResume(1)");       //使用日志 Log 类输出调试信息
30.     System.out.println("FirstActivity->onResume(1)");}
31.   protected void onPause() {         //子类重写 onPause 方法
32.     super.onPause();                 //基类或父类调用 onPause 方法
33.     Log.d(TAG, "onPause(1)");        //使用日志 Log 类输出调试信息
34.     System.out.println("FirstActivity->onPause(1)");}
35.   protected void onStop() {          //子类重写 onStop 方法
36.     super.onStop();                  //基类或父类调用 onStop 方法
37.     Log.d(TAG, "onStop(1)");         //使用日志 Log 类输出调试信息
38.     System.out.println("FirstActivity->onStop(1)");}
39.   protected void onDestroy() {       //子类重写 onDestroy 方法
40.     super.onDestroy();               //基类或父类调用 onDestroy 方法
41.     Log.d(TAG, "onDestroy(1)");      //使用日志 Log 类输出调试信息
42.     System.out.println("FirstActivity->onDestroy(1)");}
43.   protected void onRestoreInstanceState(Bundle savedInstanceState){
      //子类重写该方法
44.     super.onRestoreInstanceState(savedInstanceState);
                                  //基类调用 onRestoreInstanceState 方法
45.     Log.d(TAG, "onRestoreInstanceState(1)"); //使用日志 Log 类输出调试信息
46.     System.out.println("FirstActivity->onRestoreInstanceState(1)");}
47.   protected void onSaveInstanceState(Bundle outState) {
                                  //子类重写 onSaveInstanceState 方法
48.     super.onSaveInstanceState(outState);//基类或父类调用 onSaveInstanceState 方法
49.     Log.d(TAG, "onSaveInstanceState(1)"); //使用日志 Log 类输出调试信息
50.     System.out.println("FirstActivity->onSaveInstanceState(1)");}}
```

源代码清单 1.14   first_layout.xml

```
1. <?xml version="1.0" encoding="utf-8"?>    ← xml 版本和编码方式
2. <LinearLayout xmlns:android="http://schemas.android.com/apk/res/android"
3.     android:layout_width="fill_parent"    ← 布局宽度        ← xml 命名空间
4.     android:layout_height="fill_parent"   ← 布局宽度
5.     android:orientation="vertical"        ← 布局方向
```

```
6.      android:gravity="center_horizontal" >    ← 布局位置
7.      <Button    ← 按钮控件标签
8.          android:id="@+id/startBtn"    ← 按钮控件ID
9.          android:layout_width="wrap_content"    ← 按钮控件宽度
10.         android:layout_height="wrap_content"    ← 按钮控件高度
11.         android:gravity="center_horizontal|top"    ← 按钮控件位置
12.         android:text="@string/StartSecondActivity"/>    ← 按钮显示文本
13. </LinearLayout>    ← 线性布局标签
```

步骤2：根据1.4.3节手动创建活动的方法创建活动 SecondActivity 类和布局文件 second_layout.xml（分别见源代码清单 1.15 和 1.16）。AndroidManifest.xml 中两个活动的注册源代码清单见 1.17。

源代码清单 1.15　SecondActivity.java

```
1.  public class SecondActivity extends Activity {
                            //从基类或父类 Activity 公有派生子类 SecondActivity
2.      private static final String TAG="SecondActivity";
                            //定义私有、静态和最终(只读)字符串
3.      //该活动创建时被首先调用
4.      protected void onCreate(Bundle savedInstanceState) {
                            //子类重写 onCreate 方法
5.          super.onCreate(savedInstanceState);//基类或父类调用 onCreate 方法
6.          setContentView(R.layout.second_layout);//设置第二个活动的屏幕布局
7.          //按住 Ctrl 键，移动鼠标，出现链接线，单击此链接可跳转到屏幕布局资源定义文件
8.          Log.d(TAG, "onCreate(2)");    //使用日志 Log 类输出调试信息
9.          System.out.println("SecondActivity->onCreate(2)");
                            //采用 Java 经典方法输出调试信息
10.         Button btnSecondStart = (Button)findViewById(R.id.SecondBtnStart);
                            //通过资源 ID 绑定控件
11.         btnSecondStart.setOnClickListener(new View.OnClickListener() {
                            //设置该按钮单击监听接口
12.             public void onClick(View v) {//重写该按钮的单击回调事件处理方法
13.                 //构造意图对象 intent,第一个参数为当前上下文 SecondActivity.this,
14.                 //第二个参数为希望启动的活动类名 FirstActivity.class
15.                 Intent intent = new Intent(SecondActivity.this,First
                    Activity.class);
16.                 startActivity(intent); //启动意图设置的活动 FirstActivity}});
17.         Button btnSecondClose= (Button)findViewById(R.id.SecondBtnClose);
                            //通过资源 ID 绑定按钮
18.         btnSecondClose.setOnClickListener(new View.OnClickListener() {
                            //设置该按钮单击监听接口
19.             public void onClick(View v) {    //重写该按钮的单击回调事件处理方法
20.                 finish();//关闭该活动 SecondActivity}});}
21.     protected void onRestoreInstanceState(Bundle savedInstanceState) {
                            //子类重写方法
22.         super.onRestoreInstanceState(savedInstanceState); //基类或父类调用方法
23.         Log.d(TAG, " onRestoreInstanceState (2)"); //使用日志 Log 类输出调试信息
24.         System.out.println("SecondActivity-> onRestoreInstanceState (2)");}
25.     protected void onStart() {    //子类重写 onStart 方法
26.         super.onStart();    //基类或父类调用 onStart 方法
```

```
27.     Log.d(TAG, " onStart (2)");          //使用日志 Log 类输出调试信息
28.     System.out.println("SecondActivity-> onStart (2)");}
29. protected void onRestart() {              //子类重写 onRestart 方法
30.     super.onRestart();                    //基类或父类调用 onRestart 方法
31.     Log.d(TAG, " onRestart (2)"); System.out.println("SecondActivity->
        onRestart (2)");}
32. protected void onResume() {               //子类重写 onResume 方法
33.     super.onResume();                     //基类或父类调用 onResume 方法
34.     Log.d(TAG, " onResume (2)");          //使用日志 Log 类输出调试信息
35.     System.out.println("SecondActivity-> onResume (2)");}
36. protected void onSaveInstanceState(Bundle outState) {//子类重写方法
37.     super.onSaveInstanceState(outState);  //基类或父类调用方法
38.     Log.d(TAG, " onSaveInstanceState (2)"); //使用日志 Log 类输出调试信息
39.     System.out.println("SecondActivity-> onSaveInstanceState (2)");}
40. protected void onPause() {                //子类重写 onPause 方法
41.     super.onPause();                      //基类或父类调用 onPause 方法
42.     Log.d(TAG, " onPause (2)");           //使用日志 Log 类输出调试信息
43.     System.out.println("SecondActivity-> onPause (2)");}
44. protected void onStop() {                 //子类重写 onStop 方法
45.     super.onStop();                       //基类或父类调用 onStop 方法
46.     Log.d(TAG, " onStop (2)");            //使用日志 Log 类输出调试信息
47.     System.out.println("SecondActivity-> onStop (2)");}
48. protected void onDestroy() {              //子类重写 onDestroy 方法
49.     super.onDestroy();                    //基类或父类调用 onDestroy 方法
50.     Log.d(TAG, " onDestroy (2)");         //使用日志 Log 类输出调试信息
51.     System.out.println("SecondActivity-> onDestroy (2)");}}
```

源代码清单 1.16 second_layout.xml

```
1. <?xml version="1.0" encoding="utf-8"?>         ← xml 版本和编码方式
2. <LinearLayout xmlns:android="http://schemas.android.com/apk/res/android"
3.     android:layout_width="fill_parent"   ← 布局宽度        ← xml 命名空间
4.     android:layout_height="fill_parent"  ← 布局高度
5.     android:orientation="vertical"       ← 布局方向
6.     android:gravity="center_horizontal" > ← 布局位置
7.     <Button                              ← 按钮控件标签
8.         android:id="@+id/SecondBtnStart" ← 按钮控件 ID
9.         android:layout_width="wrap_content"  ← 按钮控件宽度
10.        android:layout_height="wrap_content" ← 按钮控件宽度
11.        android:text="@string/StartFirstActivity" />  ← 按钮显示文本
12.    <Button  ← 按钮控件标签
13.        android:id="@+id/SecondBtnClose" ← 按钮控件 ID
14.        android:layout_width="wrap_content"  ← 按钮控件宽度
15.        android:layout_height="wrap_content" ← 按钮控件高度
16.        android:text="@string/CloseSecondActivity" /> ← 按钮显示文本
17. </LinearLayout>  ← 线性布局标签
```

源代码清单 1.17 AndroidManifest.xml 中两个活动的注册

```
1. <?xml version="1.0" encoding="utf-8"?>
2. <!-- 说明 xml 版本和编码方法 -->
```

```xml
3.   <!-- 根标签为 manifest-->
4.   <!-- package="sziit.lihz.ch01_7"属性: 配置应用包名-->
5.   <!-- android:versionCode="1"属性: 配置应用程序版本代码 -->
6.   <!-- android:versionName="1.0"属性:配置应用程序名称 -->
7.   <manifest xmlns:android="http://schemas.android.com/apk/res/android"
8.       package="sziit.lihz.ch01_7" android:versionCode="1" android:versionName="1.0" >
9.       <!-- 配置应用程序所需最小 Android 系统版本 -->
10.      <uses-sdk android:minSdkVersion="17" />
11.      <!-- 应用程序配置标签为:application -->
12.      <!-- android:icon="@drawable/ic_launcher"属性: 配置应用程序的图标 -->
13.      <!-- android:label="@string/app_title"属性: 配置应用程序的标签 -->
14.      <!-- android:theme="@android:style/Theme.Light"属性: 配置应用主题风格 -->
15.      <application android:icon="@drawable/ic_launcher" android:label="@string/app_title"
16.          android:theme="@android:style/Theme.Light">
17.          <!-- 活动配置标签为: activity，注册第一个活动 FirstActivity-->
18.          <!--android:name="sziit.lihz.ch01_7.FirstActivity"属性: 配置该活动类名 -->
19.          <!--android:label="@string/first_activity_title"属性: 配置该活动的标题 -->
20.          <activity android:name="sziit.lihz.ch01_7.FirstActivity"
21.              android:label="@string/first_activity_title" >
22.              <!-- 意图过滤标签: intent-filter-->
23.              <intent-filter>
24.                  <!--该属性设置: 该活动为 android 主启项,定应用程序最先启动该 Activity，
                        是程序的活动入口 -->
25.                  <action android:name="android.intent.action.MAIN" />
26.                  <!-- 该属性: 决定应用程序是否显示在程序列表里 -->
27.                  <category android:name="android.intent.category.LAUNCHER" />
28.              </intent-filter>    </activity>
29.          <!-- 活动配置标签为: activity, 注册第二个活动 SecondActivity-->
30.          <!--android:name="sziit.lihz.ch01_7.SecondActivity"属性: 配置该活动类名 -->
31.          <!--android:label="@@string/second_activity_title"属性: 配置该活动的标题 -->
32.          <activity android:name="sziit.lihz.ch01_7.SecondActivity"
33.              android:label="@string/second_activity_title" />
34.      </application>    </manifest>
```

步骤3：根据图1.22所示的重写/实现方法对话框，分别给FirstActivity类和SecondActivity类添加并重写活动生命周期方法 onCreate()、onStart()、onResume()、onPause()、onRestart()、onStop()、onDestroy()等。

步骤4：分别给 first_layout.xml 和 second_layout.xml 布局中的按钮添加 OnClickListener 监听器，增加 onClick()事件回调方法。

步骤5：在生命周期方法 onCreate()、onResume()、onPause()、onRestart()、onStop()、onDestroy()等中增加 Log 类输出调试信息或 Java 经典方法输出调试信息。

步骤6：启动 Android 虚拟设备，运行该项目，运行界面如图1.30所示。DDMS LogCat 日志界面如图1.31所示。该程序启动出现第一个活动界面后，若有其他活动启动，此时若单击 Back 按钮，可观察到活动1停止后被恢复的过程，如图1.32所示。

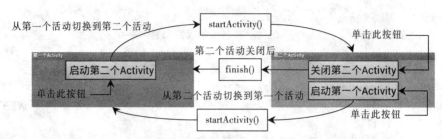

图 1.30　两个活动相互切换实训项目运行界面和切换过程

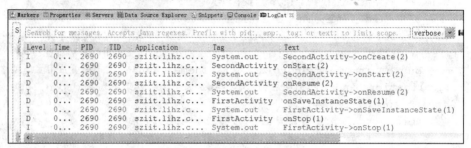

图 1.31　通过 DDMS Logcat 观察两个活动相互切换生命周期回调方法执行过程

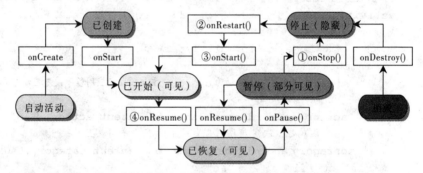

图 1.32　第一个活动停止和恢复时生命周期回调方法执行过程

（3）项目小结

① 活动的基类为 android.app.Activity，可以通过 ADT 自动向导和手动方式创建子活动，通过调用 public void setContentView(int layoutResID) 方法设置用户界面布局，提供与用户之间的交互。

② 可利用重写/实现对话框生成活动生命周期接口的方法 onCreate()（创建）、onStart()（启动）、onResume()（恢复）、onPause()（暂停）、onRestart()（重启）、onStop()（停止）、onDestroy()（销毁）等，然后通过 android.util.Log 日志类方法 Log.v()、Log.d()、Log.e()、Log.i()、Log.w() 等或 Java 经典方法 System.out.println() 输出日志信息，再利用 DDMS Logcat 工具可以观察活动栈管理活动生命周期的执行情况。

③ 可通过意图构造器 public Intent(Context packageContext, Class<?> cls) 传递两个参数构造意图对象，第一个参数为应用程序的上下文，第二个参数为活动包名，调用 public void startActivity(Intent intent) 方法启动活动，调用 void android.app.Activity.finish() 方法关闭活动。

④ 活动必须在 AndroidManifest.xml 中注册，必须掌握意图过滤标签 <intent-filter> 中设置的以下两个属性：

a. `<action android:name="android.intent.action.MAIN" />`，该属性决定该活动为 android 主启项，应用程序最先启动该 Activity，是程序的活动入口。

b. `<category android:name="android.intent.category.LAUNCHER"/>`，该属性决定应用程序是否显示在程序列表中或主屏幕桌面上。

2. 定制 Android 应用全局入口

通过本项目实训可了解到 Android 应用的真正入口为 Application 对象的 onCreate()方法，而非活动的 onCreate()方法，达到以下实训目标，熟练掌握相应的操作编程技巧。

（1）项目实训目标

① 掌握利用 ADT 创建 Android 应用项目的操作方法；
② 掌握手动创建 Android 应用类的方法；
③ 掌握手动创建活动类的方法；
④ 掌握利用重写/实现对话框生成活动生命周期接口的方法；
⑤ 掌握利用 DDMS LogCat 输出日志观察应用全局入口为 Application 的 onCreate()方法；
⑥ 掌握利用 Android 应用对象保持全局变量的方法；
⑦ 掌握从一个活动启动另一个活动的方法及其关闭活动的方法。

该项目完整源代码参见附件包：Ch01_ex08MainApplicationExam.rar。

（2）项目实施步骤

步骤 1：根据 1.3.1 节的方法，利用 ADT 创建项目 Ch01_ex08MainApplicationExam，类似扩展项目实训 1，将活动名设为 MainActivity，屏幕布局文件设置为 activity_main.xml，单击【Finish】按钮。

步骤 2：根据 1.4.3 节手动创建活动的方法创建 LoginActivity 类和布局文件 login_main.xml，并在 AndroidManifest.xml 中注册活动 LoginActivity。

步骤 3：根据图 1.22 所示的重写/实现方法对话框，分别给 MainActivity 类和 LoginActivity 类添加并重写活动生命周期方法 onCreate()、onStart()、onResume()、onPause()、onRestart()、onStop()、onDestroy()等。

步骤 4：类似 1.4.3 节手动创建活动类的方法创建 android.app.Application 类的子类 MainApp，并在 AndroidManifest.xml 中的`<application>`标签增加属性 android:name=".MainApp"，配置自定义的应用 MainApp；根据图 1.22 所示的重写/实现方法对话框，添加 onCreate()方法回调等，定义全局变量 m_isLogin 记录全局状态，见源代码清单 1.18。

源代码清单 1.18 /Ch01_ex08MainApplicationExam/src/sziit/lihz/ch01_8/MainApp.java

```
1. public class MainApp extends Application {//从应用基类 Application 派生子应用类
2.     private static final String TAG = "MainApp"; private boolean m_isLogin =
       true; //定义全局变量
3.     public void onCreate() {   //子类重写基类方法 onCreate
4.         super.onCreate();          //基类或父类调用 onCreate
5.         Log.v(TAG, "onCreate()");System.out.println("MainApp->onCreate()");}
6.     ......
7.     public void unregisterActivityLifecycleCallbacks(
8.         ActivityLifecycleCallbacks callback) {//子类重写基类方法
9.         super.unregisterActivityLifecycleCallbacks(callback);
10.        System.out.println("MainApp->unregisterActivityLifecycleCallbacks()");}}
```

步骤 5：在 MainActivity 类的 onCreate(Bundle savedInstanceState)方法中根据全局变量 m_isLogin 的状态决定程序的跳转流程，其关键代码见源代码清单 1.19。

源代码清单 1.19　\Ch01_ex08MainApplicationExam\src\sziit\lihz\ch01_8\MainActivity.java

```
1.  public class MainActivity extends Activity {
2.      private static final String TAG = "MainActivity";
3.      protected void onCreate(Bundle savedInstanceState) {
4.          super.onCreate(savedInstanceState);
5.          ......
6.          MainApp myApp = (MainApp) this.getApplicationContext();
                                                          //获取全局应用对象
7.          if(!myApp.isLogin()) {
8.              //构造意图对象 intent,第一个参数为当前上下文 FirstActivity.this,
9.              //第二个参数为希望启动的活动类名 SecondActivity.class
10.             Intent intent = new Intent(MainActivity.this,LoginActivity.class);
11.             startActivity(intent);//启动意图设置的活动 SecondActivity
12.             Log.v(TAG, "重定向到登录界面！");
13.         } else { Intent intent = new Intent(Intent.ACTION_VIEW, Uri.parse
                ("http://www.baidu.com"));
14.             intent.setClassName("com.android.browser", "com.android.browser.
                BrowserActivity");
15.             startActivity(intent);
16.             Log.v(TAG, "重定向到网页！");    }}}
```

步骤 6：在生命周期方法 onCreate()、onResume()、onPause()、onRestart()、onStop()、onDestroy() 和 MainApp 类的 onCreate()等方法中增加 Log 类输出调试信息或 Java 经典方法输出调试信息。

步骤 7：启动 Android 虚拟设备（AVD），运行该项目，运行界面如图 1.33 所示。利用 DDMS LogCat 工具可观察到程序的执行次序为："MainApp->onCreate()"→"MainActivity-> onCreate()"，先调用应用对象的 onCreate()方法，再调用活动的 onCreate()方法。

MainActivity

LoginActivity

图 1.33　定制 Android 应用全局入口项目运行界面

（3）项目小结

① 应用的基类为 android.app.Application，重写该类的 onCreate()方法，可在应用对象中保持 Android 应用程序的全局变量，便于控制程序的流程。

② 掌握定制 Android 应用对象的方法，应用对象的 onCreate()方法才是程序真正的入口函数。

③ 通过 getApplicationContext()或 getApplication()方法可获得全局应用对象的实例，从而在 Android 应用程序的不同组件中获得全局变量值，实现全局数据在不同活动组件、服务组件等之间的全局共享。

④ 掌握构造意图对象的方法和活动之间的跳转方法 startActivity(intent)。

本章小结

本章主要介绍了 Android 的基本概念和系统构架知识、安卓活动和意图的基本概念和作用、安卓应用和上下文的基本概念和作用，重点讲解了 Android 应用程序的框架模型，通过项目实例的实施让用户掌握搭建 Android 开发环境、构建 Android 应用程序、创建 Android 活动、定制 Android 应用程序的启动入口，以及运行和调试 Android 应用程序的基本方法和应用技巧。

强化练习

1. 填空题

（1）Android 四大核心组件分别是（　　）、（　　）、（　　）和（　　）。
（2）活动的四种状态分别是（　　）、（　　）、（　　）和（　　）。
（3）活动生命周期的三种类型分别是（　　）、（　　）和（　　）。
（4）活动的生命周期事件回调函数包括（　　）、（　　）、（　　）、（　　）、（　　）、（　　）和（　　）。
（5）活动的四种加载模式分别是（　　）、（　　）、（　　）和（　　）。
（6）活动运行时首先执行的生命周期事件回调函数为（　　）。
（7）活动销毁时最后执行的生命周期事件回调函数为（　　）。
（8）Dalvik 虚拟机调试监控服务的简写为（　　）。
（9）安卓开发工具的简写为（　　）。
（10）可扩展标记语言的简写为（　　）。
（11）Java 开发工具的简写为（　　）。
（12）软件开发工具包的简写为（　　）。
（13）Android 虚拟设备的简写为（　　）。
（14）Android 针对活动的管理使用的是栈机制，活动栈保存了已经启动并且没有终止的活动，其遵循的原则是（　　）。
（15）在字符串资源源文件 strings.xml 定义字符串的标签名为（　　）。
（16）在如下 AndroidManifest.xml 代码中，表示该项目会出现在 LAUNCHER 桌面上的属性设置为(　　)；表示该项目可启动运行的属性代码为(　　)；表示活动<activity>的标签属性为（　　）。

```
1. <activity
2.     android:name="sziit.lihz.ch01_6.MainActivity"
3.     android:label="@string/app_name" >
4.     <intent-filter>
5.         <action android:name="android.intent.action.MAIN" />
6.         <category android:name="android.intent.category.LAUNCHER" />
7.     </intent-filter>
8. </activity>
```

（17）资源字符串文件/Ch01_ex01_HelloWorld/res/values/strings.xml 中采用键值对定义字符串如下所示，在第 3 行代码字符串键值对中，name 为 app_name，其对应的值为（　　），在 xml 文件中引用该字符串的方式为（　　　　　）；在第 4 行代码字符串键值对中，name 为 hello_world，其对应的值为（　　　　），在 xml 文件中引用该字符串的方式为（　　　　　）。

```
1. <?xml version="1.0" encoding="utf-8"?>
2. <resources>
3.     <string name="app_name">Ch01_ex01_HelloWorld</string>
4.     <string name="hello_world">Hello world!</string>
5. </resources>
```

（18）启动活动的常用方法为（　　　　）；关闭活动的常用方法为（　　　　）。

（19）组件（　　　　）是提供支持跨应用共享数据的唯一方式。

（20）为了便于程序员运行和调试 Android 应用程序，Android 提供了（　　　　），可方便地在其上运行程序，而不需要实际的移动终端。

（21）为了支持 Java 程序运行，需要安装（　　　　）。

（22）Android 平板电脑中有很多应用，如联系人管理、多媒体和摄像等，它们都属于 Android 的（　　　　）层。

2. 单选题

（1）android.util.Log 可设置五种输出日志的级别，其中可显示所有类型的消息的调用方法为（　　）。

  A. Log.i(TAG, strings) B. Log.d(TAG, strings) C. Log.w(TAG, strings)

  D. Log.e(TAG, strings) E. Log.v(TAG, strings)

（2）活动运行时首先执行的生命周期事件回调函数为（　　）。

  A. onCreate() B. onStart() C. onRestart() D. onResume()

  E. onPause() F. onStop() G. onDestroy()

（3）活动销毁时最后执行的生命周期事件回调函数为（　　）。

  A. onCreate() B. onDestroy() C. onRestart() D. onResume()

  E. onPause() F. onStop() G. onStart()

（4）R.java 类文件在建立项目时由 Android 提供的工具自动生成，该工具名称为（　　）。

  A. 资源打包工具（Android Asset Packaging Tool，AAPT）

  B. Dalvik 虚拟机调试监控服务（Dalvik Debug Monitor Service，DDMS）

  C. 安卓开发工具（Android Development Tools，ADT）

  D. 可扩展标记语言（Extensible Markup Language，XML）

  E. Java 开发工具（Java Development Kit，JDK）

  F. 软件开发工具包（Software Development Kit，SDK）

  G. Android 虚拟设备（Android Virtual Devic，AVD）

（5）在 AndroidManifest.xml 中设置活动在桌面启动器中显示的名称的属性为（　　）。

  A. android:name="string"

  B. android:theme="resourceortheme"

  C. android:label="stringresource"

  D. android:icon="drawableresource"

  E. android:exported=["true"|"false"]

F. android:enabled=["true"|"false"]

（6）在 AndroidManifest.xml 中设置应用程序可调试的应用的属性是（　　）。

　　A. android:description="string resource"

　　B. android:enabled=["true" | "false"]

　　C. android:icon="drawable resource"

　　D. android:label="string resource"

　　E. android:debuggable=["true" | "false"]

　　F. android:name="string"

（7）下列不属于 Android 体系结构中的应用程序层的是（　　）。

　　A. 短消息（SMS，Short Message Services）程序

　　B. 日历

　　C. 电话簿

　　D. 嵌入式数据库 SQLite

（8）资源目录 res 包含了 Android 应用项目的全部资源，命名规则正确的说法为（　　）。

　　A. 可以支持数字（0~9）、下画线（_）和大小写字符（a~z，A~Z）

　　B. 只能支持数字（0~9）和大小写字母（a~z，A~Z）

　　C. 只支持数字（0~9）、下画线（_）和小写字符（a~z），第一个可以是字母和下画线（_）

　　D. 只支持数字（0~9）、下画线（_）和小写字符（a~z），第一个必须是字母

（9）Android 应用程序设计主要是调用(　　)提供的 API 实现。

　　A. 应用程序层　　　B. 应用框架层　　　C. 应用视图层　　　D. 系统库层

（10）Android 应用项目启动最先加载的是 AndroidManifest.xml，若项目中包含多个活动，则决定最先加载活动的属性为（　　）。

　　A. android.intent.category.LAUNCHER　　　B. android.intent.category.ICON

　　C. android.intent.action.MAIN　　　D. android.intent.category.ACTIVITY

（11）在活动的生命周期中，首先被调用的回调函数为（　　）。

　　A. onStart()　　　B. onResume()　　　C. onRestart()　　　D. onCreate()

（12）若要向 Android 工程添加字符串资源，应将其添加到（　　）文件。

　　A. dimens.xml　　　B. styles.xml　　　C. strings.xml　　　D. value.xml

3. 问答题、操作题

（1）什么是 Android？

（2）什么是 Android 系统构架？

（3）Android 的平台包含哪些特性？

（4）什么是 Android 应用？

（5）什么是 Android 上下文？

（6）什么是活动？

（7）什么是意图？

（8）如何创建 Android 程序？

（9）简述 Activity 的生命周期。

（10）Activity 包含哪四种状态？

（11）Activity 有哪些常用方法？

（12）如何使用 ADT 创建一个 Android 应用项目？

（13）<application>标签中属性 android:debuggable="true"起什么作用？

（14）如何修改 Android 应用程序的标题、布局中显示内容、字体和颜色？

（15）如何修改隐藏应用程序标题栏最大化屏幕、定制显示风格和设置屏幕背景图片？

（16）如何在字符串资源文件 strings.xml 中定义一个字符串 hello_world，如何在布局文件 activity_main.xml 中利用其设置文本视图控件 TextView 的 android:text 属性值？

（17）创建一个名为 MyHelloWorld 的 Android 应用项目，并简述该项目中各个组成文件的作用。

第 2 章
Android 用户界面布局管理器（Layout Manager）

【知识目标】
- 了解 Android 布局管理器（Layout Manager）的基本概念；
- 了解 Android 界面组件（Component）结构层次；
- 了解 Android 用户界面（User Interface）的一般结构；
- 掌握线性布局（LinearLayout）管理器的基本概念和常用属性；
- 掌握相对布局（RelativeLayout）管理器的基本概念和常用属性；
- 掌握表格布局（TableLayout）管理器的基本概念和常用属性；
- 掌握帧布局（FrameLayout）管理器的基本概念和常用属性；
- 掌握网格布局（GridLayout）管理器的基本概念和常用属性；
- 掌握绝对布局（AbsoluteLayout）管理器的基本概念和常用属性；
- 了解 Android 布局嵌套的应用方法；
- 了解 merge 标签和 include 标签。

【能力目标】
- 掌握线性布局（LinearLayout）属性设置方法；
- 掌握相对布局（RelativeLayout）属性设置方法；
- 掌握表格布局（TableLayout）属性设置方法；
- 掌握帧布局（FrameLayout）属性设置方法；
- 掌握网格布局（GridLayout）属性设置方法；
- 掌握绝对布局（AbsoluteLayout）属性设置方法。

【重点、难点】
- 应用线性布局（LinearLayout）实现用户界面方法和技巧；
- 应用相对布局（RelativeLayout）实现用户界面方法和技巧；
- 应用表格布局（TableLayout）实现用户界面方法和技巧；
- 应用帧布局（FrameLayout）实现用户界面方法和技巧；
- 应用网格布局（GridLayout）实现用户界面方法和技巧。

2.1 学习导入

2.1.1 什么是 Android 布局管理器（Layout Manager）

Android 应用程序的用户界面（User Interface，UI）开发主要包括：用户界面显示和事件处理两个方面的内容。UI 显示主要通过视图（View）和视图组（ViewGroup）两大类实现：ViewGroup 类负责整体布局，View 类负责控件应用。事件处理主要包括：监听事件和回调事件。

Android 应用程序中的活动必须拥有一个 XML 格式的 UI 布局文件，活动要通过 View 对象和 ViewGroup 对象将 XML 格式所描绘的用户界面绘制出来，才能与用户进行交互。

Android UI 主要包括几类布局管理器（Layout Manager）：线性布局（LinearLayout）、相对布局（RelativeLayout）、帧布局（FrameLayout）、表格布局（TableLayout）和绝对布局（AbsoluteLayout）等。Android 的这些布局管理器都是以视图组为基类派生出来的，它们之间的继承关系图 UML（Unified Modeling Language，统一建模语言）如图 2.1 所示。从该图可知，线性布局、相对布局、帧布局和绝对布局是直接继承 ViewGroup 类，表格布局是继承自 LinearLayout 类。Android 平台中的布局管理器本质上是可视组件，它既可作为容器包含其他可视组件，也可作为组件加入到其他布局管理器中，从而构成繁茂的 Android 界面组件。

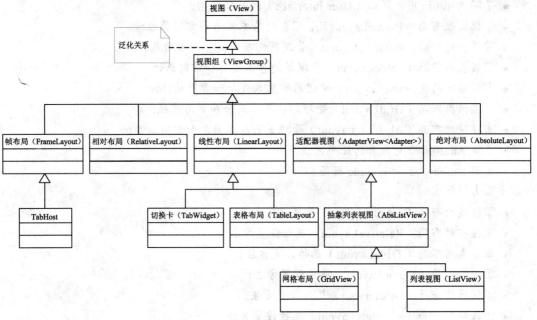

图 2.1　Android 布局管理器之间的关系图

2.1.2　什么是 Android 界面组件（Component）结构层次

在 Android 平台中，所有的可视组件都是视图的子类；视图组是 View 类的一个重要直接子类；Android 布局管理器是 ViewGroup 类的一组重要的直接或间接子类，Android 界面组件结构层次示意图如图 2.2 所示。

图 2.2 中的类在继承和归集上存在一定的交错。Android 应用程序最基本的组件为活动，每个活动实际上就是应用程序的一个界面。一个活动的用户界面是由 View 对象、ViewGroup 对象以及它们派生类的对象所组成的。因此，Android UI 的一般结构如图 2.3 所示。继承自 android.view.View 的子类控件可通过 ViewGroup 容器组织在一起，ViewGroup 类本身也是 View 的子类。

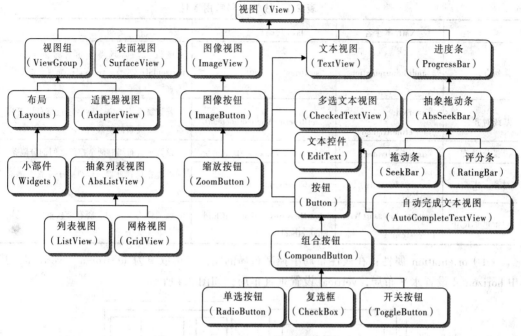

图 2.2 Android 界面组件结构层次示意图

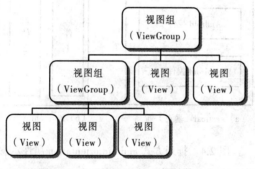

图 2.3 Android UI 的一般结构

2.2 技术准备

2.2.1 线性布局管理器

1. 线性布局作用

线性布局提供了控制水平或垂直排列的模型，会将容器中的组件一个一个排列起来。通过 android:orientation 属性，可控制组件横向或者纵向排列。线性布局中的组件不会自动换行，组件依次排列到尽头之后，越界的组件不会显示出来。

2. 线性布局常用属性

线性布局常用属性包括：排列方式、基线对齐、设置分隔条、对齐方式和权重最小尺寸等，如表 2.1 所示（XML 属性名均以 android:开头）。

表 2.1 线性布局常用属性

属 性	XML 属性名	Java 代码设置方法	功 能
排列方式	android:orientation	setOrientation(int)	设置布局管理器内组件排列方式。horizontal：水平；vertical：垂直；默认为垂直
基线对齐	android:baselineAligned	setBaselineAligned(boolean)	若该属性为 false，就会阻止该布局管理器与其子元素的基线对齐
设置分隔条	android:divider	setDividerDrawable(Drawable)	设置垂直布局时两个控件之间的分隔条
对齐方式	android:gravity	setGravity(int)	设置布局管理器内组件（子元素）的对齐方式
权重最小尺寸	android:measureWithLargestChild	setMeasureWithLargestChildEnable(boolean)	该属性为 true 时，所有带权重的子元素都会取最大子元素的最小尺寸

（1）orientation 属性：在线性布局中控制布局的朝向，可取值为 horizontal、vertical，其中 horizontal 设置水平布局，vertical 设置垂直布局，如图 2.4 所示。

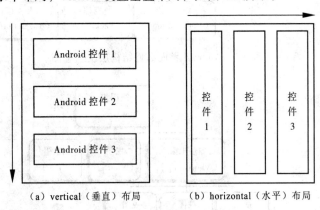

图 2.4 排列方式 android:orientation 属性示意

所谓水平放置，指的是该容器里中存放的控件或者容器只能以一行的形式出现，放置的控件只能是该行中的某个位置，两个控件或者容器之间只有左右关系没有其他方向上的关系，当水平方向满屏时不会自动换行，再放置的控件将在屏幕之外存在，无法看到。

所谓垂直放置，就相当于一列，放置的控件或者容器只能在该列中的某个位置，两个控件之间只存在上下方向的关系，不存在其他方向上的关系。当这一列放满后，再添加的控件就置于屏幕之外，无法看到。

（2）baselineAligned 属性：设置线性布局基线对齐方式。

（3）divider 属性：设置垂直布局时两个控件之间的分隔条。

（4）gravity 属性：设置布局管理器内组件（子元素）的对齐方式，其属性取值包括 top、bottom、left、right、center_vertical（垂直方向居中）、center_horizontal（水平方向居中）、fill_vertical（垂直方向拉伸）、fill_horizontal（水平方向拉伸）、center、fill、clip_vertical、clip_horizontal。可以同时指定多种对齐方式，如 left|center_vertical 左侧垂直居中。可取值常量及其含义描述如表 2.2 所示。

表 2.2　gravity 属性可取值常量及其含义描述

常　量	含　义　描　述	是否改变控件大小
top	将控件放在其容器的顶部，不改变其大小	否
bottom	将控件放在其容器的底部，不改变其大小	否
left	将控件放在其容器的左侧，不改变其大小	否
right	将控件放在其容器的右侧，不改变其大小	否
center	将控件横纵居中，不改变其大小	否
center_vertical	将控件纵向居中（垂直对齐方式：垂直方向上居中对齐）	否
center_horizontal	将对象横向居中（水平对齐方式：水平方向上居中对齐）	否
fill_vertical	必要时增加控件的纵向大小，沿垂直方向完全充满其容器	是
fill_horizontal	必要时增加控件的横向大小，沿水平方向完全充满其容器	是
fill	必要时增加控件的横纵向大小，以完全充满其容器	是
clip_vertical	垂直方向裁剪；附加选项，用于按照容器的边来剪切对象的顶部和/或底部的内容。剪切基于其纵向对齐设置：顶部对齐时，剪切底部；底部对齐时剪切顶部；除此之外剪切顶部和底部	是
clip_horizontal	水平方向裁剪；附加选项，用于按照容器的边来剪切对象的左侧和/或右侧的内容。剪切基于其横向对齐设置：左侧对齐时，剪切右侧；右侧对齐时剪切左侧；除此之外剪切左侧和右侧	是

（5）measureWithLargestChild 属性：设置权重最小尺寸。

3. 线性布局子元素控制

线性布局的子元素，即线性布局中的组件，都受到 LinearLayout.LayoutParams 控制，线性布局包含的子元素可以执行的属性如表 2.3 所示（XML 属性名均以 android:开头）。

表 2.3　线性布局子元素控制属性

常用属性	XML 属性	功　　能
对齐方式	android:layout_gravity	设置该元素在线性布局（父容器）的对齐方式，即该组件本身的对齐方式，注意要与 android:gravity 区分
所占权重	android:layout_weight	设置该元素在线性布局（父容器）中所占的权重，例如都是 1 的情况下，那个方向（android:orientation 方向）长度都是一样的
所占宽度	android:layout_width	容器的宽度，该值必须设置
所占高度	android:layout_height	容器的高度，该值必须设置
控件 ID	android:id	标识该布局，可通过 findViewById(int id)找到该布局
设置背景	android:background	设置线性布局背景
点击事件	android:clickable	设置线性布局是否可以响应点击事件
根布局属性	xmlns:android= "http://schemas.android.co m/apk/res/android"	设置该值表示可以当做一个布局文件，被 Android 系统解释使用，每个布局的根布局必须包含该属性，否则，系统将找不到布局
左内边距	android:paddingLeft	设置该布局左内边距,该值设置后，位于该布局的 View 或者 ViewGroup 均在 padding 距离内放置，边距内不能放置控件
右内边距	android:paddingRight	设置该布局的右内边距，同上
上内边距	android:paddingTop	设置该布局的上内边距，同 android:paddingLeft
下内边距	android:paddingBottom	设置该布局的下内边距，同上

续表

常用属性	XML 属性	功　　能
内边距	android:padding	设置该布局的四个方向内边距，同上
外边距	android:layout_margin	表示外边距，还有与内边距相类似的属性，分别为上、下、左、右外边距
最小高度	android:minHeigh	表示该布局的最小高度，layout_height="wrap_content"
最小宽度	android:minWidth	表示该布局的最小宽度，Layout_width="wrap_content"

4. 线性布局控制子元素排列与在父元素中排列

线性布局控制子元素排列与在父元素中排列的属性类型及说明如表 2.4 所示。

表 2.4　控制子元素排列与在父元素中排列属性类型

属性类型	功　　能
设备组件本身属性	带 layout 的属性是设置本身组件属性，如 android:layout_gravity 设置的是本身的对齐方式
设置子元素属性	不带 layout 的属性是设置其所包含的子元素，如 android:gravity 设置的是该容器子组件的对齐方式
LayoutParams 属性	所有的布局管理器都提供了相应的 LayoutParams 内部类，这些内部类用于控制该布局本身，如对齐方式 layout_gravity，所占权重 layout_weight，这些属性用于设置本元素在父容器中的对齐方式
容器属性	在 android:后面没有 layout 的属性基本都是容器属性，android:gravity 作用是指定本元素包含的子元素的对齐方式，只有容器才支持这个属性

提示：Android 中使用的常用单位，包括 px、dip/dp、sp、in、mm 和 pt 等，如表 2.5 所示。

表 2.5　Android 中使用的常用单位

单位名称	描　　述
px	像素（pixels），基本原色素及其灰度的基本编码，每个 px 对应屏幕上的一个点
dip/dp	设备独立像素（device independent pixels），该单位基于屏幕密度
sp	比例像素（scale pixels），处理字体的大小，可根据用户字体大小进行缩放
in	英寸（inch），标准长度单位，1 in=2.54 cm
mm	毫米（millimeter），标准长度单位，1000 mm 相当于 1 m
pt	磅（point），标准长度单位，1/72 英寸，等于 0.3527 mm

5. 线性布局常见用法

（1）获取 LinearLayout 的宽高

- 调用 View.getHeight()和 View.getWidth()方法可获得组件内 View 子控件高和宽。
- 调用 View.getMeasuredWidth()和 View.getMeasuredHeight()方法可从外部获得 View 对象宽高，如源代码清单 2.1 所示（完整源代码 Ch02_ex01_LinearLayout.rar）：

源代码清单 2.1　\Ch02_ex01_LinearLayout\src\sziit\lihz\ch02_ex01_linearlayout\MainActivity.java

```
1. //从上下文检索窗口快速获取LayoutInflater 实例
2. View view = getLayoutInflater().inflate(R.layout. activity_main, null);
3. LinearLayout layout = (LinearLayout) view.findViewById(R.id.linearlayout);
4. layout.measure(0, 0);
            //调用该测量方法之后才能通过getMeasuredHeight()等方法获取宽、高
5. int width = layout.getMeasuredWidth();      //获取宽度
6. int height = layout.getMeasuredHeight();    //获取高度
```

- 调用 View.getLayoutParams().width 和 View.getLayoutParams().height 从 LayoutParams（或 xml 布局文件）中获取 View 组件的宽高。

（2）在 LinearLayout 中添加分隔线
- 使用 ImageView 添加；
- 使用 LinearLayout 标签的 android:showDividers 属性添加；
- 使用代码 linearLayout.setShowDividers()设置显示分隔线样式，使用 linearLayout.setDividerDrawable()设置分隔线图片。

6. 线性布局典型案例

（1）演示 layout_gravity 和 gravity 属性区别的案例

实现要点：
- android:gravity：用来设置 view 本身的文本应该显示在 view 的什么位置，默认值是左侧；
- android:layout_gravity：是相对于包含该元素的父元素来说的，设置该元素在父元素的什么位置。
- 创建 Ch02_ex02_LinearLayout 项目，在 layout 目录中增加布局文件 linearlayout_exam1.xml（注意：文件名不允许有大写字母），其代码如清单 2.2 所示，其布局效果如图 2.5 所示。

源代码清单 2.2　\Ch02_ex01_LinearLayout\res\layout\linearlayout_exam1.xml

```
1.  <?xml version="1.0" encoding="utf-8"?>
                              //xml 文件开始声明，版本1.0，编码方式 GB2312 编码
2.  <LinearLayout xmlns:android="http://schemas.android.com/apk/res/android"
3.      android:layout_width="fill_parent"    //布局宽度：填充父控件宽度
4.      android:layout_height="fill_parent"   //布局高度：填充父控件高度
5.      android:orientation="vertical" >
6.      <Button                ← 按钮标签
7.          android:layout_width="100dip"     ← 按钮宽度
8.          android:layout_height="40dip"     ← 按钮高度
9.          android:layout_gravity="bottom|left"   ← 按钮位置
10.         android:background="#FFFF00"      ← 按钮背景色
11.         android:gravity="top|left"        //文本信息显示在控件的顶部向左对齐位置
12.         android:text="按钮控件" />         //显示的文本信息
13.     <TextView              ← 文本视图标签
14.         android:layout_width="100dip"     ← 文本视图宽度
15.         android:layout_height="40dip"     ← 文本视图高度
16.         android:layout_gravity="top|center_horizontal"   ← 文本视图位置
17.         android:background="#00FFFF"      ← 文本视图背景色
18.         android:gravity="bottom|center"   //文本信息显示在控件的底部居中位置
19.         android:text="文本视图控件" />     //显示的文本信息
20. </LinearLayout>           ← 线性布局标签
```

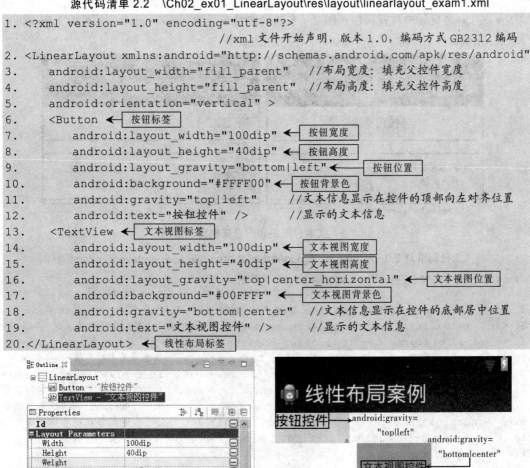

（a）界面设计　　　　　　　　　　（b）布局效果图

图 2.5　演示 layout_gravity 和 gravity 属性区别

Tips 2.1：Button 并没有按照第 9 行所设置的 android:layout_gravity="bottom|left"属性那样显示在左下方，TextView 也没有按照第 16 行所设置的 android:layout_gravity="top|center_horizontal"属性那样显示在顶部正中央，这是因为设置了线性布局的 android:orientation 属性为"vertical"。对于线性布局，若设置 android:orientation="vertical"，则 android:layout_gravity 设置的属性只在水平方向生效，如图 2.5（a）所示 Button 显示在水平方向的最左边，TextView 显示在屏幕的水平正中央；类似的，若设置 android:orientation="horizontal"，则 android:layout_gravity 属性只在垂直方向生效。

（2）文本（TextView）、按钮（Button）和图像按钮（ImageButton）排列案例

实现要点：

- 通过修改 android:gravity 属性来控制 LinearLayout 中子元素的排列情况；
- 在 Ch02_ex02_LinearLayout 项目 layout 目录中增加布局文件 linearlayout_exam2.xml，其代码如清单 2.3 所示，其布局效果见图 2.6；
- 顶部+水平居中对齐属性：图 2.6 左边的线性布局的 android:gravity 属性为 top|center_horizontal；
- 左部+垂直居中对齐属性：图 2.6 右边的线性布局的 android:gravity 属性为 left|center_vertical。

图 2.6　文本、按钮和图像按钮排列案例

源代码清单 2.3　\Ch02_ex02_LinearLayout\res\layout\linearlayout_exam2.xml

```
1. <?xml version="1.0" encoding="utf-8"?>
                             //xml 文件开始声明，版本1.0，编码方式GB2312编码
2. <!-- android:gravity="left|center_horizontal":左部+垂直居中对齐属性 -->
3. <!-- android:gravity="right|center_horizontal":右部+垂直居中对齐属性 -->
4. <!-- android:gravity="top|center_horizontal":顶部+水平居中对齐属性 -->
5. <!-- android:gravity="bottom|center_horizontal":底部+水平居中对齐属性 -->
6. <LinearLayout xmlns:android="http://schemas.android.com/apk/res/android"
7.     android:layout_width="fill_parent"         //布局宽度：填充父控件宽度
8.     android:layout_height="fill_parent"        //布局高度：填充父控件高度
9.     android:gravity="top|center_horizontal"    //子控件在顶部水平居中
10.    android:orientation="vertical" >   ←垂直方向布局  //垂直方向线性布局
11.    <Button  ← 按钮标签
12.        android:layout_width="wrap_content"//按钮控件宽度：与显示的信息宽度相同
13.        android:layout_height="wrap_content"//按钮控件高度：与显示的信息高度相同
14.        android:text="按钮控件" />  ← 按钮显示文本
15.    <TextView ← 文本视图标签
```

```
16.        android:layout_width="wrap_content"  //按钮控件宽度：与显示的信息宽度相同
17.        android:layout_height="wrap_content" //按钮控件高度：与显示的信息高度相同
18.        android:text="文本控件" />   ← 文本视图显示文本
19.    <ImageButton  ← 图像按钮标签
20.        android:layout_width="wrap_content"  //按钮控件宽度：与显示的信息宽度相同
21.        android:layout_height="wrap_content" //按钮控件高度：与显示的信息高度相同
22.        android:src="@drawable/ic_launcher"  ← 图像按钮显示图片
23.        android:text="图像控件" />
24.</LinearLayout>
```

（3）三个图像按钮（ImageButton）各自对齐案例

实现要点：

- 三个水平方向的图像按钮，分别左对齐、居中对齐、右对齐；
- 水平线性布局：最顶层线性布局的 orientation 是 horizontal（水平方向）；
- 等分三个线性布局：第二层线性布局的 orientation 是 vertical（垂直方向），并且宽度是 fill_parent，依靠权重分配宽度；
- 设置图像按钮（ImageButton）对齐方式：图像按钮的 android:layout_gravity 属性分别设置为 left、center、right，默认为 left；
- 在 Ch02_ex02_LinearLayout 项目 layout 目录中增加布局文件 linearlayout_exam3.xml，其核心代码如源代码清单 2.4 所示，其布局效果如图 2.7 所示。

图 2.7　三个图像按钮各自对齐案例

源代码清单 2.4　\Ch02_ex02_LinearLayout\res\layout\linearlayout_exam3.xml

```
1. <?xml version="1.0" encoding="utf-8"?>
                                        //xml 文件开始声明，版本 1.0，编码方式 GB2312 编码
2. <LinearLayout xmlns:android="http://schemas.android.com/apk/res/android"
3.     android:layout_width="fill_parent"     //布局宽度：填充父控件宽度
4.     android:layout_height="fill_parent"    //布局高度：填充父控件高度
5.     android:orientation="horizontal" > ← 最顶层水平线性布局  //水平方向线性布局
6.     <LinearLayout
7.         android:layout_width="fill_parent"    //布局宽度：填充父控件宽度
8.         android:layout_height="wrap_content"  //布局高度：与显示的信息高度相同
9.         android:layout_weight="1"  ← 依靠权重分配宽度
10.        android:background="#ff0000"          //背景色
11.        android:orientation="vertical" > ← 第二层垂直线性布局
12.        <ImageButton    ← 图像按钮标签
13.            android:layout_width="wrap_content"   //图片控件宽度：与显示的信息宽度相同
14.            android:layout_height="wrap_content"  //图片控件高度：与显示的信息高度相同
15.            android:layout_gravity="left"  ← 图像靠左对齐
16.            android:src="@drawable/ic_launcher" > //设置图片控件显示的图片来源
17.        </ImageButton>
```

```
18.    </LinearLayout>    ← 第二层线性布局标签
19.    ……
20.</LinearLayout>    ← 最顶层线性布局标签
```

（4）动态设置线性布局案例
- 调用 setOrientation(int)设置线性布局方向，若为1，则设置为纵向，否则设置为横向。
- 创建 Ch02_ex03_LinearLayout 项目，修改 activity_main.xml 的代码（见源代码清单 2.5），编写 MainActivity.java 代码（见源代码清单 2.6），运行结果如图 2.8 所示。

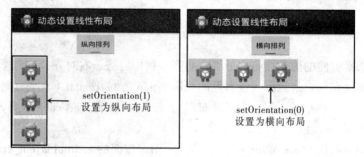

图 2.8 动态设置线性布局案例

源代码清单 2.5 \res\layout\activity_main.xml

```
1. <?xml version="1.0" encoding="utf-8"?>
                                //xml 文件开始声明，版本1.0，编码方式 GB2312 编码
2. <LinearLayout xmlns:android="http://schemas.android.com/apk/res/android"
3.     ……
4.     <ToggleButton  ← 开关按钮标签
5.         android:id="@+id/toggleButton"  ← 开关按钮 id
6.         android:layout_gravity="center"  ← 开关按钮位置
7.         android:checked="false"
8.         android:textOff="横向排列"  ← 开关按钮初始值
9.         android:textOn="纵向排列" />  ← checked 为 true 时显示文本
10.    <LinearLayout  ← 线性布局标签
11.        android:id="@+id/mLinearLayout"    //线性布局 id 号
12.        android:layout_width="fill_parent"    //布局宽度：填充父控件宽度
13.        android:layout_height="fill_parent"    //布局高度：填充父控件高度
14.        android:orientation="vertical" >  ← 设置垂直布局
15.        <ImageButton  ← 图像按钮标签
16.            android:layout_width="wrap_content"
                                //图片控件宽度：与所显示图片宽度一致
17.            android:layout_height="wrap_content"
                                //图片控件高度：与所显示图片高度一致
18.            android:src="@drawable/ic_launcher" >  ← 设置图像按钮图片
19.        </ImageButton>
20.        ……
21.    </LinearLayout>
22.</LinearLayout>
```

源代码清单 2.6 \ch02_ex03_linearlayout\MainActivity.java

```
1. public class MainActivity extends Activity { //从活动基类 Activity 派生子类
```

```
2.  protected void onCreate(Bundle savedInstanceState) {  //重写活动创建方法
3.      super.onCreate(savedInstanceState);   //调用活动基类方法
4.      setContentView(R.layout.activity_main);//设置活动屏幕布局文件资源
5.      //绑定布局资源文件中的开关按钮ToggleButton对象
6.      ToggleButton toggleButton = (ToggleButton) this.findViewById(R.id.toggleButton);
7.      //绑定布局资源文件中的线性布局LinearLayout对象
8.       final LinearLayout mLL = (LinearLayout) this.findViewById(R.id.mLinearLayout);
9.      //设置OnCheckedChangeListener监听事件并处理onCheckedChanged()回调方法
10.     toggleButton.setOnCheckedChangeListener(new OnCheckedChangeListener() {
11.          public void onCheckedChanged(CompoundButton buttonView, boolean isChecked) {
12.         /**当ToggleButton按钮在不断按下时会在true和false间相互切换,会传一个
           布尔型的值,用来判断是true还是false,然后区分是纵向还是横向*/
13.          if(isChecked){//当isChecked==true时
14.              mLL.setOrientation(1);// 1 :设置为纵向(垂直)布局
15.          }else{//当isChecked==false时
16.              mLL.setOrientation(0);// 0 :设置为横向(水平)布局
17.          } } });  }}
```

2.2.2 相对布局管理器

1. 相对布局的作用

相对布局是指利用控件之间的相对位置关系来对布局进行放置。换句话说,该容器中的控件与其他任何一个控件或者容器(包括父控件)有相对关系。在相对布局容器中,子组件的位置总是相对兄弟组件,父容器来决定的,其允许子元素指定它们相对于其父元素或兄弟元素的位置,这是相对布局中最常用的布局方式之一。相对布局灵活性大,属性多,操作难度大,属性之间产生冲突的可能性也大,使用时要多做测试。

2. 相对布局支持的常用属性

相对布局的常用属性包括:对齐方式、忽略对齐方式等,如表2.6所示(XML属性名均以android:开头)。

表2.6 相对布局常用属性

属 性	XML属性	Java代码设置方法	功 能
对齐方式	android:gravity	setGravity(int)	设置布局容器内子元素的对齐方式,注意与android:layout_ gravity区分,后者是设置组件本身元素对齐方式
忽略对齐方式	android:ignoreGravity	setIgnoreGravity(int)	设置该组件不受gravity属性影响,因为gravity属性影响容器内所有的组件的对齐方式,设置之后,该组件就可以例外

3. 相对布局LayoutParams属性

(1)第一类:只能设置boolean值的属性(true或false)。这些属性都是相对父容器的,确定是否在父容器中居中(水平,垂直),是否位于父容器的上/下/左/右端,如表2.7所示(XML属性名均以android:开头)。

表2.7 LayoutParams 只能设置 boolean 值的属性

序号	属性名	功能描述
1	android:layout_centerHorizontal	确定是否在父容器中水平居中
2	android:layout_centerVertical	确定是否在父容器中垂直居中
3	android:layout_centerInParent	确定是否在父容器中位于中央
4	android:layout_alignParentTop	确定是否在父容器中顶端对齐
5	android:layout_alignParentBottom	确定是否在父容器中底端对齐
6	android:layout_alignParentLeft	确定是否在父容器中左对齐
7	android:layout_alignParentRight	确定是否在父容器中右对齐
8	android:layout_alignWithParentIfMissing	如果对应的兄弟元素找不到,就以父元素做参照物

（2）第二类：只能设置相对其他组件 id 的属性,属性值须为 id 的引用名 "@id/idname",如表2.8所示。

表2.8 LayoutParams 只能设置相对其他组件 id 的属性

序号	属性名	功能描述
1	android:layout_toLeftOf	位于所给 id 组件左侧
2	android:layout_toRightOf	位于所给 id 组件右侧
3	android:layout_above	位于所给 id 组件的上方
4	android:layout_below	位于所给 id 组件的下方
5	android:layout_alignTop	与所给 id 组件顶部对齐
6	android:layout_alignBottom	与所给 id 组件底部对齐
7	android:layout_alignLeft	与所给 id 组件左对齐
8	android:layout_alignRight	与所给 id 组件右对齐

（3）第三类：属性值为具体的像素值（如 30 px、40 dip 等）,如图2.9所示。

表2.9 值为具体像素值的属性

序号	属性名	功能描述
1	android:layout_marginBottom	离某元素底边缘的距离
2	android:layout_marginLeft	离某元素左边缘的距离
3	android:layout_marginRight	离某元素右边缘的距离
4	android:layout_marginTop	离某元素上边缘的距离

4. 相对布局常用方法

（1）使用 getLocationOnScreen()方法获取屏幕中一个组件的位置

步骤1：创建数组。先创建一个整型数组,数组大小为 2；将这个数组传入 getLocationOnScreen()方法。

步骤2：将位置信息传入数组。调用 View.getLocationOnScreen()方法,返回的是一个数组 int[2],其中 int[0]是横坐标,int[1]是纵坐标（参见 Ch02_ex04_RelativeLayout.rar）。

（2）使用 LayoutParams 设置属性值

在 Android 中可对任何属性进行设置,利用 LayoutParams 对象可设置相对布局属性值,先使用 LayoutParams.addRule()方法设置所有组件的属性值,然后调用 View.set LayoutParams()

方法传入 LayoutParams 对象，并更新 View 的相应 Layout Params 属性值，以向容器中添加该组件。代码中动态设置布局属性的步骤如下：

步骤1：创建 LayoutParams 对象；
步骤2：调用 LayoutParams 对象的 addRule()方法设置相关属性；
步骤3：调用 View.setLayoutParams()方法将准备好的 LayoutParams 对象设置给组件；
步骤4：调用 addView()方法将 View 对象设置到布局中。如源代码清单 2.7 所示。

源代码清单 2.7　\Ch02_ex04_RelativeLayout\src\sziit\lihz\ch02_ex04_relativelayout\MainActivity.java

```
1.  public class MainActivity extends Activity {
2.     protected void onCreate(Bundle savedInstanceState) {
3.        super.onCreate(savedInstanceState);
4.        setContentView(R.layout.activity_main);
5.        //装载布局文件
6.        View view = getLayoutInflater().inflate(R.layout.activity_main, null);
7.        RelativeLayout relativeLayout = (RelativeLayout) view.findViewById
           (R.id.relayout);
8.        //装载要动态添加的布局文件
9.        TextView textView = (TextView) relativeLayout.findViewById(R.id.textview);
10.       int locations[] = new int[2];  //创建数组，将该数组传入getLoca tionOn
          Screen()方法
11.       //textView.getLocationInWindow(locations);
12.       textView.getLocationOnScreen(locations);    //获取位置信息
13.       int xCoord = locations[0];                   //获取宽度
14.       int yCoord = locations[1];                   //获取高度
15.       yCoord = locations[1]; }}
```

5. 相对布局典型案例

（1）相对布局案例 1：五个图片按钮排成梅花形状，其效果如图 2.9 所示，布局源代码清单见 2.8（项目源代码见 Ch02_ex05_RelativeLayout.rar）。

源代码清单 2.8　\Ch02_ex05_RelativeLayout\res\layout\activity_main.xml

```
1.  <?xml version="1.0" encoding="utf-8"?>    // XML 序言，版本 1.0，编码格式 GB2312
2.  <RelativeLayout xmlns:android="http://schemas.android.com/apk/res/android"
3.     android:layout_width="fill_parent"       //布局宽度：填充父控件高度
4.     android:layout_height="fill_parent" >//布局高度：填充父控件高度
5.     <ImageButton                              //图片按钮标签
6.        android:id="@+id/imgbtn1"              //图片按钮1的 id 号
7.        android:layout_width="80dp"            //图片按钮1的宽度,80（设备的独立像素
8.        android:layout_height="80dp"           //图片按钮1的高度,80（设备的独立像素
9.        android:scaleType="fitXY"              //缩放类型
10.       android:layout_centerInParent="true"//在父控件中央
11.       android:src="@drawable/flower1"/>//设置图片
12.    <ImageButton                              //图片按钮标签
13.       android:id="@+id/imgbtn2"              //图片按钮2的 id 号
14.       android:layout_width="80dp"            //图片按钮宽度:80dp（设备的独立像素）
15.       android:layout_height="80dp"           //图片按钮高度:80dp（设备的独立像素）
16.       android:scaleType="fitXY"              //把图片不按比例扩大/缩小到View的大小显示
17.       android:layout_above="@+id/imgbtn1"    //位于 imgbtn1 控件上方
```

```
18.         android:layout_alignLeft="@+id/imgbtn1"      //与控件 imgbtn1 左边对齐
19.         android:src="@drawable/flower2"/>            //图片按钮所显示图片来源
20.     <ImageButton                                     //图片按钮标签
21.         android:id="@+id/imgbtn3"                    //图片按钮3的id号
22.         android:layout_width="80dp"
23.         android:layout_height="80dp"
24.         android:scaleType="fitXY"
25.         android:layout_alignLeft="@+id/imgbtn1"      //与控件 imgbtn1 向左对齐
26.         android:layout_below="@+id/imgbtn1"          //位于 imgbtn1 控件下方
27.         android:layout_centerInParent="true"         //置于父控件居中
28.         android:src="@drawable/flower3"/>            //图片按钮所显示图片来源
29.     <ImageButton                                     //图片按钮标签
30.         android:id="@+id/imgbtn4"                    //图片按钮4的id号
31.         android:layout_width="80dp"
32.         android:layout_height="80dp"
33.         android:scaleType="fitXY"
34.         android:layout_alignTop="@+id/imgbtn1"       //与控件 imgbtn1 顶部对齐
35.         android:layout_centerInParent="true"         //置于父控件居中
36.         android:layout_toLeftOf="@+id/imgbtn1"       //位于控件 imgbtn1 左边
37.         android:src="@drawable/flower4"/>            //图片按钮所显示图片来源
38.     <ImageButton                                     //图片按钮标签
39.         android:id="@+id/imgbtn5"                    //图片按钮5的id号
40.         android:layout_width="80dp"
41.         android:layout_height="80dp"
42.         android:scaleType="fitXY"
43.         android:layout_alignTop="@+id/imgbtn1"       //与控件 imgbtn1 顶部对齐
44.         android:layout_centerInParent="true"         //置于父控件居中
45.         android:layout_toRightOf="@+id/imgbtn1"      //位于控件 imgbtn1 右边
46.         android:src="@drawable/flower5"/>            //图片按钮所显示图片来源
47. </RelativeLayout>
```

要点：注意每个组件的属性，先要确定方位，再进行对齐，组件左边界对齐，组件上边界对齐；两个按钮，若只有 android:layout_above="@+id/button1"，则布局如图 2.10（a）所示；若加上 android:layout_alignLeft="@+id/button1"，则布局如图 2.10（b）所示。

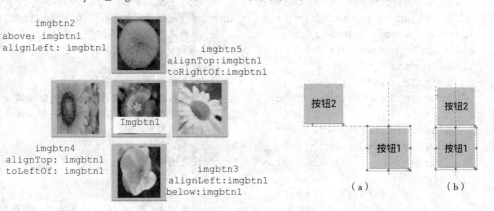

图 2.9　五个图片按钮排成梅花形状　　　　图 2.10　两个按钮布局情形

（2）相对布局案例 2：自定义简单文本信息输入表单界面布局，其效果如图 2.11 所示，

relative_layout_1.xml 布局代码见源代码清单 2.9（项目源代码见 Ch02_ex05_RelativeLayout.rar）。

（3）相对布局案例3：利用 RelativeLayout 的属性控制三个按钮的布局位置，其效果如图 2.12 所示，relative_layout_2.xml 布局代码见源代码清单 2.10（项目源代码见 Ch02_ex05_Relative Layout.rar）。

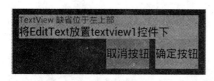
图 2.11　相对布局输入表单案例

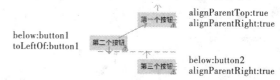
图 2.12　相对布局控制三个按钮

源代码清单 2.9　\Ch02_ex05_RelativeLayout\res\layout\relative_layout_1.xml

```xml
1. <?xml version="1.0" encoding="utf-8"?>
2. <!-- 利用相对布局创建一个输入表单案例 -->
3. <RelativeLayout xmlns:android="http://schemas.android.com/apk/res/android"
4.     android:layout_width="match_parent"       //布局宽度：与父控件宽度相匹配
5.     android:layout_height="wrap_content"      //布局高度：与内容自适应
6.     android:background="#0000FF"              //布局背景色：蓝色
7.     android:padding="10dip" >//布局控件与边界的距离:10dip（设备的独立像素）
8.     <!-- textview1 控件 默认位于左上部 -->
9.     <TextView   //文本控件标签
10.        android:id="@+id/textview1"           //文本控件id号：+ 表示新增id号
11.        android:layout_width="match_parent"   //文本控件宽度：与父控件宽度相匹配
12.        android:layout_height="wrap_content"  //文本控件高度：与内容自适应
13.        android:background="#00FF00"          //文本控件背景色：绿色
14.        android:text="TextView 默认位于左上部" />  //文本控件显示信息
15.    <!-- 将 EditText 放在 TextView1 控件下 -->
16.    <EditText   //编辑框控件标签
17.        android:id="@+id/edittext1"           //编辑框控件id号
18.        android:layout_width="match_parent"   //编辑框控件宽度：与父控件宽度相匹配
19.        android:layout_height="wrap_content"  //编辑框控件高度：与内容自适应
20.        android:layout_below="@id/textview1"  //编辑框控件位于控件 TextView1 下方
21.        android:background="#aaaa00"          //编辑框控件背景色：灰色
22.        android:text="将 EditText 放置 textview1 控件下" />  //编辑框控件显示内容
23.    <!-- ok_button 按钮位于 edittext1 控件下，向父控件右边对齐 -->
24.    <Button                                   //按钮控件标签
25.        android:id="@+id/ok_button"           //按钮控件id号
26.        android:layout_width="wrap_content"   //按钮控件宽度：根据内容宽度自动调整
27.        android:layout_height="wrap_content"  //按钮控件高度：根据内容高度自动调整
28.        android:layout_alignParentRight="true"//按钮控件与父控件右边对齐
29.        android:layout_below="@id/edittext1"  //按钮控件位于 EditText1 控件下方
30.        android:layout_marginLeft="10dip"     //按钮控件与左边距离: 10dip
31.        android:background="#00cccc"          //按钮控件背景色
32.        android:text="确定按钮" />            //按钮控件显示信息
33.    <!--cancel_button 按钮向 ok_button 的上部对齐，且位于其左边10dip 位置处-->
34.    <Button   //按钮控件标签
35.        android:id="@+id/cancel_button"       //按钮控件id号
36.        android:layout_width="wrap_content"   //按钮控件宽度：根据内容宽度自动调整
37.        android:layout_height="wrap_content"  //按钮控件高度：根据内容高度自动调整
38.        android:layout_alignTop="@id/ok_button"  //按钮控件与 ok_button 控件顶部对齐
```

```
39.        android:layout_toLeftOf="@id/ok_button"  //按钮控件与 ok_button 控件左边对齐
40.        android:background="#00aaaa"             //按钮控件背景色
41.        android:text="取消按钮" />                //按钮控件显示信息
42.    </RelativeLayout>                            //相对布局标签
```

源代码清单 2.10   \Ch02_ex05_RelativeLayout\res\layout\relative_layout_2.xml

```
1.  <?xml version="1.0" encoding="utf-8"?>
2.  <!-- 每个 XML 文档都由 XML 序言开始 -->
3.  <!-- xml version="1.0" :这行代码表示按照 1.0 版本的 XML 规则进行解析 -->
4.  <!-- encoding = "utf-8":表示此 xml 文件采用 utf-8 的编码格式。编码格式也可以是 GB2312 -->
5.  <!-- RelativeLayout:表示采用相对布局管理器 -->
6.  <RelativeLayout xmlns:android="http://schemas.android.com/apk/res/android"
7.      android:layout_width="match_parent"        //布局宽度：与父控件宽度相匹配
8.      android:layout_height="match_parent">      //布局高度：与父控件高度相匹配
9.      <Button                                    //按钮控件标签
10.         android:id="@+id/button1"              //按钮控件 id 号
11.         android:layout_width="wrap_content"    //按钮控件宽度：根据内容宽度自动调整
12.         android:layout_height="wrap_content"   //按钮控件高度：根据内容高度自动调整
13.         android:layout_alignParentRight="true" //按钮控件与父控件右边对齐
14.         android:layout_alignParentTop="true"   //按钮控件与父控件顶部对齐
15.         android:layout_marginLeft="20dp"       //按钮控件与左间距 20dp
16.         android:layout_marginTop="20dp"        //按钮控件与上间距 20dp
17.         android:gravity="center_vertical|left" //按钮控件显示文本左对齐|垂直居中
18.         android:text="第一个按钮" />             //按钮控件显示内容
19.     <Button                                    //按钮控件标签
20.         android:id="@+id/button2"              //按钮控件 id 号
21.         android:layout_width="wrap_content"    //按钮控件宽度：根据内容宽度自动调整
22.         android:layout_height="wrap_content"   //按钮控件高度：根据内容高度自动调整
23.         android:layout_below="@+id/button1"    //位置在第一个按钮 button1 的下方
24.         android:layout_marginTop="15dp"        //设置上间距 15dp
25.         android:layout_toLeftOf="@+id/button1" //与第一个按钮 button1 的左边对齐
26.         android:gravity="center_vertical|left" //按钮控件显示文本左对齐|垂直居中
27.         android:text="第二个按钮" />             //按钮控件显示内容
28.     <Button                                    //按钮控件标签
29.         android:id="@+id/button3"              //按钮控件 id 号
30.         android:layout_width="wrap_content"    //宽度匹配内容
31.         android:layout_height="wrap_content"   //高度匹配内容
32.         android:layout_below="@+id/button2"    //位置在第二个按钮 button2 的下面
33.         android:layout_marginTop="15dp"        //设置上间距 15dp
34.         android:layout_alignParentRight="true" //贴紧父元素右边
35.         android:gravity="center_vertical|left" //按钮控件显示文本左对齐|垂直居中
36.         android:text="第三个按钮"/>              //按钮控件显示内容
37.     </RelativeLayout>                          //相对布局标签
```

2.2.3　帧布局管理器

1. 帧布局的作用

帧布局也称框架布局，为每个组件创建一个空白区域，一个区域称为一帧，这些帧会根据 FrameLayout 中定义的 android:layout_gravity 属性自动对齐。对于放置前后的关系，在没有设置其他属性之前，Android 系统采用的是叠放原则，即后加入结点的层叠在上面。设置属性

android:bringToFront="true|false"将前面放置的控件提到最前面可见。

2. 帧布局支持的常用属性

帧布局支持的常用属性如表2.10所示。

表2.10 帧布局支持的常用属性（XML属性名均以android:开头）

XML属性	Java方法	功能描述
android:foreground	setForeground(Drawable)	设置该帧布局容器的前景图像
android:foregroundGravity	setForegroundGravity(int)	设置绘制前景图像的gravity属性

所谓前景图像，即永远处于帧布局最顶层，直接面对用户的图像，也就是不会被覆盖的图像。

3. 帧布局典型案例

（1）三个TextView帧布局的简单案例

实现要点：

- FrameLayout：xml文件中的帧布局标签。
- 当向帧布局中增加组件时，所有组件默认都会置于帧布局的左上角，其大小由子控件中最大的子控件决定；若组件大小相同，则同时只能看到最上面的那个组件。
- android:layout_gravity：该属性可设置帧布局中组件的对齐方式。
- 创建Ch02_ex06_FrameLayout项目，修改布局文件activity_main.xml为FrameLayout布局，并增加三个文本控件，如源代码清单2.11所示，其布局效果如图2.13（a）所示。将控件的android:layout_gravity属性分别设置为center_horizontal、right和left后，其布局效果如图2.13（b）所示。

源代码清单2.11　\Ch02_ex06_FrameLayout\res\layout\activity_main.xml

```
1.  <?xml version="1.0" encoding="utf-8"?>       //xml版本，编码方式"utf-8"
2.  <FrameLayout xmlns:android="http://schemas.android.com/apk/res/android"
3.      android:id="@+id/framelayout01"          //帧布局组件id号
4.      android:layout_width="fill_parent"       //帧布局宽度:填充父组件（屏幕）宽度
5.      android:layout_height="fill_parent"      //帧布局高度:填充父组件（屏幕）高度
6.      android:background="#F0F0F0">            //帧布局背景色
7.      <TextView                                //文本视图组件标签
8.          android:id="@+id/textview1"          //TextView控件id号
9.          android:layout_width="wrap_content"  //控件布局宽度：根据显示内容自动调整
10.         android:layout_height="wrap_content" //控件布局高度：根据显示内容自动调整
11.         android:text="红色大字体"             //控件显示文本信息
12.         android:textColor="#FFFF0000"        //控件显示文本颜色
13.         android:textSize="60px" >            //控件显示文本大小（60像素）
14.     </TextView>
15.     <TextView                                //文本视图组件标签
16.         android:id="@+id/textview2"          //TextView控件id号
17.         android:layout_width="wrap_content"  //控件布局宽度：根据显示内容自动调整
18.         android:layout_height="wrap_content" //控件布局高度：根据显示内容自动调整
19.         android:text="绿色中字体"             //控件显示文本信息
20.         android:textColor="#FF00FF00"        //控件显示文本颜色
21.         android:textSize="40px" >            //控件显示文本大小（40像素）
22.     </TextView>
```

```
23.    <TextView                                    //文本视图组件标签
24.        android:id="@+id/TextView03"             //TextView 控件 id 号
25.        android:layout_width="wrap_content"      //控件布局宽度：根据显示内容自动调整
26.        android:layout_height="wrap_content"     //控件布局高度：根据显示内容自动调整
27.        android:text="蓝色小字体"                 //控件显示文本信息
28.        android:textColor="#FF0000FF"            //控件显示文本颜色
29.        android:textSize="20px" >                //控件显示文本大小（20 像素）
30.    </TextView>
31.</FrameLayout>                                   //帧布局标签
```

（a）默认 layout_gravity 属性　　　　　　　　（b）定制 layout_gravity 属性

图 2.13　三个文本视图帧布局的简单案例

类似的，按钮在水平方向上分别左对齐、居中对齐、右对齐的源代码清单如 2.12 所示，其布局效果如图 2.14（a）所示，若未分别设置三个按钮的 android:layout_gravity 属性，则只能见到最后一个按钮，如图 2.14（b）所示。

（a）三个按钮分别左对齐、居中对齐、右对齐　　　　（b）未设置 android:layout_gravity 属性

图 2.14　三个按钮帧布局的简单案例

源代码清单 2.12　\Ch02_ex06_FrameLayout\res\layout\three_button_framelayout.xml

```
1. <?xml version="1.0" encoding="utf-8"?>          //xml 版本，编码方式"utf-8"
2. <FrameLayout xmlns:android="http://schemas.android.com/apk/res/android"
3.     android:layout_width="match_parent"         //帧布局宽度：匹配父组件（屏幕）宽度
4.     android:layout_height="match_parent"        //帧布局高度：匹配父组件（屏幕）高度
5.     android:background="#FFFFFF">               //帧布局背景色
6.     <Button                                     //按钮控件标签
7.         android:layout_width="wrap_content"     //按钮控件宽度：根据显示内容自动调整
8.         android:layout_height="wrap_content"    //按钮控件高度：根据显示内容自动调整
9.         android:layout_gravity="left"           //按钮控件向父控件左边对齐
10.        android:textColor="#f00"                //按钮控件显示文本颜色
11.        android:text="红色按钮 1" />             //按钮控件显示文本信息
12.    <Button                                     //按钮控件标签
13.        android:layout_width="wrap_content"     //按钮控件宽度：根据显示内容自动调整
14.        android:layout_height="wrap_content"    //按钮控件高度：根据显示内容自动调整
15.        android:layout_gravity="center_horizontal"//按钮控件向父控件顶部居中对齐
16.        android:textColor="#0f0"                //按钮控件显示文本颜色
17.        android:text="绿色按钮 2" />             //按钮控件显示文本信息
18.    <Button                                     //按钮控件标签
19.        android:layout_width="wrap_content"     //按钮控件宽度：根据显示内容自动调整
20.        android:layout_height="wrap_content"    //按钮控件高度：根据显示内容自动调整
```

```
21.        android:layout_gravity="right"          //按钮控件向父控件右边对齐
22.        android:textColor="#00f"                 //按钮控件显示文本颜色
23.        android:text="蓝色按钮 3" />              //按钮控件显示文本信息
24.</FrameLayout>                                    //帧布局标签
```

（2）帧动画简单案例（项目源代码见 Ch02_ex07_FrameLayout.rar）

实现要点：

- 主要实现流程：步骤1，定义名为picframe的空帧布局，将前景图像的位置android:foregroundGravity 设置为中央位置（center），android:layout_gravity 属性设置为 center_horizontal；步骤2，在活动中利用 findViewById()方法获取该 FrameLayout 对象；步骤3，新建一个句柄（Handler）对象，重写消息处理方法 handlerMessage()，调用图像更新的方法；步骤4，自定义 move()方法，通过 setForeground()方法动态设置前景图片显示的位图；步骤5，在 onCreate()方法中新建一个计时器对象 Timer，重写 run()方法，每隔 200 毫秒向 Handler 发送空信息 MSG_CTRL_COLOR。
- 加载资源图片文件方法：getResources().getDrawable(int)。
- 设置前景图片显示位图的方法：setForeground(Drawable)。
- 定时器控制 Handler：创建 Handler 对象，实现 handleMessage()方法。
- 定时器任务：创建一个定时器对象 Timer，实现定时器任务调度 schedule()，利用句柄对象定时发送消息 handler.sendEmptyMessage(MSG_CTRL_COLOR)。

帧布局见源代码清单 2.13，核心源代码清单见 2.14，运行结果如图 2.15 所示。

源代码清单 2.13   \Ch02_ex07_FrameLayout\res\layout\framelayout_ex1.xml

```
1. <FrameLayout xmlns:android="http://schemas.android.com/apk/res/android"
2.     xmlns:tools="http://schemas.android.com/tools"
3.     android:id="@+id/picframe"                       //设置帧布局的id号
4.     android:layout_width="wrap_content"              //帧布局宽度：自适应内容
5.     android:layout_height="wrap_content"             //帧布局宽度：自适应内容
6.     android:layout_gravity="center_horizontal"       //帧布局位置：水平居中
7.     android:foregroundGravity="center" >             //帧布局前景位置：居中
8. </FrameLayout>                                        //帧布局标签
```

源代码清单 2.14   \Ch02_ex07_framelayout\MainActivity.java

```
1. public class MainActivity extends Activity {//从基类 Activity 派生子类 MainActivity
2.     protected void onCreate(Bundle savedInstanceState) {//重写 onCreate 方法
3.         super.onCreate(savedInstanceState);//基类调用 onCreate 方法
4.         setContentView(R.layout.activity_main);  //加载布局文件
5.         picframe = (FrameLayout) findViewById(R.id.picframe);//绑定帧布局对象
6.         // 初始化 Drawable 资源数组,通过 getResources().getDrawable()
7.         // 来获取布局中的 Drawable 资源
8.         for (int i = 0; i < names.length; i++) {pics[i] = getResources().
getDrawable(names[i]);}
9.         new Timer().schedule(new TimerTask() {
                              //定义一个定时器对象,定时发送信息给 handler
10.        public void run() { //发送一条空信息来通知系统改变前景图片
11.            handler.sendEmptyMessage(MSG_CTRL_COLOR);}//发送空消息
12.        }, 0, 200);}
13.    private final int MSG_CTRL_COLOR = 0x123456;
```

```
14.    FrameLayout picframe = null;//初始化变量，帧布局
15.    //Drawable 资源数组对应的资源 id
16.    final int[] names = new int[] { R.drawable.p1, R.drawable.s_2,
17.         R.drawable.s_3, R.drawable.s_4, R.drawable.s_5, R.drawable.s_6,
18.         R.drawable.s_7, R.drawable.s_8 };
19.    Drawable[] pics = new Drawable[names.length];
                               //在 Activity 中创建 Drawable 组件数组
20.    //自定义一个用于定时更新 UI 界面的 handler 类对象
21.    Handler handler = new Handler() {//定义一个 handler 对象来刷新帧布局的前景图像
22.        int i = 0;
23.        public void handleMessage(Message msg) {//重写句柄对象处理消息方法
24.            if (msg.what == MSG_CTRL_COLOR) {  //判断信息是否为本应用发出的
25.                i++; move(i % 8);}             //有 8 个图片作为动画的素材
26.            super.handleMessage(msg);}};       //父对象调用基类方法
27.    //定义走路时切换图片的方法
28.    void move(int i) {// 通过 setForeground 来设置前景图像
29.        picframe.setForeground(pics[i]);}}    //设置前景图片
```

图 2.15　帧动画简单案例运行结果

（3）帧布局霓虹灯案例（项目源代码见 Ch02_ex08_FrameLayout.rar）

实现要点：

- 利用 FrameLayout 和按钮控件绘制霓虹灯效果的层叠布局（见源代码清单 2.15）。
- 后挡前：后面的视图组件会遮挡前面的视图组件，越在前面，被遮挡的概率越大。
- 界面居中：将所有的按钮组件的对齐方式 android:layout_gravity 设置为 center。
- 正方形：所有 Button 组件都设置 android:layou.t_width 和 android:layout_height 属性，将 Button 的宽和高设置为相等，后面的控件比前面的边长依次少 40。
- 颜色：每个 Button 组件的背景色均不同。
- 颜色资源：在根结点<resources>下创建<color>子结点，color 属性标签 name 自定义，子文本为颜色代码（见源代码清单 2.16）。也可直接在代码中定义颜色数组：

```
    final int[] colors = new int[] { Color.MAGENTA, Color.BLUE, Color.GREEN,
Color.GRAY, Color.YELLOW, Color.RED };
```

- 文本：每个 Button 组件的显示文本设置为其次序编号；文本显示对齐方式 android:gravity 设置为 left。
- 定时器控制句柄 handler：创建 Handler 对象，实现 handlerMessage()方法，通过定时器定时发送消息，句柄收到消息后调用 setBackgroundResource()方法更新背景色，实现霓虹灯动态效果（见源代码清单 2.17），运行效果如图 2.16 所示。

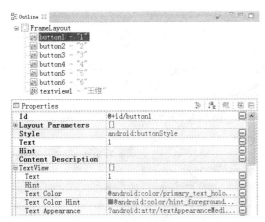
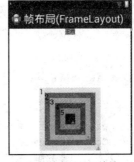

（a）界面设计　　　　　　　　（b）帧布局界面效果

图 2.16　帧布局霓虹灯案例运行结果

源代码清单 2.15　\Ch02_ex08_FrameLayout\res\layout\framelayout_ex2.xml

```
1. <?xml version="1.0" encoding="utf-8"?>      //xml 版本，编码方式"utf-8"
2. <FrameLayout xmlns:android="http://schemas.android.com/apk/res/android"
3.     android:layout_width="match_parent"     //帧布局宽度:匹配父组件（屏幕）宽度
4.     android:layout_height="match_parent">   //帧布局高度:匹配父组件（屏幕）高度
5.     <Button                                  //按钮控件标签
6.         android:id="@+id/button1"            //按钮控件 id 号
7.         android:layout_width="240px"         //按钮控件宽度: 240 px
8.         android:layout_height="240px"        //按钮控件高度: 240 px
9.         android:layout_gravity="center"      //按钮控件布局位置: 居中
10.        android:background="#ff0"            //按钮控件背景色
11.        android:gravity="left"               //按钮控件显示文本位置: 左对齐
12.        android:text="1" />                  //按钮控件显示文本信息
13.    ……
14.    <Button                                  //按钮控件标签
15.        android:id="@+id/button6"            //按钮控件 id 号
16.        android:layout_width="40px"          //按钮控件宽度: 40 px
17.        android:layout_height="40px"         //按钮控件高度: 40 px
18.        android:layout_gravity="center"      //按钮控件布局位置: 居中
19.        android:background="#00f"            //按钮控件背景色
20.        android:gravity="left"               //按钮控件显示文本位置: 左对齐
21.        android:text="6" />                  //按钮控件显示文本信息
22.    <TextView                                //文本视图控件标签
23.        android:id="@+id/textview1"          //文本视图控件 id 号
24.        android:layout_width="wrap_content"  //文本视图控件宽度:与显示内容自适应
25.        android:layout_height="wrap_content" //文本视图控件高度:与显示内容自适应
26.        android:layout_gravity="top|center"  //文本视图控件布局位置: 顶部居中
27.        android:background="#0f0"            //文本视图控件背景色
28.        android:text="王维《春眠》"           //文本视图控件显示内容
29.        android:textSize="20px" />           //文本视图控件显示字体大小
30.</FrameLayout>                               //帧布局标签
```

源代码清单 2.16　\Ch02_ex08_FrameLayout\res\values\colors.xml

```
1. <?xml version="1.0" encoding="utf-8"?>
```

```
2.    <resources>    //资源标签
3.      <color name = "color1">#0ff</color>    //定义颜色键值对,名为"color1",值为#0ff
4.      <color name = "color2">#f0f</color>    //color: 颜色标签
5.      <color name = "color3">#ff0</color>    <color name = "color4">#f00</color>
6.      <color name = "color5">#0f0</color>    <color name = "color6">#00f</color>
7.    </resources>
```

源代码清单 2.17   \Ch02_ex08_framelayout\MainActivity.java

```
1.  public class MainActivity extends Activity {
                                         //从基类 Activity 派生子类 MainActivity
2.    protected void onCreate(Bundle savedInstanceState) {//重写 onCreate 方法
3.        super.onCreate(savedInstanceState);    //调用基类方法
4.        setContentView(R.layout.activity_main); //加载布局文件
5.        //初始化组件数组,通过 findViewById()来获取布局中的按钮组件
6.        for (int i = 0; i < names.length; i++) {
7.            buttons[i] = (Button) findViewById(names[i]);//遍历绑定按钮控件对象
8.        }
9.        textview1 = (TextView) findViewById(R.id.textview1);//绑定 TextView 对象
10.       Resources res = getResources();          //获得资源
11.       jueju = res.getStringArray(R.array.jueju); //获得字符串数组
12.       // 定义一个线程周期性地改变 currentColor 变量值,每隔 0.4 秒更换一次颜色
13.       new Timer().schedule(new TimerTask() {   //新建定时器,定时调度回调任务
14.           public void run() {
              //发送 MSG_CTRL_COLOR 消息通知系统改变 6 个 Button 组件的背景色
15.               handler.sendEmptyMessage(MSG_CTRL_COLOR);//发送空消息
16.           }
17.       }, 0, 400);                //400 毫秒执行一次
18.   }
19.   private int curColor = 0; //该变量用来控制霓虹灯颜色变化
20.   private final int MSG_CTRL_COLOR = 0x123456;//定义消息常量
21.   //颜色对应的资源 id
22.   final int[] colors = new int[] { R.color.color1, R.color.color2,
23.       R.color.color3, R.color.color4, R.color.color5, R.color.color6 };
24.   //View 组件对应的资源 id
25.   final int[] names = new int[] { R.id.button1, R.id.button2, R.id.button3,
26.       R.id.button4, R.id.button5, R.id.button6 };
27.   Button[] buttons = new Button[names.length];
                                       //在 Activity 中创建 Button 组件数组
28.   String[] jueju = new String[names.length];//字符串数组
29.   TextView textview1;         //定义文本视图对象
30.   //定义该 Handler,以便在定时器中固定调用 handleMessage 方法
31.   //Handler 主要接收定时器子线程发送的数据,并用此数据配合主线程更新 UI
32.   Handler handler = new Handler() {        //构建句柄对象
33.       public void handleMessage(Message msg) {//重写处理消息方法
34.           if(msg.what == MSG_CTRL_COLOR) {
                              //消息来自本程序 sendEmptyMessage 所发送
35.               for(int i = 0; i < names.length; i++) {//循环设置 Button 背景颜色
36.                   buttons[i].setBackgroundResource(colors[(i + curColor)
37.                       % names.length]);         //设置按钮背景色
38.               }
39.               textview1.setText(jueju[curColor % names.length]);//显示文本信息
```

```
40.            textview1.setBackgroundColor(colors[(curColor) % names.length]);
                                                //设置背景色
41.            curColor++;            //当前颜色计数器加 1
42.        }
43.        super.handleMessage(msg);  //父对象调用基类方法
44.    };};};}
```

2.2.4 表格布局管理器

1. 表格布局的作用

TableLayout 类以行和列的形式对控件进行管理，每一行为一个 TableRow 对象或一个 View 控件。当为 TableRow 对象时，可在 TableRow 下添加子控件，默认情况下，每个子控件占据一列；当为 View 时，该 View 将独占一行。

2. 表格布局的一些概念

- 继承关系：表格布局继承了 LinearLayout，其本质为线性布局管理器。
- 控制组件：表格布局采用行列形式管理 UI 组件，但并不需要声明有多少行列，只需要添加 TableRow 和其他组件就可控制表格的行数和列数，而网格布局（GridLayout）需指定行列数。
- 增加行的方法：可采用 TableRow 增加行列和组件增加行两种方法。TableRow 增加行列：向表格布局中添加一个 TableRow，一个 TableRow 就是一个表格行，同时 TableRow 也是容器，可向其中添加子元素，每添加一个组件，就增加了一列。组件增加行：直接向 TableLayout 中添加组件，就相当于添加了一行。
- 列宽：表格布局中，列的宽度由该列最宽的单元格决定，整个表格的宽度默认充满父容器本身。

3. 单元格行为方式

（1）行为方式概念

- 隐藏（Collapsed）：若某列被设置成 Collapsed，则该列所有单元格会被隐藏。
- 拉伸（Stretchable）：若某列被设为 Stretchable，则该列所有单元格的宽度可被拉伸，保证表格能完全填满表格剩余空间。
- 收缩（Shrinkable）：若某列被设为 Shrinkable，则该列所有单元格宽度可被收缩，保证表格能适应父容器的宽度。

（2）行为方式属性和单元格属性

行为方式属性称 TableLayout 全局属性，也即列属性，如表 2.11 所示。

表 2.11 TableLayout 的行为方式属性（XML 属性名均以 android:开头）

名 称	XML 属性	设 置 方 法	功 能
隐藏	android:collapsed Columns	setColumnCollapsed (int, boolean)	设置需要被隐藏的列的序号，在 xml 文件中，如果隐藏多列，多列序号间用逗号隔开
拉伸	android:stretch Columns	setStretchAllColumns (boolean)	设置允许被拉伸的列的序列号，xml 文件中多个序列号之间用逗号隔开
收缩	android:shrinkable Columns	setShrinkableAll Columns(boolean)	设置允许收缩的列的序号，xml 文件中多个序号之间可以用逗号隔开

说明：列可同时具备 stretchColumns 及 shrinkColumns 属性，若这样，当该列的内容很多时，则系统根据需要自动调节该行的 layout_height，从而多行显示其内容。

TableLayout 的单元格属性如表 2.12 所示。

表 2.12　TableLayout 的单元格属性（XML 属性名均以 android:开头）

XML 属性	功　能　描　述
android:layout_column	指定该单元格在第几列显示
android:layout_span	指定该单元格占据的列数（未指定时为 1）

表 2.13 中列出了 TableLayout 行为方式属性的一些示例代码。

表 2.13　TableLayout 行为方式属性的示例代码

序　号	示　例　代　码	功　能　描　述
1	android:collapseColumns="*"	隐藏所有行
2	android:collapseColumns="1"	隐藏第 2 列（从 0 开始计算），若有多列要隐藏，则用","隔开
3	android:stretchColumns="1,2"	设置列 1,2 为可伸展列
4	android:stretchColumns="0"	第 0 列可伸展
5	android:shrinkColumns="1,2"	第 1,2 列皆可收缩
6	android:layout_column="1"	该控件显示在第 1 列
7	android:layout_span="2"	该控件占据 2 列

4. 表格布局典型案例

（1）表格布局案例 1（项目源代码见 Ch02_ex09_TableLayout.rar）

实现要点：

- 列属性：设置 collapseColumns、shrinkColumns 和 stretchColumns 等属性。
- 单元格属性：设置 layout_column、layout_span 等属性。
- 宽度属性：设置布局宽度 layout_width 属性，如表 2.14 所示。

代码清单见 2.18，运行效果见图 2.17。

表 2.14　表格布局案例 1 主要属性设置

序　号	设　置　属　性	功　能　描　述
1	android:collapseColumns="0"	第 0 列被隐藏：该列被隐藏
2	android:shrinkColumns="1"	第 1 列可收缩：该列向列方向收缩，可以很深
3	android:stretchColumns="2"	第 2 列可伸展：该列向行方向伸展，可以很长
4	android:layout_column="0"	该控件被指定在第 0 列
5	layout_column="2",layout_span="4"	该控件被指定在第 2 列，可跨 4 列
6	layout_column="1",layout_span="4"	该控件被指定在第 1 列，可跨 4 列
7	android:layout_width="1dip"	设置表格宽度属性

源代码清单 2.18　\Ch02_ex09_TableLayout\res\layout\tablelayout_ex1.xml

```
1. <?xml version="1.0" encoding="utf-8"?>
2. <LinearLayout xmlns:android="http://schemas.android.com/apk/res/android"
3.     android:layout_width="fill_parent"      //布局宽度：填充父控件宽度
4.     android:layout_height="wrap_content"    //布局高度：根据内容高度自动调整高度
5.     android:background="#00f"               //布局背景色
6.     android:orientation="vertical"          //线性布局方向：垂直排列
```

```
7.      android:padding="0dip">              //与父控件之间的间隔
8.      <TextView                             //文本控件标签，表格布局自动增加一行
9.          android:layout_width="fill_parent"   //文本控件宽度：填充父控件
10.         android:layout_height="wrap_content"
                                              //文本控件高度：根据内容高度自动调整高度
11.         android:background="#7f00ffff"    //文本控件背景色
12.         android:text="表1：演示 列属性设置"  //文本控件显示内容
13.         android:textSize="15sp" />        //文本控件显示字体大小
14.     <!-- 第1个TableLayout，描述表中的列属性：第0列被隐藏，第1列可收缩，第2列可
        伸展 -->
15.     <TableLayout                          //表格布局标签
16.         android:id="@+id/tab1"            //表格布局id号
17.         android:layout_width="fill_parent"   //表格布局宽度：填充父控件宽度
18.         android:layout_height="wrap_content"
                                              //表格布局高度：根据内容高度自动调整高度
19.         android:collapseColumns="0"       //设置表格布局第0列被隐藏属性
20.         android:padding="1dip"            //表格布局单元间隔距离
21.         android:shrinkColumns="1"         //设置表格布局第1列可收缩属性
22.         android:stretchColumns="2" >      //设置表格布局第2列可伸展属性
23.         <TableRow>                        //表格列标签
24.             <Button android:text="第0列被隐藏了" />  //自动创建布局高度第0列
25.             <Button android:text="第1列可收缩" />    //自动创建布局高度第1列
26.             <Button android:text="第2列可伸展" />    //自动创建布局高度第2列
27.         </TableRow>
28.         <TableRow>                        //表格列标签
29.             <Button android:text="行路难·其一" />  <!-- 该列被隐藏了 -->
30.             <Button android:text="多歧路，今安在？" />  <!-- 该列向列方向收缩，可
                以很深 -->
31.             <Button android:text="长风破浪会有时，直挂云帆济沧海。" />
32.             <!-- 该列向行方向伸展,可以很长 -->
33.         </TableRow>
34.     </TableLayout>
35.     <TextView                             //文本控件标签，表格布局自动增加一行
36.         android:layout_width="fill_parent"   //文本控件宽度：填充父控件
37.         android:layout_height="wrap_content"
                                              //文本控件高度：根据内容高度自动调整高度
38.         android:background="#7f00ffff"    //文本控件背景色
39.         android:text="表2:演示单元格属性设置"  //文本控件显示内容
40.         android:textSize="15sp" />        //文本控件显示字体大小
41.     <!-- 第2个TableLayout，描述表中单元格的属性，包括：android:layout_column及
        android:layout_span -->
42.     <TableLayout                          //表格布局标签
43.         android:id="@+id/tab2"            //表格布局id号
44.         android:layout_width="wrap_content"
                                              //表格布局宽度：根据内容宽度自动调整宽度
45.         android:layout_height="wrap_content"
                                              //表格布局高度：根据内容高度自动调整高度
46.         android:padding="0dip" >          //表格布局单元间隔距离
47.         <TableRow>                        //表格列标签
48.             <Button android:text="第0列" />  //自动创建布局高度第0列
```

```xml
49.        <Button android:text="第1列" />      //自动创建布局高度第1列
50.        <Button android:text="第2列" />      //自动创建布局高度第2列
51.        <Button android:text="第3列" />      //自动创建布局高度第3列
52.        <Button android:text="第4列" />      //自动创建布局高度第4列
53.    </TableRow>
54.    <TableRow>                              //表格列标签
55.        <TextView android:layout_column="0"  //该控件指定到第0列
56.            android:text="金樽清酒斗十千，" />  <!-- 该控件被指定在第0列 -->
57.        <TextView
58.            android:layout_column="2" android:layout_span="4"
59.            android:text="玉盘珍羞直万钱。" />  <!-- 该控件被指定在第2列，可
                跨4列 -->
60.    </TableRow>
61.    <TableRow>                              //表格列标签
62.        <TextView
63.            android:layout_column="1" android:layout_span="4"
64.            android:text="停杯投箸不能食，拔剑四顾心茫然。" />
65.        <!-- 该控件被指定在第1列，可跨4列 -->
66.    </TableRow>
67.</TableLayout>
68.<TextView                                   //文本视图标签
69.    android:layout_width="wrap_content" android:layout_height="wrap_content"
70.    android:background="#7f00ffff"
71.    android:text="表3：演示可伸展特性，默认非均匀布局"
72.    android:textSize="15sp" />  <!-- 第3个TableLayout，使用可伸展特性布局
         -->
73.<TableLayout                                //表格布局标签
74.    android:id="@+id/tab3"                  //表格布局id号
75.    android:layout_width="fill_parent"   android:layout_height="wrap_
        content"
76.    android:padding="0dip"  android:stretchColumns="*" >
                                                //表格布局所有列均可伸展
77.    <TableRow>  //表格列标签，宽度根据内容自适应
78.        <Button android:text="江雪" />       //自动创建布局高度第0列
79.        <Button android:text="柳宗元" />     //自动创建布局高度第1列
80.        <Button android:text="千山鸟飞绝" />  //自动创建布局高度第2列
81.    </TableRow>
82.</TableLayout>
83.<TextView
84.    android:layout_width="wrap_content" android:layout_height="wrap_content"
85.    android:background="#7f00ffff"
86.    android:text="表4：演示可伸展特性，设置各控件宽度一致" android:textSize="15sp" />
87.<!-- 第4个TableLayout，设置可伸展特性，且指定每个控件宽度一致，如1dip -->
88.<TableLayout                                //表格布局标签
89.    android:id="@+id/tab4"                  //表格布局id号
90.    android:layout_width="fill_parent" android:layout_height="wrap_content"
91.    android:padding="0dip" android:stretchColumns="*" >
                                                //表格布局所有列均可伸展
92.    <TableRow>                              //表格列标签，设置宽度均相等
93.        <Button                             //按钮控件标签
```

```
94.            android:layout_width="1dip"         //设置宽度均相等
95.            android:text="江雪" />               //按钮控件显示文本信息
96.        <Button                                 //按钮控件标签
97.            android:layout_width="1dip"         //设置宽度均相等
98.            android:text="柳宗元" />             //按钮控件显示文本信息
99.        <Button                                 //按钮控件标签
100.           android:layout_width="1dip"         //设置宽度均相等
101.           android:text="孤舟蓑笠翁" />         //按钮控件显示文本信息
102.       </TableRow>
103.   </TableLayout>
104.</LinearLayout>
```

（2）表格布局案例2（项目源代码见 Ch02_ex10_TableLayout.rar）

实现要点：

- 独自一行按钮：向表格布局中添加按钮，该按钮就会独自占据一行。
- 收缩按钮：设置 android:stretchable 属性标签，属性值是要收缩的列（从0编号）。
- 拉伸按钮：设置 android:shrinkable 属性标签，属性值是要拉伸的列。

源代码清单见2.19，运行效果如图2.18所示。

图2.17 表格布局案例1界面

图2.18 表格布局案例2界面

源代码清单2.19 \Ch02_ex10_TableLayout\res\layout\tablelayout_ex2.xml

```
1. <LinearLayout xmlns:android="http://schemas.android.com/apk/res/android"
2.     xmlns:tools="http://schemas.android.com/tools"
3.     android:layout_width="match_parent" android:layout_height="match_parent"
       //宽高属性
4.     android:background="#0f0" android:orientation="vertical" >
                                             //设置背景色和方向属性
5.     <!-- LinearLayout 默认是水平的，这里设置其方向为垂直 -->
6.     <!-- 表格布局，第1列允许收缩，第2列允许拉伸，注意这里行列的计数都是从0开始的 -->
7.     <TableLayout                          //表格布局标签
8.         android:layout_width="fill_parent" android:layout_height="wrap_content"
9.         android:shrinkColumns="1"         //设置第1列允许收缩属性
10.        android:stretchColumns="2">       //设置第2列允许拉伸属性
11.        <!-- 向 TableLayout 中直接添加 TextView 组件，独占一行 -->
12.        <TextView    //向TableLayout增加文本视图控件自动增加一行表格
13.            android:layout_width="fill_parent" android:layout_height="wrap_
               content"
14.            android:text="独自一行的文本视图" />
```

```
15.     <TableRow>   //表格列标签,3个按钮各占一列（共3列）
16.        <Button      android:layout_width="wrap_content"
17.            android:layout_height="wrap_content" android:text="普通的按钮" />
18.        <Button android:layout_width="wrap_content"
19.            android:layout_height="wrap_content"
20.            android:text="收缩的按钮" />//该列会向高度方向扩展，宽度方向收缩
21.        <Button      android:layout_width="wrap_content"
22.            android:layout_height="wrap_content"
23.            android:text="拉伸的按钮"  />//该列会根据文本内容尽可能伸展宽度
24.     </TableRow>
25. </TableLayout>
26. <!-- 第1列按钮会隐藏掉 -->
27. <TableLayout  //表格布局标签
28.     android:layout_width="fill_parent" android:layout_height="wrap_content"
29.     android:collapseColumns="1">//设置第1列按钮隐藏属性
30.     <TableRow>
31.        <Button android:layout_width="wrap_content"
32.            android:layout_height="wrap_content" android:text="普通按钮0"/>
33.        <Button android:layout_width="wrap_content"
34.            android:layout_height="wrap_content"
35.            android:text="普通按钮1(被隐藏)" />//android:collapseColumns="1"
36.        <Button android:layout_width="wrap_content"
37.            android:layout_height="wrap_content" android:text="普通按钮2" />
38.     </TableRow>
39. </TableLayout>
40. <!--设置第1列和第2列可被拉伸 -->
41. <TableLayout
42.     android:layout_width="fill_parent" android:layout_height="wrap_content"
43.     android:stretchColumns="1,2" >//设置第1列和第2列可被拉伸属性
44.     <TableRow>
45.        <Button android:layout_width="wrap_content"
46.            android:layout_height="wrap_content" android:text="静夜思" />
47.        <Button android:layout_width="wrap_content"
48.            android:layout_height="wrap_content" android:text="李白" />
49.        <Button android:layout_width="wrap_content"
50.            android:layout_height="wrap_content"
51.            android:text="床前明月光，疑是地上霜。"/>
52.     </TableRow>
53.     <TableRow>
54.        <TextView android:layout_width="wrap_content"
55.            android:layout_height="wrap_content" android:text="静夜思" />
56.        <TextView android:layout_width="wrap_content"
57.            android:layout_height="wrap_content"
58.            android:text="举头望明月，低头思故乡。" />
59.     </TableRow>    </TableLayout>   </LinearLayout>
```

2.2.5 网格布局管理器

1. 网格布局的作用

网格布局将整个容器划分成 rows×columns 个网格，每个网格可放置一个组件，还可设置一个组件横跨多少列、多少行（跨多行和多列），不存在一个网格放多个组件的情况。类似线性布局，可设置容器中组件的对齐方式。

2. 网格布局的常用属性

网格布局的常用属性如表 2.15 所示。

表 2.15 网格布局的常用属性（XML 属性名均以 android:开头）

常用属性	XML 属性	设置方法	作用
对齐模式	android:alignmentMode	setAlignmentMode(int)	设置网格布局管理器的对齐模式
列数	android:columnCount	setColumnCount(int)	设置网格布局的最大列数
是否保留列序列号	android:columnOrderPreserved	setColumnOrderPreserved(boolean)	设置网格容器是否保留列序列号
行数	android:rowCount	setRowCount(int)	设置该网格的最大行数
是否保留行序列号	android:rowOrderPreserved	setRowOrderPreserved(int)	设置该网格容器是否保留行序列号
页边距	android:useDefaultMargins	setUseDefaultMargins(boolean)	设置该布局是否使用默认的页边距

3. 网格布局的 LayoutParams 属性

网格布局的 LayoutParams 属性如表 2.16 所示。

表 2.16 网格布局的 LayoutParams 属性（XML 属性名均以 android:开头）

LayoutParams 属性	XML 属性	作用
设置位置列	android:layout_column	设置子组件在 GridLayout 的哪一列
设置横向跨列	android:layout_columnSpan	设置该子组件在 GridLayout 中横向跨几列
设置占据空间方式（组件排列方式）	android:layout_gravity	setGravity(int)，设置该组件采用何种方式占据该网格的空间（center,left,right,buttom）
设置行位置	android:layout_row	设置该子组件在 GridLayout 的第几行
设置横跨行数	android:layout_rowSpan	设置该子组件在 GridLayout 纵向横跨几行
设置组件排列方式	android:orientation	vertical（竖直,默认）或者 horizontal（水平）
设置布局为几行	android:rowCount="3"	设置网格布局有 3 行
设置布局为几列	android:columnCount="3"	设置网格布局有 3 列

4. 网格布局典型案例

（1）利用网格布局实现表单（项目源文件见 Ch02_ex11_GridLayout.rar）

实现要点：

- 设置网格布局的列数：利用 android:columnCount="4"属性设置 GridLayout 的列数。
- 设置网格布局位置列：利用 android:layout_column="0"属性设置 GridLayout 始于 0 列。
- 设置横跨列数：利用 android:layout_columnSpan="4"属性设置该控件横跨 4 列。

该案例源代码清单见 2.20，布局视图如图 2.19 所示。

源代码清单 2.20　\Ch02_ex11_GridLayout\res\layout\grid_layout_1.xml

```xml
1. <?xml version="1.0" encoding="utf-8"?>
2. <GridLayout xmlns:android="http://schemas.android.com/apk/res/android"
3.     android:layout_width="match_parent"      //网格布局宽度：与父控件相匹配
4.     android:layout_height="wrap_content"     //网格布局高度：与控件内容相适应
5.     android:background="#0f0"                //网格布局背景色
6.     android:columnCount="4"                  //网格布局的列数
7.     android:padding="10dip" >                //网格布局间隔距离10dip（设备无关像素）
8.     <TextView android:text="请在下输入信息:" />  //该文本视图独占网格布局1行4列
9.     <EditText                                //编辑控件标签
10.        android:layout_column="0"             //编辑控件始于网格布局0列
11.        android:layout_columnSpan="4"         //编辑控件横跨网格布局4列
12.        android:text="编辑框输入的信息"         //编辑控件显示信息
13.        android:layout_gravity="fill_horizontal" />//编辑控件布局位置：水平填充
14.    <Button                                   //按钮控件标签
15.        android:layout_column="2"             //按钮控件始于网格布局2列
16.        android:text="取消" />                //按钮控件显示信息
17.    <Button                                   //按钮控件标签
18.        android:layout_marginLeft="10dip"     //按钮控件左边距离10dip
19.        android:text="确定" />                //按钮控件显示信息
20. </GridLayout>                                //网格布局标签
```

图 2.19　网格布局实现表单案例

（2）利用网格布局实现计算器界面（Ch02_ex12_ GridLayout.rar）

实现要点：

- 设置网格布局行：android:rowCount="6"，设置 GridLayout 为 6 行。
- 设置网格布局列：android:columnCount="4"，设置 GridLayout 为 4 列。
- 设置控件横跨 4 列：android:layout_columnSpan="4"，TextView 横跨 4 列，独占一行。
- 设置按钮横跨列 2 列：android:layout_columnSpan=4"，Button 横跨 4 列，两个占一行。
- 设置 textView 中的文本与边框有 4 像素间隔：android:padding = "4px"。

将组件设置给网格布局的流程：

步骤 1．设置组件所在行属性：GridLayout.SpecrowSpec = GridLayout.spec(int)。

步骤 2．设置组件所在列属性：GridLayout.Spec colSpec= GridLayout.spec(int)。

步骤 3．创建 LayoutParams 对象：LayoutParams lp=new LayoutParams(rowSpec, colSpec)。

步骤 4．设置组件占满容器：lp.setGravity(Gravity.FILL)。

步骤 5．将组件添加到布局中：gridlayout.addView(view, params)。

布局源代码清单见 2.21；Java 源代码清单见 2.22；运行界面如图 2.20 所示。

源代码清单 2.21　\Ch02_ex12_GridLayout\res\layout\caculator_layout.xml

```xml
1.<GridLayout xmlns:android="http://schemas.android.com/apk/res/android"
```

```
2.     xmlns:tools="http://schemas.android.com/tools"
3.     android:id="@+id/gridlayout"              //定义网格布局的id号
4.     android:layout_width="match_parent"      //网格布局的宽度：填充父控件
5.     android:layout_height="wrap_content"
                                                //网格布局的高度：根据控件内容高度自动调整高度
6.     android:background="#00f"                //设置网格布局的背景色
7.     android:columnCount="4"                  //设置GridLayout布局的列为4
8.     android:rowCount="6" >                   //设置GridLayout布局的行为6
9. <!-- 定义6行 * 4列 GridLayout，如下文本视图和按钮组件都横跨4列，独占一行 -->
10.    <TextView   //文本视图标签
11.        android:layout_width="match_parent"//文本视图布局宽度：填充父控件宽度
12.        android:layout_height="wrap_content"
                                                //文本视图高度：根据控件内容高度自动调整
13.        android:layout_columnSpan="4"        //文本视图横跨4列，独占网格布局一行
14.        android:layout_marginLeft="4px"      //文本视图左边距：4个像素
15.        android:layout_marginRight="4px"     //文本视图右边距：4个像素
16.        android:background="#eee"            //文本视图背景色
17.        android:gravity="right"              //文本视图显示内容向右对齐
18.        android:padding="5px"                //文本视图布局间间隔
19.        android:text="0123456789"            //文本视图显示信息
20.        android:textColor="#0f0"             //文本视图字体颜色
21.        android:textSize="50sp" />           //文本视图字体大小
22.    <Button     //按钮控件标签
23.        android:layout_width="match_parent"  //按钮布局宽度：填充父控件宽度
24.        android:layout_height="wrap_content"
                                                //按钮布局高度：根据控件内容高度自动调整
25.        android:layout_columnSpan="4"        //按钮横跨4列，独占网格布局一行
26.        android:background="#ccc"            //按钮背景色
27.        android:text="清除" />                //按钮显示信息
28.</GridLayout>                                //网格布局标签
```

源代码清单2.22　\Ch02_ex12_GridLayout\src\lihz\sziit\ch02_ex12_gridlayout\MainActivity.java

```
1. public class MainActivity extends Activity {
                                    //从基类Activity公有派生子类MainActivity
2.     private GridLayout gridlayout;  //声明私有网格布局（GridLayout）对象gridlayout
3.     String uiName[] = new String[] { "7", "8", "9", "/", "4", "5", "6", "*",
4.         "1", "2", "3", "-", ".", "0", "=", "+" };// 定义需要放到按钮上的字符串
5.     protected void onCreate(Bundle savedInstanceState) {
                                                //子类重写onCreate方法
6.         super.onCreate(savedInstanceState);   //父对象调用基类方法
7.         setContentView(R.layout.caculator_layout); //设置计算器布局界面
8.         DisplayMetrics dm = new DisplayMetrics();
9.         getWindowManager().getDefaultDisplay().getMetrics(dm);//获取屏幕信息
10.        int screenWidth = dm.widthPixels;     //获得屏幕宽度
11.        gridlayout = (GridLayout) findViewById(R.id.gridlayout);
                                                //绑定布局对象
12.        for (int i = 0; i < uiName.length; i++) { //遍历数组
13.            Button button = new Button(this);  //构建按钮对象
14.            button.setText(uiName[i]);         //设置按钮显示信息
15.            button.setTextSize(40);            //设置按钮字体大小
16.            button.setWidth(screenWidth/4);    //设置按钮宽度
```

```
17.        Spec rowSpec = GridLayout.spec(i / 4 + 2);  //指定组件所在行
18.        Spec colSpec = GridLayout.spec(i % 4);      //指定组件所在列
19.        LayoutParams lp = new LayoutParams(rowSpec, colSpec);
                                                      //构造 LayoutParams 对象
20.        lp.setGravity(Gravity.FILL);               //设置组件充满网格布局属性
21.        gridlayout.addView(button, lp);            //将按钮组件添加到 GridLayout
22.     }}
```

也可直接只用 xml 布局实现计算器界面,其源代码清单如 2.23,界面效果如图 2.21 所示。

源代码清单 2.23   \Ch02_ex12_GridLayout\res\layout\gridlayout_calculator.xml

```
1. <GridLayout xmlns:android="http://schemas.android.com/apk/res/android"
2.     xmlns:tools="http://schemas.android.com/tools"
3.     android:id="@+id/gridlayout1"            //网格布局id号
4.     android:layout_width="wrap_content"      //网格布局宽度: 与子控件宽度相适应
5.     android:layout_height="wrap_content"     //网格布局高度: 与子控件高度相适应
6.     android:columnCount="4"                  //网格布局的列数
7.     android:orientation="horizontal"         //网格布局方向: 水平布局
8.     android:rowCount="6" >                   //网格布局的行数
9.     <TextView                                //文本视图控件标签
10.        android:layout_columnSpan="4"        //文本视图控件横跨4列,独占网格布局一行
11.        android:layout_gravity="right"       //文本视图控件显示内容对齐方式: 右对齐
12.        android:layout_marginLeft="5dp"      //文本视图与左边控件间隔
13.        android:layout_marginRight="5dp"     //文本视图与右边控件间隔
14.        android:padding="4px"                //文本视图中文本与边框的间隔
15.        android:background="#eee"            //文本视图背景色
16.        android:text="0"                     //文本视图显示信息
17.        android:textColor="#000"             //文本视图字体颜色
18.        android:textSize="50sp" />           //文本视图字体大小
19.    <Button                                  //按钮控件标签
20.        android:layout_columnSpan="2"        //按钮控件横跨2列
21.        android:layout_gravity="fill"        //按钮控件填充2列单元格
22.        android:text="回退" />                //按钮控件显示文本
23.    <Button                                  //按钮控件标签
24.        android:layout_columnSpan="2"        //按钮控件横跨2列
25.        android:layout_gravity="fill"        //按钮控件填充2列单元格
26.        android:text="清空" />                //按钮控件显示文本
27.    <Button android:text="+" />              //沿水平方向布局按钮,每个占1列,4个占一行
28.    <Button android:text="1" />              //网格布局1个单元格,显示"1"
29.    <Button android:text="2" />              //网格布局1个单元格,显示"2"
30.    <Button android:text="3" />              //网格布局1个单元格,显示"3"
31.    <Button android:text="-" />              //网格布局1个单元格,显示"-"
32.    <Button android:text="4" />              //网格布局1个单元格,显示"4"
33.    <Button android:text="5" />              //网格布局1个单元格,显示"5"
34.    <Button android:text="6" />              //网格布局1个单元格,显示"6"
35.    <Button android:text="*" />              //网格布局1个单元格,显示"*"
36.    <Button android:text="7" />              //网格布局1个单元格,显示"7"
37.    <Button android:text="8" />              //网格布局1个单元格,显示"8"
38.    <Button android:text="9" />              //网格布局1个单元格,显示"9"
39.    <Button android:text="/" />              //网格布局1个单元格,显示"/"
40.    <Button android:layout_width="wrap_content" android:text="." />
```

```
41.    <Button android:text="0" />      //网格布局1个单元格,显示"0"
42.    <Button android:text="=" />      //网格布局1个单元格,显示"="
43.</GridLayout>                         //网格布局标签
```

图 2.20 网格布局和编码实现计算器界面

图 2.21 网格布局计算器界面

2.2.6 绝对布局管理器

1. 绝对布局的作用

绝对布局管理器采用坐标进行定位,屏幕是二维结构,绝对布局管理器就按照 x 和 y 坐标进行定位,坐标的原点位于屏幕左上角。在绝对布局中,组件位置通过 x、y 坐标(或 top、left 等属性)来控制,布局容器不再管理组件位置和大小,这些均可由用户自定义。但这种布局不能适配不同的分辨率、屏幕大小,已经过时。若仅为一种特定设备开发布局,可选择绝对布局方式。

2. 绝对布局的属性

绝对布局的属性如表 2.17 所示。

表 2.17 绝对布局的属性

序号	XML 属性	功能
1	android:layout_x	设置组件的 x 坐标值
2	android:layout_y	设置组件的 y 坐标值
3	android:layout_width	设置宽度是否充满父容器,或者仅仅包含子控件
4	android:layout_height	设置高度是否充满父容器,或者仅仅包含子控件
5	android:width	设置组件的宽度,可指定一个数字+单位,如 20dp 或 20px
6	android:height	设置组件的高度,可指定一个数字+单位,如 20dp 或 20px

3. 绝对布局典型案例

实现要点:
- 设置组件的 x 坐标值:利用 android:layout_x="100px"设置按钮的 x 坐标属性等。
- 设置组件的 y 坐标值:利用 android:layout_y="24dp"设置图像按钮的 y 坐标属性等。

该案例的源代码清单见 2.24(项目源代码见 Ch02_ex13_AbsoluteLayout.rar),效果如图 2.22 所示。

源代码清单 2.24 \Ch02_ex13_AbsoluteLayout\res\layout\absolutelayout_ex.xml

```
1. <?xml version="1.0" encoding="utf-8"?>  //xml 文件头,xml 解析版本和编码方式
```

```
2. <AbsoluteLayout xmlns:android="http://schemas.android.com/apk/res/android"
3.     android:layout_width="fill_parent"           //布局宽度：填充父控件宽度
4.     android:layout_height="fill_parent" >        //布局高度：填充父控件高度
5.     <Button                                      //按钮控件标签
6.         android:layout_width="wrap_content"      //按钮控件宽度：与内容宽度相适应
7.         android:layout_height="wrap_content"     //按钮控件高度：与内容高度相适应
8.         android:layout_x="0px"                   //按钮控件的 x 坐标
9.         android:layout_y="0px"                   //按钮控件的 y 坐标
10.        android:text="购物" />                   //按钮控件显示信息
11.    <Button                                      //按钮控件标签
12.        android:layout_width="wrap_content"      //按钮控件宽度：与内容宽度相适应
13.        android:layout_height="wrap_content"     //按钮控件高度：与内容高度相适应
14.        android:layout_x="100px"                 //按钮控件的 x 坐标
15.        android:layout_y="0px"                   //按钮控件的 y 坐标
16.        android:text="团购" />                   //按钮控件显示信息
17.    <Button                                      //按钮控件标签
18.        android:layout_width="wrap_content"      //按钮控件宽度：与内容宽度相适应
19.        android:layout_height="wrap_content"     //按钮控件高度：与内容高度相适应
20.        android:layout_x="0px"                   //按钮控件的 x 坐标
21.        android:layout_y="80px"                  //按钮控件的 y 坐标
22.        android:text="音乐" />                   //按钮控件显示信息
23.    <Button                                      //按钮控件标签
24.        android:layout_width="wrap_content"      //按钮控件宽度：与内容宽度相适应
25.        android:layout_height="wrap_content"     //按钮控件高度：与内容高度相适应
26.        android:layout_x="100px"                 //按钮控件的 x 坐标
27.        android:layout_y="80px"                  //按钮控件的 y 坐标
28.        android:text="动漫" />                   //按钮控件显示信息
29.    <TextView                                    //文本视图控件标签
30.        android:id="@+id/textView"               //文本视图控件 id 号
31.        android:layout_width="wrap_content"      //文本视图控件宽度：与内容宽度相适应
32.        android:layout_height="wrap_content"     //文本视图控件高度：与内容高度相适应
33.        android:layout_x="140dp"                 //文本视图控件的 x 坐标
34.        android:layout_y="0dp"                   //文本视图控件的 y 坐标
35.        android:textSize="28px"                  //文本视图控件显示字体大小
36.        android:textColor="#0f0"                 //文本视图控件字体颜色
37.        android:text="教书启智，育人铸魂" />     //文本视图控件显示信息
38.    <ImageView                                   //图像视图控件标签
39.        android:id="@+id/imageView"              //图像视图控件 id 号
40.        android:layout_width="wrap_content"      //图像视图宽度：与图片宽度相适应
41.        android:layout_height="wrap_content"     //图像视图高度：与图片高度相适应
42.        android:layout_x="200dp"                 //图像视图的 x 坐标
43.        android:layout_y="24dp"                  //图像视图的 y 坐标
44.        android:src="@drawable/ic_launcher" />   //图像视图的图片来源
45.    <TextView                                    //文本视图控件标签
46.        android:id="@+id/textView"               //文本视图控件 id 号
47.        android:layout_width="wrap_content"      //文本视图控件宽度：与内容宽度相适应
48.        android:layout_height="wrap_content"     //文本视图控件高度：与内容高度相适应
49.        android:layout_x="145dp"                 //文本视图控件的 x 坐标
50.        android:layout_y="70dp"                  //文本视图控件的 y 坐标
51.        android:textSize="35px"                  //文本视图控件显示字体大小
```

```
52.        android:textColor="#00f"            //文本视图控件字体颜
53.        android:text="世界因您而美丽" />     //文本视图控件显示信息
54.</AbsoluteLayout>                            //绝大布局标签
```

图 2.22 绝对布局典型案例

2.3 案例实施

2.3.1 利用线性布局显示贺知章《回乡偶书》

1. 运行结果

四个文本框控件以垂直方向线性布局居中显示贺知章的《回乡偶书》，控件背景颜色依次为红色、绿色、蓝色和黄色。本例运行结果如图 2.23 所示。

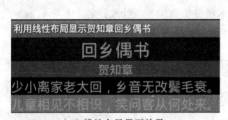 

（a）线性布局界面效果　　　　　　　　　　（b）界面设计轮廓

图 2.23 利用线性布局显示贺知章《回乡偶书》

2. 实训目的

（1）掌握线性布局标签及其属性设置方法；
（2）了解文本框控件标签及其属性设置方法；
（3）了解 TextView 对象方法；
（4）了解 findViewById()方法。

3. 实训过程

步骤 1：利用 Android Project 项目向导生成名为 Ch02_ex14_LinearLayout 的应用程序。

步骤 2：打开字符串资源文件 strings.xml，利用 Android 可视化资源编辑器，修改并定义相关字符串，如源代码清单 2.25 所示。

源代码清单 2.25　\Ch02_ex14_LinearLayout\res\values\strings.xml

```
1. <?xml version="1.0" encoding="utf-8"?>  //xml 文件头，xml 解析版本和编码方式
2. <resources>  //资源标签
3.     <string name="strTitle">回乡偶书</string>
                                    //定义字符串键值对(strTitle,回乡偶书)
```

```
4.    <string name="app_name">利用线性布局显示贺知章回乡偶书</string>
                                       //定义字符串键值对
5.    <string name="strAuthor">贺知章</string>
                                       //定义字符串键值对(strAuthor,贺知章)
6.    <string name="strContent1">少小离家老大回，乡音无改鬓毛衰。</string>
7.    <string name="strContent2">儿童相见不相识，笑问客从何处来。</string>
8. </resources>
```

步骤3：修改主布局文件 main_linearlayout.xml，如源代码清单 2.26 所示。

源代码清单 2.26　\Ch02_ex14_LinearLayout\res\layout\main_linearlayout.xml

```
1. <?xml version="1.0" encoding="utf-8"?>
2. <LinearLayout xmlns:android="http://schemas.android.com/apk/res/android"
3.     android:orientation="vertical"            //线性布局方向：垂直方向
4.     android:layout_width="fill_parent"        //线性布局宽度：填充父控件宽度
5.     android:layout_height="50pt">             //线性布局高度：50pt (磅，point)
6. <LinearLayout                                 //线性布局标签
7.     android:orientation="vertical"            //线性布局方向：垂直方向
8.     android:layout_width="fill_parent"        //线性布局宽度：填充父控件宽度
9.     android:layout_height="40pt"              //线性布局高度：40pt
10.    android:layout_weight="1">                //线性布局权重
11.    <TextView                                 //文本视图标签
12.        android:id="@+id/tvTitle"             //文本视图 id 号
13.        android:text="@string/strTitle"
                                       //文本视图显示信息，引入自定义字符串 strTitle
14.        android:textSize="12pt"               //文本视图字体大小：12pt
15.        android:background="#aa0000"          //文本视图背景色
16.        android:gravity="center_horizontal"   //文本视图中内容显示位置：水平居中
17.        android:layout_width="fill_parent"    //文本视图宽度：填充父控件宽度
18.        android:layout_height="wrap_content"  //文本视图高度：与内容高度相适应
19.        android:layout_weight="1"/>           //文本视图权重
20.    <TextView                                 //文本视图标签
21.        android:id="@+id/tvAuthor"            //文本视图 id 号
22.        android:text="@string/strAuthor"
                                       //文本视图显示信息，引入自定义字符串 strAuthor
23.        android:textSize="8pt"                //文本视图字体大小：8pt
24.        android:gravity="center_horizontal"   //文本视图中内容显示位置：水平居中
25.        android:background="#00aa00"          //文本视图背景色
26.        android:layout_width="fill_parent"    //文本视图宽度：填充父控件宽度
27.        android:layout_height="wrap_content"  //文本视图高度：与内容高度相适应
28.        android:layout_weight="1"/>           //文本视图权重
29.    <TextView                                 //文本视图标签
30.        android:id="@+id/tvContent1"          //文本视图 id 号
31.        android:text="@string/strContent1"
                                       //文本视图显示信息，引入自定义字符串 strContent1
32.        android:textSize="9pt"                //文本视图字体大小：9pt
33.        android:gravity="center_horizontal"   //文本视图中内容显示位置：水平居中
34.        android:background="#0000aa"          //文本视图背景色
35.        android:layout_width="fill_parent"    //文本视图宽度：填充父控件宽度
36.        android:layout_height="wrap_content"  //文本视图高度：与内容高度相适应
37.        android:layout_weight="1"/>           //文本视图权重
38.    <TextView                                 //文本视图标签
```

```
39.        android:id="@+id/tvContent2"              //文本视图id号
40.        android:text="@string/strContent2"
                                                     //文本视图显示信息,引入自定义字符串strContent2
41.        android:textSize="9pt"                    //文本视图字体大小:9pt
42.        android:gravity="center_horizontal"       //文本视图中内容显示位置:水平居中
43.        android:background="#aaaa00"              //文本视图背景色
44.        android:layout_width="fill_parent"        //文本视图宽度:填充父控件宽度
45.
46.        android:layout_height="wrap_content"      //文本视图高度:与内容高度相适应
47.        android:layout_weight="1"/>               //文本视图权重
48.    </LinearLayout>                               //线性布局标签
49.</LinearLayout>                                   //线性布局标签
```

步骤4:打开Ch02_01_Activity.java源代码文件,确保在onCreate()方法中通过setContentView()方法加载了屏幕布局文件main_linearlayout.xml,如源代码清单2.27所示。

源代码清单2.27　\Ch02_ex14_LinearLayout\src\sziit\lihz\Ch02_ex14_Activity.java (硬编码方式)

```
1. package sziit.lihz.ch02_ex14_LinearLayout; /*定义用户包*/
2. import android.app.Activity; /*引入Activity包*/
3. import android.os.Bundle; /*引入Bundle包*/
4. import android.widget.TextView; /*引入TextView包*/
5. public class Ch02_ex14_Activity extends Activity {
                                    //从活动基类Activity派生子类
6.   /**当活动第一次创建时调用.*/
7.   public void onCreate(Bundle savedInstanceState) {
                                    //子类重写基类onCreate方法
8.      super.onCreate(savedInstanceState); /*调用父类同名方法*/
9.      setContentView(R.layout.main_linearlayout); /*设置活动(Activity)的
        屏幕布局*/
10.     TextView tvTitle=(TextView)findViewById(R.id.tvTitle); /*查找标
        题文本框控件*/
11.     tvTitle.setText("回乡偶书"); //设置该文本视图对象显示信息
12.     TextView tvAuthor=(TextView)findViewById(R.id.tvAuthor);
                                    //绑定该控件对象
13.     tvAuthor.setText("贺知章"); //设置该文本视图对象显示信息
14.     TextView tvContent1=(TextView)findViewById(R.id.tvContent1);
                                    //绑定该控件对象
15.     tvContent1.setText("少小离家老大回,乡音无改鬓毛衰。");
                                    //设置该文本视图显示信息
16.     TextView tvContent2=(TextView)findViewById(R.id.tvContent2);
                                    //绑定该控件对象
17.     tvContent2.setText("儿童相见不相识,笑问客从何处来。");
                                    //设置该文本视图显示信息
18.   }}
```

- 代码行10:利用方法findViewById(),根据屏幕布局文件main_linearlayout.xml中定义的资源id名android:id="@+id/tvTitle",查找该id对应的文本视图控件对象。
- 代码行11:利用setText()方式直接以硬编码方式设置标题信息为"回乡偶书"。
- 代码行12:利用方法findViewById(),根据屏幕布局文件main_linearlayout.xml中定义的资源id名android:id="@+id/tvAuthor",查找该id对应的文本视图控件对象。

- 代码行 13：利用 setText()方式直接以硬编码方式设置作者信息为"贺知章"。
- 代码行 14：利用方法 findViewById()，根据屏幕布局文件 main_linearlayout.xml 中定义的资源 id 名 android:id="@+id/tvContent1"，查找内容首行文本编辑控件对象。
- 代码行 15：利用 setText()方式直接以硬编码方式设置内容首行信息为"少小离家老大回，乡音无改鬓毛衰。"。
- 代码行 16：利用方法 findViewById()，根据屏幕布局文件 main_linearlayout.xml 中定义的资源 id 名 android:id="@+id/tvContent1"，查找内容第二行文本编辑控件对象。
- 代码行 17：利用 setText()方式直接以硬编码方式设置内容第二行信息为"儿童相见不相识，笑问客从何处来。"。

步骤 5：将步骤 4 的硬编码设置文本信息 setText()方法调用改成从资源文件读取字符串信息，如源代码清单 2.28 所示。

源代码清单 2.28 \Ch02_ex14_LinearLayout\Ch02_ex14_Activity.java (资源引用方式)

```
1. public class Ch02_ex14_Activity extends Activity {
                                            //从活动基类Activity派生子类
2.     /**当活动第一次创建时调用.*/
3.     public void onCreate(Bundle savedInstanceState) {
                                            //子类重写基类onCreate方法
4.         super.onCreate(savedInstanceState); /*调用父类同名方法*/
5.         setContentView(R.layout.main_linearlayout); /*设置活动(Activity)的
           屏幕布局*/
6.         TextView tvTitle=(TextView)findViewById(R.id.tvTitle); /*查找标题文
           本框控件*/
7.         tvTitle.setText(R.string.strTitle);    //设置该文本视图显示信息
8.         TextView tvAuthor=(TextView)findViewById(R.id.tvAuthor);
                                            //通过文本视图id绑定对象
9.         tvAuthor.setText(R.string.strAuthor);  //设置该文本视图显示信息
10.        TextView tvContent1=(TextView)findViewById(R.id.tvContent1);
                                            //通过id绑定对象
11.        tvContent1.setText(R.string. strContent1); //设置该文本视图显示信息
12.        TextView tvContent2=(TextView)findViewById(R.id.tvContent2);
                                            //通过id绑定对象
13.        tvContent2.setText(R.string.strTitle); //设置该文本视图显示信息
14.    }}
```

- 代码行 7：根据字符串资源文件 strings.xml 中定义的字符串名<string name="strTitle">回乡偶书</string>，通过资源引用 R.string.strTitle，利用 setText()方法设置标题信息。
- 代码行 9：根据字符串资源文件 strings.xml 中定义的字符串名<string name="strAuthor">贺知章</string>，通过资源引用 R.string.strAuthor，利用 setText()方法设置作者信息。
- 代码行 11：根据字符串资源文件 strings.xml 中定义的字符串名<string name= "strContent1">少小离家老大回，乡音无改鬓毛衰。</string>，通过资源引用 R.string. strContent1，利用 setText()方法设置首行内容信息。
- 代码行 13：根据字符串资源文件 strings.xml 中定义的字符串名<string name= "strContent2">儿童相见不相识，笑问客从何处来。</string>，通过资源引用 R.string.strTitle，利用 setText()方法设置第二行内容信息。

TextView 控件的典型对象方法及其 XML 属性对应关系如表 2.18 所示。

表 2.18　TextView 的对象方法与 XML 属性对应关系

序　号	对象方法	XML 属性	描　　述
1	setText()	android:text	文本信息
2	setTextColor()	android:textColor	文本色
3	setBackgroundColor()	android:background	文本背景色
4	setWidth()	android:width	文本宽度
5	setHeight()	android:height	文本高度
6	setText()	android:textSize	文本尺寸
7	setGravity()	android:gravity	对齐方式
8	setHint()	android:hint	提示信息
9	setHintTextColor	android:textColorHint	提示信息颜色

2.3.2　利用相对布局实现密码验证界面

1. 运行结果

将一个文本视图控件、一个编辑控件和两个按钮控件，利用相对布局实现密码验证界面。本例运行结果如图 2.24 所示。

（a）界面设计

（b）密码验证界面效果

图 2.24　显示密码验证的相对布局范例

2. 实训目的

（1）掌握相对布局标签及其属性设置方法；
（2）了解 RelativeLayout.LayoutParams 参数设置方法；
（3）了解文本视图对象方法；
（4）了解编辑控件对象方法；
（5）了解按钮控件对象方法；
（6）掌握控件查找方法 findViewById()。

3. 实训过程

步骤 1：利用 Android Project 项目向导生成名为 Ch02_ex15_RelativeLayout 的应用程序。
步骤 2：修改屏幕布局文件 main_relativelayout.xml，如源代码清单 2.29 所示。

源代码清单 2.29　\Ch02_ex15_RelativeLayout\res\layout\main_relativelayout.xml

```
1.<?xml version="1.0" encoding="utf-8"?> //xml 文件头，定义 xml 版本和编码方法
```

```
2.  <RelativeLayout xmlns:android="http://schemas.android.com/apk/res/android"
3.      android:layout_width="fill_parent"         //线性布局宽度：填充父控件宽度
4.      android:layout_height="fill_parent">       //线性布局高度：填充父控件高度
5.      <TextView                                  //文本视图控件标签
6.          android:id="@+id/tvLabel"              //文本视图控件 id 号
7.          android:layout_width="fill_parent"     //文本视图控件宽度：填充父控件宽度
8.          android:layout_height="wrap_content"
                                                   //文本视图控件高度：与显示内容高度相适应
9.          android:text="请输入密码:"/>           //文本视图控件信息信息
10.     <EditText                                  //文本编辑控件标签
11.         android:id="@+id/etPassword"           //文本编辑控件 id 号
12.         android:layout_width="fill_parent"     //文本编辑控件宽度：填充父控件宽度
13.         android:layout_height="wrap_content"
                                                   //文本编辑控件高度：与显示内容高度相适应
14.         android:background="@android:drawable/editbox_background"
                                                   //文本编辑控件背景色
15.         android:layout_below="@id/tvLabel"
                                        //文本编辑控件位于文本视图控件 tvLabel 下面
16.         android:password="true"/>              //启动密码属性
17.     <Button                                    //按钮控件标签
18.         android:id="@+id/btnOk"                //按钮控件 id 号
19.         android:layout_width="wrap_content"    //按钮控件宽度：与显示内容宽度相适应
20.         android:layout_height="wrap_content"   //按钮控件高度：与显示内容高度相适应
21.         android:layout_below="@id/etPassword"
                                        //按钮控件位于编辑控件 etPassword 下面
22.         android:layout_alignParentLeft="true"     //启动与父控件左对齐属性
23.         android:layout_marginLeft="10dip"
                                        //与左控件左边距为 10dip（设备无关像素）
24.         android:text="确认" />                 //按钮控件显示信息
25.     <Button                                    //按钮控件标签
26.         android:id="@+id/btnCancel"            //按钮控件 id 号
27.         android:layout_width="wrap_content"    //按钮控件宽度：与显示内容宽度相适应
28.         android:layout_height="wrap_content"   //按钮控件高度：与显示内容高度相适应
29.         android:layout_toRightOf="@id/btnOk"   //该按钮位于 btnOk 控件右边
30.         android:layout_alignTop="@id/btnOk"    //该按钮与 btnOk 控件顶部对齐
31.         android:text="取消" />                 //按钮控件显示信息
32. </RelativeLayout>                              //相对布局标签
```

Tips 2.2：android:layout_*属性，例如 layout_below, layout_alignParentRight, and layout_toLeftOf、android:layout_alignParentLeft、android:layout_marginLeft。当使用屏幕相对布局标签 RelativeLayout 时，这些 android:layout_*属性定义了各个控件之间的相对位置关系。有些属性使用子控件的资源 ID 定义其相对位置。详细请参见相对布局 LayoutParams 属性表。

2.3.3 利用表格布局实现菜单

1. 运行结果

本例运行结果如图 2.25 所示。以灰色背景、黑色字体显示菜单，右边为菜单的快捷键。

图 2.25　显示菜单的表格布局范例

2．实训目的

（1）掌握表格布局(TableLayout)标签及其属性设置方法；

（2）掌握 TableRow 标签及其属性设置方法；

（3）了解文本视图控件标签及其属性设置方法。

3．实训过程

步骤1：创建名为 Ch02_ex16_TableLayout 的 Android 应用程序。

步骤2：编辑屏幕布局文件 main_tablelayout.xml，其代码清单如 2.30 所示。

代码清单 2.30　\Ch02_ex16_TableLayout\res\layout\main_tablelayout.xml

```
1.  <?xml version="1.0" encoding="utf-8"?>             //xml 文本头，定义 xml 解析版本和编码方式
2.  <TableLayout xmlns:android="http://schemas.android.com/apk/res/android"
3.       android:layout_width="fill_parent"            //表格布局宽度：填充父控件宽度
4.       android:layout_height="fill_parent"           //表格布局高度：填充父控件高度
5.       android:background="#009999"                  //表格布局背景色
6.       android:stretchColumns="1">                   //设置表格布局第1列伸展属性
7.       <TableRow>                                    //表格行标签
8.           <TextView                                 //文本视图标签
9.               android:id="@+id/tvNew"               //文本视图 id 号
10.              android:layout_column="1"             //指定该文本视图到第1列
11.              android:text="新建"                   //文本视图显示信息
12.              android:padding="3dip" />             //文本控件显示内容与边框间距
13.          <TextView
14.              android:text="ALT+SHIFT+N"            //文本视图显示信息
15.              android:gravity="right"               //文本视图显示信息与右边界对齐
16.              android:padding="3dip" />             //文本控件显示内容与边框间距
17.      </TableRow>                                   //表格行标签结束标记
18.      <TableRow>                                    //表格行标签
19.          <TextView                                 //文本视图标签
20.              android:id="@+id/tvOpen"              //文本视图 id 号
21.              android:layout_column="1"             //指定该文本视图到第1列
22.              android:text="打开文件..."            //文本视图显示信息
23.              android:padding="3dip" />             //文本控件显示内容与边框间距
24.          <TextView                                 //文本视图标签
25.              android:text="Ctrl+O"                 //文本视图显示信息
26.              android:gravity="right"               //文本视图显示信息与右边界对齐
27.              android:padding="3dip" />             //文本控件显示内容与边框间距
28.      </TableRow>
```

```
29.    <View                                            //视图标签,独占一行,用于显示菜单分隔线
30.        android:layout_height="2dip"                 //视图高度
31.        android:background="#FF909090" />            //视图背景色
32.    <TableRow>                                       //表格行标签
33.        <TextView                                    //文本视图标签
34.            android:id="@+id/tvClose"                //文本视图 id 号
35.            android:layout_column="1"                //指定该文本视图到第 1 列
36.            android:text="关闭"                      //文本视图显示信息
37.            android:padding="3dip" />                //文本控件显示内容与边框间距
38.        <TextView                                    //文本视图标签
39.            android:text="Ctrl+W"                    //文本视图显示信息
40.            android:gravity="right"                  //文本视图显示信息与右边界对齐
41.            android:padding="3dip" />                //文本控件显示内容与边框间距
42.    </TableRow>
43.    <TableRow>                                       //表格行标签
44.        <TextView                                    //文本视图标签
45.            android:id="@+id/tvCloseAll"             //文本视图 id 号
46.            android:layout_column="1"                //指定该文本视图到第 1 列
47.            android:text="全部关闭"                  //文本视图显示信息
48.            android:padding="3dip" />                //文本控件显示内容与边框间距
49.        <TextView                                    //文本视图标签
50.            android:text="Ctrl+SHIFT+W"              //文本视图显示信息
51.            android:gravity="right"                  //文本视图显示信息与右边界对齐
52.            android:padding="3dip" />                //文本控件显示内容与边框间距
53.    </TableRow>
54.    <View                                            //视图标签,独占一行,用于显示菜单分隔线
55.        android:layout_height="2dip"                 //视图高度
56.        android:background="#FF909090" />            //视图背景色
57.    <TableRow>                                       //表格行标签
58.        <TextView                                    //文本视图标签
59.            android:text="X"                         //文本视图显示信息
60.            android:padding="3dip" />                //文本控件显示内容与边框间距
61.        <TextView                                    //文本视图标签
62.            android:id="@+id/tvImport"               //文本视图 id 号
63.            android:text="导入..."                   //文本视图显示信息
64.            android:padding="3dip" />                //文本控件显示内容与边框间距
65.         <TextView                                   //文本视图标签
66.            android:text="Ctrl+I"                    //文本视图显示信息
67.            android:gravity="right"                  //文本视图显示信息与右边界对齐
68.            android:padding="3dip" />                //文本控件显示内容与边框间距
69.    </TableRow>
70.    <TableRow>                                       //表格行标签
71.        <TextView                                    //文本视图标签
72.            android:text="X"                         //文本视图显示信息
73.            android:padding="3dip" />                //文本控件显示内容与边框间距
74.        <TextView                                    //文本视图标签
75.            android:id="@+id/tvExport"               //文本视图 id 号
76.            android:text="导出..."                   //文本视图显示信息
77.            android:padding="3dip" />                //文本控件显示内容与边框间距
78.        <TextView                                    //文本视图标签
79.            android:text="Ctrl+E"                    //文本视图显示信息
80.            android:gravity="right"                  //文本视图显示信息与右边界对齐
```

```
81.            android:padding="3dip" />         //文本控件显示内容与边框间距
82.        </TableRow>
83.        <View                                 //视图标签,独占一行,用于显示菜单分隔线
84.            android:layout_height="2dip"      //视图高度
85.            android:background="#FF909090" /> //视图背景色
86.        <TableRow>                            //表格行标签
87.            <TextView                         //文本视图标签
88.                android:id="@+id/tvQuit"      //文本视图id号
89.                android:layout_column="1"     //指定该文本视图到第1列
90.                android:text="退出"           //文本视图显示信息
91.                android:padding="3dip" />     //文本控件显示内容与边框间距
92.        </TableRow>    </TableLayout>         //表格布局标签
```

步骤3:编辑 Ch02_ex16_Activity.java 代码,如源代码清单 2.31 所示。

源代码清单 2.31   \Ch02_ex16_TableLayout\src\sziit\lihz\
Ch02_ex16_TableLayout\Ch02_ex16_Activity.java

```
1. public class Ch02_ex16_Activity extends Activity {//从活动基类派生子类
2.    /** 当活动第一次创建时被调用. */
3.    public void onCreate(Bundle savedInstanceState) {//子类重写基类 onCreate 方法
4.        super.onCreate(savedInstanceState);  //调用父对象方法
5.        setContentView(R.layout.main_tablelayout);}}//设置活动的显示视图
```

第 5 行代码:用 setContentView(int)方法为活动加载由资源 ID(R.layout.main_tablelayout)引用的\res\layout\main_tablelayout.xml 布局文件。

2.3.4 利用网格视图布局浏览图片

1. 运行结果

本例界面设计与运行结果如图 2.26 所示。

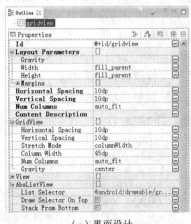 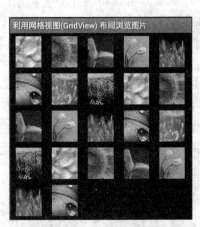

(a)界面设计   (b)显示图片界面效果

图 2.26   显示图片网格视图布局范例

2. 实训目的

(1)掌握网格视图布局标签及其属性设置方法;
(2)了解图片视图标签及其属性设置方法;
(3)了解适配器视图选项单击监听器 AdapterView.OnItemClickListener;

（4）了解视图组组件（ViewGroup）；

（5）了解列表适配器（ListAdapter）；

（6）了解视图组件；

（7）了解适配器基类（BaseAdapter）；

（8）了解 Toast 组件。

3. 实训过程

步骤 1：建立名为 Ch02_ex17_GridView 的 Android 应用程序项目。

步骤 2：将要在网格视图中显示的 JPG 图片，按高分辨率图片、中分辨率图片和低分辨率图片分别放在 /res/drawable-hdpi、/res/drawable-mdpi 和 /res/drawable-ldpi 目录，图片命名为 sample_0.jpg ~ sample_5.jpg。

步骤 3：编辑屏幕布局文件 main_gridview.xml，如源代码清单 2.32 所示。

源代码清单 2.32　\Ch02_ex17_GridView\res\layout\main_gridview.xml

```xml
1. <?xml version="1.0" encoding="utf-8"?>
2. <GridView xmlns:android="http://schemas.android.com/apk/res/android"
3.     android:id="@+id/gridview"              //网格视图布局 id
4.     android:layout_width="fill_parent"      //网格视图布局宽度：填充父控件宽度
5.     android:layout_height="fill_parent"     //网格视图布局高度：填充父控件高度
6.     android:columnWidth="45dp"              //网格视图布局列宽
7.     android:numColumns="auto_fit"           //网格视图列数：自动匹配
8.     android:verticalSpacing="10dp"          //网格视图垂直间距
9.     android:horizontalSpacing="10dp"        //网格视图水平间距
10.    android:stretchMode="columnWidth"       //网格视图伸展模式
11.    android:gravity="center"                //网格视图图像居中放置
12./>
```

步骤 4：从基类 BaseAdapter 派生名为 ImageAdapter 的子类。在包浏览器 Package Explorer 中选择 Ch02_ex17_GridView 项目，展开源代码 src 目录，选中 sziit.lihz.Ch02_ex17_GridView，右击，选择【New】→【Class】命令，打开【New Java Class】对话框。输入类名 ImgAdapter，基类（Superclass）选择 BaseAdapter；选中 Constructors from superclasss 复选框，如图 2.27 所示。单击 Finish 按钮，生成 ImgAdapter 类 Java 源代码文件模板，完善 ImgAdapter.java 子类功能后，按【Ctrl+Shift+O】组合键引入未添加的包，如源代码清单 2.33 所示。

源代码清单 2.33　\Ch02_ex17_GridView\src\sziit\lihz\Ch02_ex17_GridView\ImgAdapter.java

```java
1. package sziit.lihz.ch02_ex17_GridView;/*定义用户包*/
2. import android.content.Context;                //引入上下文包
3. import android.view.View;                     //引入视图包
4. import android.view.ViewGroup;                //引入视图组包
5. import android.widget.BaseAdapter;            //引入适配器基类
6. import android.widget.GridView;               //引入网格视图布局包
7. import android.widget.ImageView;              //引入图像视图包
8. public class ImgAdapter extends BaseAdapter {//从基类 BaseAdapter 派生子类
9.     private Context mContext;                 // 定义私有上下文对象
10.    public ImgAdapter(Context c) {            // 构造器
11.        mContext = c;
12.    }
13.    public int getCount() {                   //重写基类方法,获得容器包含子控件数
14.        return mThumbIds.length;
```

```
15.    }
16.    public Object getItem(int position) {  //重写基类方法,获得指定位置子控件对象
17.        return null;
18.    }
19.    public long getItemId(int position) {  //重写基类方法,获得指定位置子控件 id 号
20.        return 0;
21.    }
22.    //获得指定位置的视图对象
23.    public View getView(int position, View convertView, ViewGroup parent) {//重写基类方法
24.        ImageView imageView;                    //定义图像视图对象
25.        if (convertView == null) {
26.            imageView = new ImageView(mContext);  //构造图像视图对象
27.            imageView.setLayoutParams(new GridView.LayoutParams(80, 80));
                                                    //设置图像视图大小
28.            imageView.setScaleType(ImageView.ScaleType.CENTER_CROP);
                                                    //设置比例类型
29.            imageView.setPadding(6, 6, 6, 6);
30.        } else {
31.            imageView = (ImageView) convertView;
32.        }
33.        imageView.setImageResource(mThumbIds[position]);
                                                    // 设置图像视图显示的图片
34.        return imageView;
35.    }
36.    // 引用外部图像
37.    private Integer[] mThumbIds = {R.drawable.sample_2, R.drawable.sample_3,
38.            R.drawable.sample_4, R.drawable.sample_5, R.drawable.sample_6,
39.            R.drawable.sample_7, R.drawable.sample_0, R.drawable.sample_1,
40.            R.drawable.sample_2, R.drawable.sample_3, R.drawable.sample_4,
41.            R.drawable.sample_5, R.drawable.sample_6, R.drawable.sample_7,
42.            R.drawable.sample_0, R.drawable.sample_1, R.drawable.sample_2,
43.            R.drawable.sample_3, R.drawable.sample_4, R.drawable.sample_5,
44.            R.drawable.sample_6, R.drawable.sample_7 };}
```

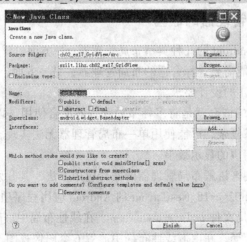

图 2.27　新建 Java 类 ImgAdapter 对话框

步骤 5：完善 Ch02_ex17_Activity.java 源代码文件，按【Ctrl+Shift+O】组合键补全未引入的包，如源代码清单 2.34 所示。

源代码清单 2.34　\Ch02_ex17_GridView\src\sziit\lihz\ch02_ex17_GridView\Ch02_ex17_Activity.java

```java
1. package sziit.lihz.ch02_ex17_GridView;/*定义用户包*/
2. import android.app.Activity;//引入Activity包*/
3. import android.os.Bundle;/*引入Bundle包*/
4. import android.view.View;        //引入视图包
5. import android.widget.AdapterView;
6. import android.widget.AdapterView.OnItemClickListener;
                                    //引入网格布局项单击监听器包
7. import android.widget.GridView;  //引入网格视图布局包
8. import android.widget.Toast;     //引入Toast包
9. public class Ch02_ex17_Activity extends Activity {
                                    //从活动基类Activity派生子类
10.    /** 当活动第一次创建时调用. */
11.    public void onCreate(Bundle savedInstanceState) {//子类重写基类onCreate方法
12.        super.onCreate(savedInstanceState);/* 调用父类同名方法 */
13.        setContentView(R.layout.main_gridview);/* 设置活动(Activity)的屏幕布局 */
14.        GridView gridview = (GridView) findViewById(R.id.gridview);
                                    //绑定GridView对象
15.        gridview.setAdapter(new ImgAdapter(this));//设置网格布局适配器
16.        //设置网格布局项单击监听器
17.        gridview.setOnItemClickListener(new OnItemClickListener() {
                                    //网格布局项单击事件回调
18.            public void onItemClick(AdapterView<?> parent, View v,
19.                    int position, long id) {
20.                Toast.makeText(Ch02_ex17_Activity.this, "" + position,
21.                    Toast.LENGTH_SHORT).show();
22.            }
23.        }); }}
```

2.4　知识拓展

2.4.1　Android 布局管理器的嵌套

前面已详细讲解了 Android 的几种典型布局管理器，一方面，开发人员在设计 Android 应用程序 UI 时，可根据功能需要选择合适的布局管理器，另一方面，Android 布局管理器还可相互嵌套，以满足复杂 UI 的设计要求。

Android 布局管理器的嵌套就是将多种布局管理器混合使用，以达到复杂布局的排版效果。若一个布局界面效果复杂，可能使用单一布局管理器无法实现，则就需要将多种布局管理器嵌套起来，以达到期望的显示效果。

本小节通过一个实例来讲解布局管理器的嵌套应用方法，其开发步骤如下：

步骤 1：建立名为 Ch02_ex18_NestLayout 的 Android 应用程序项目。

步骤 2：编辑屏幕布局文件 main_gridview.xml，在线性布局中嵌套一个文本视图控件、一个线性布局和一个表格布局；在内嵌线性布局中放置 6 个图像视图，在表格布局中水平放置一个文本编辑控件和 3 个按钮，如源代码清单 2.35 所示。

源代码清单 2.35　\Ch02_ex18_NestLayout\res\layout\nestedlayout_main.xml

```
1.  <?xml version="1.0" encoding="utf-8"?>       //xml文件头,定义xml版本和编码方式
2.  <LinearLayout xmlns:android="http://schemas.android.com/apk/res/android"
    //线性布局标签
3.      android:layout_width="fill_parent"       //布局宽度:填充父控件宽度
4.      android:layout_height="fill_parent"      //布局高度:填充父控件高度
5.      android:orientation="vertical" >         //布局方向:垂直方向
6.      <TextView                                //文本视图标签
7.          android:id="@+id/textView1"          //文本视图id号
8.          android:layout_width="fill_parent"   //文本视图宽度:填充父控件宽度
9.          android:layout_height="wrap_content" //文本视图高度:与显示内容高度相适应
10.         android:gravity="center"             //文本视图内容位置:居中显示
11.         android:text="知之者不如好之者，好之者不如乐之者" />//文本视图显示内容
12.     <LinearLayout                            //线性布局标签
13.         android:layout_width="fill_parent"   //布局宽度:填充父控件宽度
14.         android:layout_height="wrap_content" //布局高度:填充父控件高度
15.         android:gravity="center"             //布局位置:居中位置
16.         android:orientation="horizontal" >   //布局方向:水平方向
17.         <ImageView                           //图像视图标签
18.             android:id="@+id/imageView1"     //图像视图id号
19.             android:layout_width="wrap_content" //图像视图宽度:填充父控件宽度
20.             android:layout_height="wrap_content"//图像视图高度:填充父控件高度
21.             android:src="@drawable/ic_launcher" />  //图像视图显示的图片
22.         <ImageView                           //图像视图标签
23.             android:id="@+id/imageView2"     //图像视图id号
24.             android:layout_width="wrap_content" //图像视图宽度:填充父控件宽度
25.             android:layout_height="wrap_content"   //图像视图高度:填充父控件高度
26.             android:src="@drawable/ic_launcher" />//图像视图显示的图片
27.         ……
28.         <ImageView                           //图像视图标签
29.             android:id="@+id/imageView6"     //图像视图id号
30.             android:layout_width="wrap_content" //图像视图宽度:填充父控件宽度
31.             android:layout_height="wrap_content" //图像视图高度:填充父控件高度
32.             android:src="@drawable/ic_launcher" />//图像视图显示的图片
33.     </LinearLayout>
34.     <TableLayout                             //表格布局标签
35.         android:layout_width="fill_parent"   //布局宽度:填充父控件宽度
36.         android:layout_height="wrap_content" //布局高度:与显示内容高度相适应
37.         android:gravity="center"             //布局位置:居中位置
38.         android:orientation="horizontal" >   //布局方向:水平方向
39.         <TableRow                            //表格行标签
40.             <EditText                        //文本编辑控件标签
41.                 android:id="@+id/editText1"  //文本编辑控件id号
42.                 android:layout_width="wrap_content"   //文本编辑控件宽度
43.                 android:layout_height="wrap_content"  //文本编辑控件高度
44.                 android:text="搜索关键字" />  //文本编辑控件显示内容
45.             <Button                          //按钮控件标签
46.                 android:id="@+id/button1"    //按钮控件id号
```

```
47.          android:layout_width="wrap_content"   //按钮控件宽度
48.          android:layout_height="wrap_content"  //按钮控件高度
49.          android:text="搜索" />                 //按钮控件显示内容
50.        <Button                                 //按钮控件标签
51.          android:id="@+id/button2"             //按钮控件id号
52.          android:layout_width="wrap_content"   //按钮控件宽度
53.          android:layout_height="wrap_content"  //按钮控件高度
54.          android:text="收藏" />                 //按钮控件显示内容
55.        <Button                                 //按钮控件标签
56.          android:id="@+id/button3"             //按钮控件id号
57.          android:layout_width="wrap_content"   //按钮控件宽度
58.          android:layout_height="wrap_content"  //按钮控件高度
59.          android:text="截图" />                 //按钮控件显示内容
60.       </TableRow>
61.    </TableLayout>
62.</LinearLayout>
```

步骤 3：该案例界面设计如图 2.28（a）所示，运行结果如图 2.28（b）所示。

（a）界面设计　　　　　　　　　　　　（b）嵌套布局界面效果

图 2.28　嵌套布局管理器案例界面

2.4.2　Android 抽象布局标签

Android 包括三种抽象布局标签<include />、<merge />、<ViewStub />，用于布局优化，去除不必要的嵌套和视图结点，减少不必要的 infalte 操作。

1. <include>布局标签

<include>标签常用于将布局中的公共部分提取出来供其他布局使用，以实现布局模块化，这为布局编写提供了很大便利。<include>标签唯一需要的属性是 layout 属性，其指定所需包含的布局文件，可设置 android:id 和 android:layout_*属性来覆盖被引入布局根结点的对应属性值。注意，重新设置 android:id 后，子布局的顶结点 i 将发生变化。

<include/>标签能够重用布局文件，其作用就是提高代码的重用性。下面通过一个实例来讲解<include>布局标签应用方法，其开发步骤如下：

步骤 1：建立名为 Ch02_ex19_include 的 Android 应用程序项目。

步骤 2：建立可重用按钮布局文件 reuse_button.xml，如源代码清单 2.36 所示。

源代码清单 2.36　\Ch02_ex19_include\res\layout\reuse_button.xml

```
1. <?xml version="1.0" encoding="utf-8"?>   //xml 头文件，定义 xml 版本和编码方式
```

```
2.  <LinearLayout xmlns:android="http://schemas.android.com/apk/res/android"
                                                    //线性布局标签
3.      android:layout_width="fill_parent"          //布局宽度：填充父控件宽度
4.      android:layout_height="wrap_content"        //布局高度：与内容高度相匹配
5.      android:orientation="vertical" >            //布局方向：垂直方向
6.      <Button                                     //按钮控件标签
7.          android:id="@+id/subButton"             //按钮控件id号
8.          android:layout_width="wrap_content"     //按钮控件宽度：与内容宽度相匹配
9.          android:layout_height="wrap_content"    //按钮控件高度：与内容高度相匹配
10.         android:layout_gravity="center_horizontal|center_vertical"
                                                    //按钮控件布局位置
11.         android:layout_marginTop="4dp"          //按钮控件布局顶部边距
12.         android:background="#ff0000ff"          //按钮控件背景色
13.         android:clickable="true"                //按钮控件可单击
14.         android:gravity="center_horizontal|center_vertical"
                                                    //按钮控件显示内容位置
15.         android:onClick="onClick"               //按钮控件单击事件
16.         android:paddingLeft="20dp"              //按钮控件左边距
17.         android:paddingRight="20dp"             //按钮控件右边距
18.         android:text="include 标签的监听事件处理 "  //按钮控件显示内容
19.         android:textColor="#ffffff"             //按钮控件显示文本颜色
20.         android:textSize="18sp" />              //按钮控件显示字体大小
21. </LinearLayout>
```

步骤 3：修改 include1.xml 文件，利用<include>布局标签重用 reuse_button.xml 布局，如源代码清单 2.37 所示。

源代码清单 2.37　\Ch02_ex19_include\res\layout\include1.xml

```
1.  <LinearLayout xmlns:android="http://schemas.android.com/apk/res/android"
                                                    //线性布局标签
2.      xmlns:tools="http://schemas.android.com/tools"
3.      android:layout_width="fill_parent"          //线性布局宽度：填充父控件宽度
4.      android:layout_height="fill_parent"         //线性布局高度：填充父控件高度
5.      android:orientation="vertical" >            //线性布局方向：垂直方向
6.      <include                                    //利用 include 标签重用 reuse_button 布局
7.          android:id="@+id/btn1"                  //重定义布局 id 号
8.          layout="@layout/reuse_button" />        //设置重用的布局 reuse_button
9.      <include                                    //利用 include 标签重用 reuse_button 布局
10.         android:id="@+id/btn2"                  //重定义布局 id 号
11.         layout="@layout/reuse_button" />        //设置重用的布局 reuse_button
12.     <include                                    //利用 include 标签重用 reuse_button 布局
13.         android:id="@+id/btn3"                  //重定义布局 id 号
14.         layout="@layout/reuse_button" />        //设置重用的布局 reuse_button
15.     <include                                    //利用 include 标签重用 reuse_button 布局
16.         android:id="@+id/btn4"                  //重定义布局 id 号
17.         layout="@layout/reuse_button" />        //设置重用的布局 reuse_button
18. </LinearLayout>
```

步骤 4：修改 MainActivity.java，增加各个 include 标签的监听事件处理代码，如源代码清单 2.38 所示。

源代码清单 2.38 　\Ch02_ex19_include\src\sziit\lihz\ch02_ex19_include\MainActivity.java

```java
1.  public class MainActivity extends Activity {
                                    //从基类Activity公有派生子类MainActivity
2.      private final String TAG = "MainActivity";//定义私有只读字符串TAG
3.      private int[] id = { R.id.btn1, R.id.btn2, R.id.btn3, R.id.btn4 };
                                    //定义重用布局的id号数组
4.      protected void onCreate(Bundle savedInstanceState) {//重写onCreate方法
5.          super.onCreate(savedInstanceState);    //调用父对象方法onCreate
6.          setContentView(R.layout.include1);     //设置活动界面布局
7.          for (int i = 0; i < id.length; i++) {//遍历利用include所用的布局对象
8.              View v = findViewById(id[i]);     //动态绑定布局对象
9.              Button tv = (Button) v.findViewById(R.id.subButton);
                                    //动态绑定布局对象中的按钮对象
10.             tv.setId(i);                       //设置按钮的id号
11.             tv.setText(tv.getText().toString()+ tv.getId());//设置按钮显示文本
12.         } }
13.     public void onClick(View v) {        //实现include标签的监听事件处理功能
14.         Log.v(TAG, "subTextViewId:" + v.getId());//记录日志信息
15.         switch (v.getId()) {                   //根据按钮对象id处理不同的按钮
16.         case 0: case 1: case 2:    case 3:
17.             Toast.makeText(getApplicationContext(), ((Button) v).getText().
18.                 toString() + v.getId(),Toast.LENGTH_LONG).show();
19.             break;
20.         default: break;
21.         } }}
```

步骤 5：运行 Ch02_ex19_include，其结果如图 2.29 所示。

图 2.29 　<include>布局标签案例界面

2. <merge>布局标签

<merge>标签在优化 UI 结构方面起着非常重要的作用，它可删减多余的层级，提升 UI 性能。通常在使用 include 后可能导致布局嵌套过多，产生不必要的多余 layout 结点，导致解析变慢，使用<merge>标签可有效排除把一个布局插入到另一个布局时产生的多余的视图组。

<merge>多用于替换 FrameLayout，或者当一个布局包含另一个时，<merge>标签可消除视图层次结构中多余的视图组。merge 标签的典型应用情况包括：

（1）若布局顶结点是帧布局且不需要设置 background 或 padding 等属性，可用 merge 代替 FrameLayout，因为 Activity 内容视图的父视图就是个 FrameLayout，从而可用 merge 消除多余的 FrameLayout 结点。

（2）当某布局作为子布局被其他布局包含时，用 merge 作为该布局的顶结点，其在被包含时顶结点会自动被忽略，从而将该子结点全部合并到主布局中。

（3）为避免插入冗余的 ViewGroup，可用<merge>标签作为可复用布局根结点，例如：

```
<merge xmlns:android="http://schemas.android.com/apk/res/android">
  <Button   android:layout_width="fill_parent"
     android:layout_height="wrap_content"   android:text="添加"/>
  <Button   android:layout_width="fill_parent"
     android:layout_height="wrap_content"   android:text="删除"/>  </merge>
```

使用 include 标签将以上布局包含到另一个布局时，系统会忽略 merge 标签，直接把两个 Button 替换到 include 标签的位置。下面通过一个实例来讲解<merge>布局标签的应用方法，其开发步骤如下：

步骤 1：建立名为 Ch02_ex20_MergeLayout 的 Android 应用程序项目。

步骤 2：利用<merge>标签代替<FrameLayout>，消除多余的 FrameLayout 结点，如源代码清单 2.39 所示。

源代码清单 2.39   \Ch02_ex20_MergeLayout\res\layout\activity_main.xml

```
1. <merge xmlns:android="http://schemas.android.com/apk/res/android"
                                                   //用 merge 代替 FrameLayout
2.     android:layout_width="fill_parent"          //布局宽度：填充父控件宽度
3.     android:layout_height="fill_parent">        //布局高度：填充父控件高度
4.     <ImageView                                  //图像视图控件标签
5.        android:layout_width="wrap_content"      //图像视图控件宽度：由内容宽度决定
6.        android:layout_height="wrap_content"     //图像视图控件高度：由内容高度决定
7.        android:layout_gravity="center_horizontal|top"  //图像视图控件位置
8.        android:scaleType="center"               //图像视图控件比例类型
9.        android:src="@drawable/bg" />            //图像视图控件显示图片来源
10.    <TextView                                   //文本视图控件标签
11.       android:layout_width="wrap_content"      //文本视图控件宽度：由内容宽度决定
12.       android:layout_height="wrap_content"     //文本视图控件高度：由内容高度决定
13.       android:layout_gravity="center_horizontal|top"   //文本视图控件位置
14.       android:layout_marginBottom="20dip"      //文本视图控件与底部距离
15.       android:background="#AA000000"           //文本视图控件背景色（半透明）
16.       android:padding="20dip"                  //文本视图控件间距
17.       android:text="android 背景图片"           //文本视图控件显示信息
18.       android:textColor="#ffffffff" />         //文本视图控件字体颜色
19. </merge>
```

步骤 3：运行 Ch02_ex20_MergeLayout，其结果如图 2.30 所示。

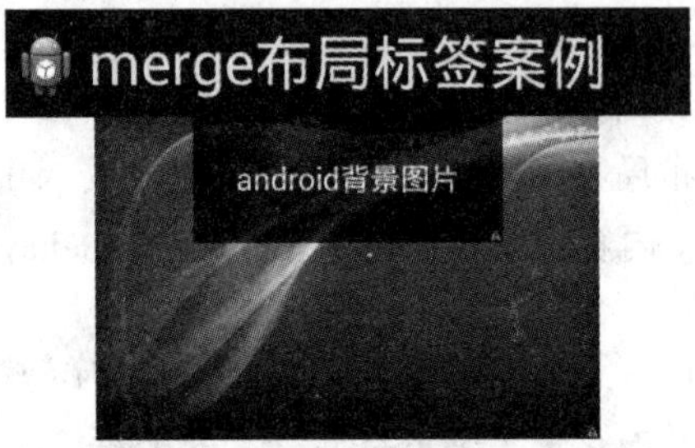

图 2.30 <merge>布局标签案例界面

本章小结

本章首先简要介绍了 Android 布局管理器的基本概念、Android 界面组件结构层次和 Android 用户界面的一般结构；其次，重点讲解了线性布局、相对布局、表格布局、帧布局、网格布局和绝对布局 6 种布局管理器的基本概念和常用属性；最后，通过典型案例帮助读者掌握这些布局管理器的使用方法和技巧。

强化练习

1. 填空题

（1）Android 中的 6 种典型布局管理器是（   ）、（   ）、（   ）、（   ）、（   ）和（   ）。

（2）Android 布局管理器均是以（   ）为基类派生出来的。

（3）Android 用户界面显示主要通过（   ）和（   ）两大类实现，（   ）类负责整体布局，（   ）负责控件应用。

（4）Android 事件处理主要包括（   ）和（   ）。

（5）Android 平台中的布局管理器本质上是（   ），它既可作为容器包含其他可视组件，也可作为组件加入到其他布局管理器中，从而构成繁茂的 Android 界面组件。

（6）（   ）提供了控制水平或垂直排列的模型，会将容器中的组件一个一个排列起来，通过 android:orientation 属性，可控制组件横向或者纵向排列。

（7）（   ）容器中，子组件的位置总是相对兄弟组件、父容器来决定的，其允许子元素指定它们相对于其父元素或兄弟元素的位置。

（8）（   ）也称框架布局，为每个组件创建一个空白区域，一个区域称为一帧。

（9）（   ）类以行和列的形式对控件进行管理，每一行为一个 TableRow 对象或一个 View 控件。

（10）（   ）将整个容器划分成 rows×columns 个网格，每个网格可放置一个组件，还可设置一个组件横跨多少列、多少行（跨多行和多列）。

（11）（   ）管理器采用坐标进行定位，屏幕是二维结构，绝对布局管理器就按照 x 和 y 坐标进行定位，坐标的原点位于屏幕左上角。

（12）统一建模语言的简写为（   ）。

（13）线性布局（   ）的 android:orientation 属性设置布局管理器内组件排列方式，

可取值为（　　　　）和（　　　　）。

（14）设备独立像素的简写为（　　　　）。

（15）（　　　　）用来设置View本身的文本应该显示在View的什么位置；而android:layout_gravity是相对于包含改元素的父元素来说的，设置该元素在父元素的什么位置。

（16）android:layout_centerHorizontal属性确定是否在父容器中（　　　　）；android:layout_alignParentLeft属性确定是否在父容器中（　　　　）。

（17）android:foreground属性设置帧布局容器的（　　　　）；android:foregroundGravity属性设置绘制（　　　　）的gravity属性。

（18）android:layout_column属性指定该单元格在（　　　　）显示；android:layout_span属性指定该单元格占据的（　　　　）。

（19）若某列被设置成Collapsed，则该列所有单元格会被（　　　　）；若某列被设为Stretchable，则该列所有单元格的宽度可被（　　　　）；若某列被设为Shrinkable，则该列所有单元格宽度可被（　　　　）。

（20）android:alignmentMode属性设置网格布局管理器的（　　　　）；android:columnCount属性设置网格布局的（　　　　）。

（21）线性布局、相对布局、帧布局和绝对布局是直接继承类为（　　　　）。

（22）在Android平台中，所有的可视组件都是（　　　　）的子类；（　　　　）是View类的一个重要直接子类；Android布局管理器是（　　　　）类的一组重要的直接或间接子类。

2．单选题

（1）在线性布局中，设置布局管理器内组件排列方式的XML属性名为（　　　　）。

　　A．android:orientation　　　　　　B．android:baselineAligned

　　C．android:divider　　　　　　　　D．android:gravity

（2）下面相对布局LayoutParams属性中只能设置boolean值的属性为（　　　　）。

　　A．android:layout_centerHorizontal　B．android:layout_toLeftOf

　　C．android:layout_above　　　　　　D．android:layout_alignLeft

（3）Android用户界面布局xml文件中的帧布局标签是（　　　　）。

　　A．LinearLayout　　B．FrameLayout　　C．RelativeLayout

　　D．TableLayout　　E．GridLayout　　　F．AbsoluteLayout

（4）相对布局LayoutParams属性中确定是否在父容器中位于中央的属性名为（　　　　）。

　　A．android:layout_centerHorizontal　B．android:layout_centerVertical

　　C．android:layout_centerInParent　　D．android:layout_alignParentLeft

（5）相对布局LayoutParams属性中设置需要被隐藏的列的序号的XML属性名为（　　　　）。

　　A．android:collapsedColumns　　　　B．android:stretchColumns

　　C．android:shrinkableColumns　　　　D．android:layout_column

（6）在网格布局中设置该网格行数的XML属性为（　　　　）。

　　A．android:alignmentMode　　　　　B．android:columnCount

　　C．android:columnOrderPreserved　　D．android:rowCount

（7）在网格布局中设置该网格列数的 XML 属性为（　　）。

A．android:alignmentMode　　　　B．android:columnCount

C．android:columnOrderPreserved　　D．android:rowCount

（8）网格布局的 LayoutParams 属性中设置子组件在 GridLayout 的哪一列的 XML 属性名为（　　）。

A．android:layout_column　　　　B．android:layout_columnSpan

C．android:layout_gravity　　　　D．android:layout_row。

（9）网格布局的 LayoutParams 属性中设置该子组件在 GridLayout 纵向横跨几行的 XML 属性名为（　　）。

A．android:layout_column　　　　B．android:layout_columnSpan

C．android:layout_row　　　　　　D．android:layout_rowSpan

（10）绝对布局中设置组件的 y 坐标值的 XML 属性名为（　　）。

A．android:layout_x　　　　　　　B．android:layout_y

C．android:layout_width　　　　　D．android:layout_height

3．问答题

（1）简述 Android 中的 6 种典型布局管理器。

（2）什么是 Android 布局管理器？

（3）什么是 Android 界面组件结构层次？

4．编程题

（1）在 Android 项目中采用线性布局管理器实现图 2.31 和图 2.32 所示的用户界面。

图 2.31　垂直线性布局界面　　　图 2.32　水平线性布局界面

（2）在 Android 项目中利用相对布局属性 android:layout_centerInParent、android:layout_above、android:layout_alignLeft、android:layout_alignRight、android:layout_below、android:layout_alignTop、android:layout_toLeftOf、android:layout_alignBottom 和 android:layout_toRightOf 在屏幕中间位置实现如图 2.33 所示的用户界面。

（a）界面设计　　　　　　　　　　（b）五个按钮布局效果图

图 2.33　五个按钮的相对布局界面

（3）在 Android 项目中利用相对布局属性 android:layout_below、android:layout_alignParentRight、android:layout_below、android:layout_marginLeft、android:layout_alignTop 和 android:layout_toLeftOf 在屏幕顶部放置 TextView 控件，将 EditText 控件放置在 TextView 下面，将 okButton 按钮放置在 EditText 控件下面，且与父控件的右边对齐，okButton 的左边间距为 10dip，cancelButton 按钮位于 okButton 左边且与其顶部保存对齐等实现如图 2.34 所示的用户界面。

（a）界面设计　　　　　　　　　（b）信息输入界面效果图

图 2.34　信息输入的相对布局界面

（4）在 Android 项目中利用相对布局属性 android:layout_alignParentLeft="true"将 TextView 控件 textView1 放置在屏幕左边，设置 android:layout_alignParentRight="true"属性将 TextView 控件 textView2 放置在屏幕右边，设置 android:layout_toLeftOf="@id/textView2 和 android:layout_toRightOf="@id/textView1"属性使得 TextView 控件 textView3 在水平方向在 textView1 和 textView2 之间扩展，实现图 2.35 所示的用户界面。

（5）在 Android 项目中利用相对布局属性 android:layout_alignParentTop="true"将 TextView 控件 textView1 放置在屏幕顶部，利用属性 android:layout_alignParentBottom="true"将 TextView 控件 textView2 放置在屏幕底部，利用 android:layout_above="@id/textView2"和 android:layout_below="@id/textView1"将 TextView 控件 textView3 放置在 textView2 上面和 textView1 下面，实现图 2.36 所示的用户界面。

图 2.35　个文本视图水平相对布局界面　　图 2.36　垂直方向相对布局

（6）在 Android 项目中利用表格布局管理器实现如图 2.37 所示的菜单用户界面。

（7）在 Android 项目中利用表格布局管理器实现图 2.38 所示的用户登录界面。

（8）在帧布局中放置 ImageView、TextView 和 Button 等三个控件，实现图 2.39 所示的系统界面。

 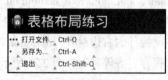

（a）界面设计　　　　　　　　　　（b）菜单界面效果

图 2.37　利用表格布局实现菜单界面

（a）界面设计　　　　　　　　　　（b）登录界面效果

图 2.38　利用表格布局实现登录界面

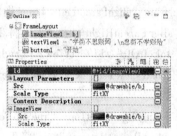 

（a）界面设计　　　　　　　　　　（b）系统界面效果

图 2.39　利用帧布局实现系统界面

（9）在 Android 项目中利用网格布局管理器实现图 2.40 所示的邮件安装界面。

（a）界面设计　　　　　　　　　　（b）邮件安装界面效果

图 2.40　利用网格布局实现邮件安装界面

（10）利用 TextView 标签分别定义两种布局文件 include1.xml 和 include2.xml，在 activity_main.xml 布局文件中，使用<include>布局标签重用这两种布局，实现图 2.41 所示的用户界面（参考源代码见 Ch02_Coding11_include.rar）。

（a）界面设计　　　　　　　　　　　（b）界面效果

图 2.41　使用<include>布局标签重用两种布局

（11）定义一个包含 Button 和 TextView 两个控件的布局文件 item.xml，在 activity_main.xml 布局文件中，使用<include>布局标签包含两个布局文件 item.xml，并添加对布局文件中按钮的事件回调处理，实现图 2.42 所示的用户界面。

 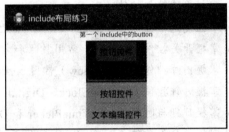

（a）界面设计　　　　　　　　　　（b）程序运行界面效果

图 2.42　<include>布局中控件事件处理方法

第 3 章 Android 常用控件和高级控件使用方法

【知识目标】
- 掌握文本控件（TextView 和 EditText）常用属性和方法；
- 掌握按钮控件（Button、ImageButton 和 ToggleButton）常用属性和方法；
- 掌握单选按钮控件（RadioButton）和单选组（RadioGroup）常用属性和方法；
- 掌握复选框控件（CheckBox）常用属性和方法；
- 掌握开关控件（Switch）常用属性和方法；
- 掌握图片控件（ImageView）常用属性和方法；
- 掌握时钟控件（AnalogClock、DigitalClock、Chronometer）常用属性和方法；
- 掌握日期与时间控件（TimePicker 和 DatePicker）常用属性和方法；
- 掌握评分条控件（RatingBar）常用属性和方法；
- 掌握自动完成文本框控件（AutoCompleteTextView、MultiAutoCompleteTextView）常用属性和方法；
- 掌握下拉列表控件（Spinner）常用属性和方法；
- 掌握滚动视图控件（ScrollView）常用属性和方法；
- 掌握列表视图控件（ListView）常用属性和方法；
- 掌握网格视图控件（GridView）常用属性和方法；
- 掌握进度条与滑块控件（ProgressBar 和 SeekBar）常用属性和方法；
- 掌握选项卡控件（TabHost）常用属性和方法；
- 掌握画廊控件（Gallery）常用属性和方法。

【能力目标】
- 掌握 Android 常用控件（TextView、Button、ImageButton、ToggleButton、RadioButton、CheckBox、Switch 和 Chronometer 等）XML 属性设置方法和 Java 编程技巧；
- 掌握 Android 高级控件（AutoCompleteTextView、Spinner、ScrollView、ListView、GridView、ProgressBar 和 Gallery 等）的 XML 属性设置方法和 Java 编程技巧。

【重点、难点】
- 应用 Android 常用控件实现用户界面方法和技巧；
- 应用 Android 高级控件实现用户界面方法和技巧。

3.1 学习导入

3.1.1 什么是 Android 视图类

Android 的 View（视图）类为所有可视化控件的基类，主要提供了控件绘制和事件处理的方法。创建用户界面所使用的控件都继承自 View，如 TextView、Button、CheckBox 等。

关于 View 及其子类的相关属性，既可以在布局 XML 文件中进行设置，也可以通过成员方法在代码中动态设置。View 类常用的属性及其对应方法如表 3.1 所示。View 类的这些 XML 属性值可方便地在 xml 代码文件中手动输入 android:后，在弹出的下拉列表中选中所需属性名，或在 Eclipse 中可视化属性页面中单击输入，如图 3.1 所示。

表 3.1 视图类常用属性及其对应方法

XML 属性名	Java 代码设置方法	功 能
android:background	setBackgroundResource(int)	设置视图背景色/背景图片；设置透明背景的两种方法为 "@android:color/transparent" 和 "@null"
android:clickable	setClickable(boolean)	设置视图是否响应单击事件
android:focusable	setFocusable(boolean)	设置视图是否获得焦点
android:focusableInTouchMode	setFocusableInTouchMode(boolean)	设置视图是否可通过触摸获得焦点
android:id	setId(int id)	设置视图控件 id 号
android:longClickable	setLongClickable(boolean)	设置视图是否响应长单击事件
android:soundEffectsEnabled	setSoundEffectsEnabled(boolean)	设置视图触发单击等事件时是否播放音效
android:minHeight	setMinHeight(int)	设置视图最小高度
android:minWidth	setMinWidth(int)	设置视图最小宽度
android:padding	setPadding (int left, int top, int right, int bottom)	设置视图左、上、右、下边距
android:paddingLeft		设置视图左边距
android:paddingTop		设置视图上边距
android:paddingRight		设置视图右边距
android:paddingBottom		设置视图下边距
android:scrollbarSize	setScrollBarSize(int)	设置滚动条的宽度
android:scrollbarStyle	setScrollBarStyle(int)	设置滚动条的风格
android:saveEnabled	setSaveEnabled(boolean)	设置视图被冻结时是否保存其状态
android:tag	setTag(Object)	设置一个文本标签
android:visibility	setVisibility(int)	设置是否显示视图控件
android:fitsSystemWindows		设置布局时是否考虑系统窗口布局
android:nextFocusLeft	setNextFocusLeftId(int)	设置向左搜索时应获得焦点的视图
android:nextFocusRight	setNextFocusRightId(int)	设置向右搜索时应获得焦点的视图
android:nextFocusDown	setNextFocusDownId(int)	设置向下搜索时应获得焦点的视图
android:nextFocusUp	setNextFocusUpId(int)	设置向上搜索时应获得焦点的视图
android:onClick		单击时从上下文中调用指定的方法

Tips：任何继承自视图的子类均拥有 View 类的属性及对应方法。

图 3.1 设置视图控件对象的属性

3.1.2 什么是 Android 视图组类

　　Android 视图组类（ViewGroup）是视图类的子类，也可充当其他控件的容器。ViewGroup 的子控件既可以是普通的 View，也可以是 ViewGroup，实际上，其使用了组合模式（Composite Pattern），也称为部分-整体模式。Android 中的一些高级控件如画廊（Gallery）、网格视图（GridView）、滚动视图（ScrollView）等都继承自 ViewGroup。通过 ViewGroup.LayoutParams 定义 View 的布局参数（见表 3.2），其 Eclipse 中的可视设置页如图 3.2 所示。LayoutParams 相当于一个 Layout 的信息包，它封装了 Layout 的位置、高、宽等信息，也可通过调用 setLayoutParams(ViewGroup.LayoutParams)方法来设置。

表 3.2 View 布局参数属性

XML 属性名	功　　能
android:layout_width	控件的宽度属性
android:layout_height	控件的高度属性
android:layout_weight	控件权重
android:layout_gravity	控件相对于其所在容器的位置
android:layout_margin	设置 View 的上下左右边框的额外空间

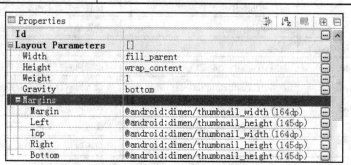

图 3.2 设置 View 布局参数（Layout Parameters）属性

上述属性中，所有 UI 控件都继承的部分重要属性和方法详细描述如下：

（1）android:id

该属性定义了视图控件的唯一标识，例如，android:id="@+id/view1"，表示该控件的 ID 为"view1"，其中"+"表示该 ID 为新增加的。

（2）android:layout_width

该属性定义了视图控件的宽度，一般可设置为"fill_parent"或"match_parent"或"wrap_content"。其中，"match_parent"表示该控件的宽度要扩展到其父控件的宽度。"wrap_content"表示该控件的宽度根据需要显示的内容进行调整，显示的内容多，则控件比较宽，显示的内容少，则宽度比较窄。"fill_parent"为 Android 2.2 以前版本的属性，其含义与"match_parent"相同。

（3）android:layout_height

该属性定义了视图控件的高度，其用法与 android:layout_width 类似。

（4）android:background

该属性设置视图控件的背景颜色或背景图片。

（5）void setId(int id)

功能：设置该视图控件的 ID 号。

参数：id 为控件的 ID 号，该参数必须为正整数。可调用 getId()获得该 ID 号或调用 findViewById(int)查找该控件对象引用。

（6）View findViewById(int id)

功能：根据给定的 ID 号查找子视图对象。若找到，则返回该视图对象的引用。

参数：id 给定搜索的 ID 号。

（7）void setBackgroundResource(int resid)

功能：设置该视图控件的背景图片资源。

参数：resid 为已经被添加到 Android 项目的背景图片资源的 ID 号。

3.2　技术准备

3.2.1　文本视图控件

1. TextView 控件作用

TextView（文本视图）的结构为：

```
java.lang.Object
   android.view.View
      android.widget.TextView
```

直接子类：Button，CheckedTextView，Chronometer，DigitalClock，EditText。

间接子类：AutoCompleteTextView，CheckBox，CompoundButton，ExtractEditText，MultiAutoCompleteTextView，RadioButton，ToggleButton。

从上可知 TextView 类继承自 View 类，TextView 控件的功能是向用户显示文本的内容，但不允许编辑，而其子类 EditView 允许用户进行编辑。在 Eclipse 图形布局（Graphical Layout）的属性页中，将其属性分为：布局参数（Layout Parameters）、视图（View）、TextView 和已过时（Deprecated）等属性。其中，所有控件公用视图属性和布局参数分别如表 3.1 和表 3.2 所示。

2. TextView 常用属性及对应方法

TextView 常用属性及对应方法如表 3.3 所示（XML 属性名均以 android:开头），这些 XML 属性值可方便地在 xml 代码文件中手动输入 android:后，在弹出的下拉列表中选中所需属性名，或在 Eclipse 可视化属性页面单击输入，如图 3.3 所示。

表 3.3　TextView 常用属性及对应方法

XML 属性名	Java 代码设置方法	功　　能
android:autoLink	setAutoLinkMask(int)	设置是否将指定格式文本转化成可单击的超链接显示
android:gravity	setGravity(int)	设置 TextView 在 x 和 y 轴方向上的显示方式
android:height	setHeight(int)	设置 TextView 的高度（以像素为单位）
android:minHeight	setMinHeight(int)	设置 TextView 的最小高度（以像素为单位）
android:maxHeight	setMaxHeight(int)	设置 TextView 的最大高度（以像素为单位）
android:width	setWidth(int)	设置 TextView 的宽度（以像素为单位）
android:minWidth	setMinWidth(int)	设置 TextView 的最小宽度（以像素为单位）
android:maxWidth	setMaxWidth(int)	设置 TextView 的最大宽度（以像素为单位）
android:hint	setHint(int)	若 TextView 中显示内容为空，显示该文本
android:text	setText(CharSequence)	设置 TextView 显示的内容
android:textColor	setTextColor(ColorStateList)	设置 TextView 的文本颜色
android:textSize	setTextSize(float)	设置 TextView 的文本大小
android:typeface	setTypeface(Typeface)	设置 TextView 的文本字体
android:ellipsize	setEllipsize(TextUtils.TruncateAt)	若设置该属性，则当 TextView 中显示的内容超过其最大长度时，会省略超出内容，可取值有 start、middle、end 和 marquee
android:textColorHighlight	setHighlightColor(int)	设置被选中文字的底色，默认为蓝色
android:textColorHint	setHintTextColor(ColorStateList)	设置提示信息文字的颜色，默认为灰色
android:textColorLink	setLinkTextColor	设置文字链接的颜色
android:singleLine	setSingleLine(boolean)	设置是否单行显示文本
android:shadowColor	setShadowLayer(float radius, float dx, float dy, int color)	指定文本阴影的颜色，需要与 shadowRadius 一起使用
android:shadowRadius		设置阴影的半径。设置为 0.1 会变成字体颜色，一般设置为 3.0 效果比较好
android:shadowDx		设置阴影横向坐标开始位置
android:shadowDy		设置阴影纵向坐标开始位置
android:freezesText	setFreezesText(boolean)	设置保存文本的内容以及光标的位置
android:ems	setEms(int)	设置 TextView 的宽度为 N 个字符的宽度
android:maxEms	setMaxEms(int)	设置 TextView 的宽度最长为 N 个字符的宽度
android:minEms	setMinEms(int)	设置 TextView 的宽度最短为 N 个字符的宽度

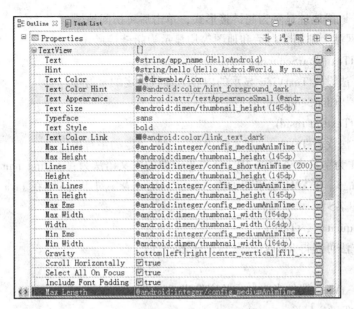

图 3.3 设置 TextView 控件属性

该控件的重要属性和方法详细描述如下：

（1）android:text

该属性设置 TextView 显示的内容。既可直接设置为某个字符串，例如 android:text="送人玫瑰，手留余香"；也可预先在 strings.xml 文件中定义好字符串资源，例如<string name="hello_world">欢迎使用 Android 开发应用程序!</string>，然后直接引用该字符串"hello_world"，android:text="@string/ hello_world "。

（2）android:textColor

该属性设置 TextView 的文本颜色。一个颜色值包含了 Alpha（透明度信息）和 RGB 值，其值总是以字符"#"开头，然后是 Alpha 和 RGB 值的字符串。其中，R 表示红色（Red）分量，G 表示绿色（Green）分量，B 表示蓝色（Blue）分量。常见的表示形式包括：#RGB、#ARGB、#RRGGBB、或#AARRGGBB。例如，android:textColor="#0f0"。颜色值资源的定义语法为：<color name="颜色名">#颜色值</color>。在 XML 代码中引用方法为：@color/<颜色名>；在 Java 中引用方法为：R.color.<颜色名>。

（3）android:textSize

该属性设置 TextView 的文本大小。Android 平台支持 6 种单位：像素（pixel, px）、设备无关像素（device independent pixel, dip）、比例无关像素（scale-independent pixel, sp）、点（point, pt）、毫米（millimeter, mm）或英寸（inches, in）。例如，android:textSize="30sp"。

（4）android:autoLink

该属性设置是否将某些文本显示为超链接形式。可选类型包括：

① "none"：所有文字均显示为普通文本形式，没有超链接。

② "web"：网站 URL 会显示为超链接形式，单击后可访问该网页。

③ "email"：邮箱地址会显示为超链接形式，单击后可发送邮件。

④ "phone"：电话显示为超链接形式，单击后可拨打电话。

⑤ "map"：地图地址显示为超链接形式，单击后可访问地图。
⑥ "all"：网站、邮箱、电话和地图等均显示为超链接形式。
（5）void setText(CharSequence text)
功能：该方法设置 TextView 显示的内容为参数给定的字符串。
参数：text 为 CharSequence 类型字符串。
（6）void setText(int resid)
功能：该方法设置 TextView 显示的内容为参数指定的字符串资源。
参数：resid 为字符串资源的 ID。字符串资源的定义语法为：<string name="字符串名">字符串值 </string>。在 XML 中引用的方法为："@string/<字符串名>"；在 Java 中引用方法为：R.string.<字符串名>。
（7）CharSequence getText()
功能：该方法可获得 TextView 显示的文本内容。
参数：无。
返回值：TextView 控件当前显示的字符串。
（8）void setTextColor(int color)
功能：该方法设置 TextView 的文本颜色。
参数：color 为自定义颜色值或系统预定义颜色常量，如 Color.BLUE 等。
示例代码如下：

```
1. TextView tv1=(TextView)this.findViewById(R.id.textView1);
                                    //引用布局文件中定义 TextView 控件
2. tv1.setText(R.string. strPhone);   //设置 TextView 显示内容为字符串资源信息
3. tv1.setTextColor(Color.GREEN);     //设置字体颜色为绿色
4. String  str=tv1.getText().toString();         //获得 TextView 显示的文本内容
5. tv1.setText("没有比人更高的山,没有比脚更长的路");//设置 TextView 显示的内容为字符串
6. tv1. setText("既然选择了远方，便只顾风雨兼程");
```

3. TextView 控件应用案例

用户既可在 XML 布局文件中声明及设置 TextView，也可在 Java 代码中生成 TextView 控件对象。TextView 控件典型应用案例实现要点及其步骤如下所述。

实现要点：

- TextView 中链接手机号码/网页/邮件/地图：使用 android:autoLink（可选值为 none/web/email/phone/map/all）属性设置一个链接，可以点击访问。例如，拨打手机号码：android:autoLink="phone"，android:text="拨电话：13760382993"，访问网页：android:autoLink="web"，android:text="百度搜索：http://www.baidu.com"；发送邮件：android:autoLink="email"，android:text="发邮件：lihz@sziit.com.cn "。
- TextView 中显示图片:通过设置背景的方式实现:android:background="@drawable/icon"。
- 设置图片在 TextView 的锚点位置：android:drawableBottom="@drawable/icon"，android:drawableTop="@drawable/icon"，android:drawableLeft="@drawable/icon"，android:drawableRight="@drawable/icon"。
- 创建 Ch03_Ex01_TextView 项目，修改布局文件 textviewlayout1.xml，如源代码清单 3.1 所示，其字符串资源文件如源代码清单 3.2 所示，其布局效果如图 3.4 所示。

源代码清单 3.1　\Ch03_Ex01_TextView\res\layout\textviewlayout1.xml

```xml
1.  <RelativeLayout xmlns:android="http://schemas.android.com/apk/res/android"
                                                                //相对布局
2.      xmlns:tools="http://schemas.android.com/tools"
3.      android:id="@+id/relayout"
4.      android:layout_width="match_parent"     //布局宽度：与父控件匹配
5.      android:layout_height="match_parent"    //布局高度：与父控件匹配
6.      android:paddingBottom="@dimen/activity_vertical_margin"
                                                //与父控件底部间距
7.      android:paddingLeft="@dimen/activity_horizontal_margin"
                                                //与父控件左边间距
8.      android:paddingRight="@dimen/activity_horizontal_margin"
                                                //与父控件右边间距
9.      android:paddingTop="@dimen/activity_vertical_margin"
                                                //与父控件顶部间距
10.     tools:context=".MainActivity" >
11.     <!-- 设置 autoLink 类型为 phone -->
12.     <TextView                               //文本视图标签
13.         android:id="@+id/textView1"         //文本视图 id 号
14.         android:layout_width="wrap_content" //文本视图宽度：与显示内容相互适应
15.         android:layout_height="wrap_content"//文本视图高度：与显示内容相互适应
16.         android:layout_centerHorizontal="true" //文本视图位于水平居中位置
17.         android:layout_centerVertical="true"   //文本视图位于垂直居中位置
18.         android:autoLink="phone"            //设置 autoLink 类型为 phone（电话）
19.         android:background="#00f"           //文本视图背景色或背景图片
20.         android:gravity="top|center_horizontal"//文本视图显示内容的位置
21.         android:text="@string/strPhone"     //文本视图显示内容
22.         android:textColor="#0f0"            //文本视图字体颜色
23.         android:textSize="18sp" />          //文本视图字体大小
24.     <!-- 设置 autoLink 类型为 web -->
25.     <TextView                               //文本视图标签
26.         android:id="@+id/textView2"         //文本视图 id 号
27.         android:layout_width="wrap_content" //文本视图宽度：与显示内容相互适应
28.         android:layout_height="wrap_content"//文本视图高度：与显示内容相互适应
29.         android:layout_above="@id/textView1"//文本视图位于控件 textView1 之上
30.         android:layout_centerHorizontal="true"//文本视图水平居中
31.         android:autoLink="web"              //设置 autoLink 类型为 web
32.         android:background="#fff"           //文本视图背景色或背景图片
33.         android:gravity="top|center_horizontal"//文本视图显示内容的位置
34.         android:text="@string/strWebAddr"   //文本视图显示内容
35.         android:textColor="#00f"            //文本视图字体颜色
36.         android:textSize="18sp" />          //文本视图字体大小
37.     <!-- 设置 autoLink 类型为 email -->
38.     <TextView                               //文本视图标签
39.         android:id="@+id/textView3"         //文本视图 id 号
40.         android:layout_width="wrap_content" //文本视图宽度：与显示内容相互适应
41.         android:layout_height="wrap_content"//文本视图高度：与显示内容相互适应
42.         android:layout_above="@id/textView2"//文本视图位于控件 textView2 之上
43.         android:layout_centerHorizontal="true" //文本视图水平居中
44.         android:autoLink="email"            //设置 autoLink 类型为 email
```

```
45.        android:background="#0f0"                //文本视图背景色或背景图片
46.        android:gravity="top|center_horizontal"  //文本视图显示内容的位置
47.        android:text="@string/strEmailAddr"      //文本视图显示内容
48.        android:textColor="#fff"                 //文本视图字体颜色
49.        android:textSize="18sp" />               //文本视图字体大小
50.    <!-- 设置字符阴影效果 -->
51.    <TextView                                    //文本视图标签
52.        android:id="@+id/textView4"              //文本视图id号
53.        android:layout_width="wrap_content"      //文本视图宽度：与显示内容相互适应
54.        android:layout_height="wrap_content"     //文本视图高度：与显示内容相互适应
55.        android:layout_above="@id/textView3"     //文本视图位于控件textView3之上
56.        android:layout_centerHorizontal="true"   //文本视图水平居中
57.        android:background="#f00"                //文本视图背景色或背景图片
58.        android:gravity="top|center_horizontal"  //文本视图显示内容的位置
59.        android:shadowColor="#00f"               //设置字符阴影效果
60.        android:text="@string/strShadowColor"    //文本视图显示内容
61.        android:textColor="#fff"                 //文本视图字体颜色
62.        android:textSize="18sp" />               //文本视图字体大小
63.    <!-- 设置textStyle风格为bold -->
64.    <TextView                                    //文本视图标签
65.        android:id="@+id/textView5"              //文本视图id号
66.        android:layout_width="wrap_content"      //文本视图宽度：与显示内容相互适应
67.        android:layout_height="wrap_content"     //文本视图高度：与显示内容相互适应
68.        android:layout_below="@id/textView1"     //文本视图位于控件textView1之下
69.        android:layout_centerHorizontal="true"   //文本视图水平居中
70.        android:background="#fff"                //文本视图背景色或背景图片
71.        android:gravity="top|center_horizontal"  //文本视图显示内容的位置
72.        android:text="@string/strTextStyle"      //文本视图显示内容
73.        android:textColor="#0f0"                 //文本视图字体颜色
74.        android:textSize="18sp"                  //文本视图字体大小
75.        android:textStyle="bold" />              //设置textStyle风格为bold
76.    <!-- 设置ellipsize为跑马灯效果 -->
77.    <TextView                                    //文本视图标签
78.        android:id="@+id/textView6"              //文本视图id号
79.        android:layout_width="wrap_content"      //文本视图宽度：与显示内容相互适应
80.        android:layout_height="wrap_content"     //文本视图高度：与显示内容相互适应
81.        android:layout_below="@id/textView5"     //文本视图位于控件textView5之下
82.        android:layout_centerHorizontal="true"   //文本视图水平居中
83.        android:background="#fff"                //文本视图背景色或背景图片
84.        android:ellipsize="marquee"              //设置ellipsize为跑马灯效果
85.        android:focusable="true"                 //设置视图可获得焦点
86.        android:focusableInTouchMode="true"      //设置视图可通过触摸获得焦点
87.        android:gravity="top|center_horizontal"  //文本视图显示内容的位置
88.        android:scrollHorizontally="true"        //设置为水平滚动
89.        android:singleLine="true"                //设置单行显示文本
90.        android:text="@string/strEllipsize"      //文本视图显示内容
91.        android:textColor="#0f0"                 //文本视图字体颜色
92.        android:textSize="18sp"                  //文本视图字体大小
93.        android:typeface="serif" />              //设置TextView的文本字体
94.    <!-- 在TextView中，通过设置背景的方式显示图片 -->
```

```
95.    <TextView                                    //文本视图标签
96.        android:id="@+id/textView7"              //文本视图 id 号
97.        android:layout_width="wrap_content"      //文本视图宽度：与显示内容相互适应
98.        android:layout_height="wrap_content"     //文本视图高度：与显示内容相互适应
99.        android:layout_below="@id/textView6"     //文本视图位于控件 textView6 之下
100.       android:layout_centerHorizontal="true"   //文本视图水平居中
101.       android:background="@drawable/ic_launcher" />
                                                    //通过设置背景的方式显示图片
102.   <LinearLayout                                //线性布局标签
103.       android:layout_width="fill_parent"       //布局宽度：填充父控件
104.       android:layout_height="fill_parent"      //布局高度：填充父控件
105.       android:layout_above="@id/textView4"     //位于控件 textView4 之上
106.       android:orientation="horizontal" >       //线性布局方向：水平方向
107.       <!-- 设置图片在 textView 的锚点位置 -->
108.       <TextView                                //文本视图标签
109.           android:id="@+id/textView8"          //文本视图 id 号
110.           android:layout_width="wrap_content"
                                                    //文本视图宽度：与显示内容相互适应
111.           android:layout_height="wrap_content"
                                                    //文本视图高度：与显示内容相互适应
112.           android:layout_alignParentLeft="true"//与父控件左边对齐
113.           android:layout_centerVertical="true" //垂直居中
114.           android:layout_gravity="left|center_vertical"//布局位置
115.           android:drawableLeft="@drawable/ic_launcher"//设置图片的锚点位置
116.           android:gravity="left|center_vertical"//文本视图显示内容的位置
117.           android:text="@string/strdrawableLeft"/>//文本视图显示内容
118.       <TextView                                //文本视图标签
119.           android:id="@+id/textView9"          //文本视图 id 号
120.           android:layout_width="wrap_content"
                                                    //文本视图宽度：与显示内容相互适应
121.           android:layout_height="wrap_content"
                                                    //文本视图高度：与显示内容相互适应
122.           android:layout_alignParentTop="true"//与父控件顶部对齐
123.           android:layout_centerHorizontal="true"//水平居中
124.           android:drawableBottom="@drawable/ic_launcher"
                                                    //设置图片的锚点位置
125.           android:text="@string/strdrawableBottom" />  //文本视图显示内容
126.       <TextView                                //文本视图标签
127.           android:id="@+id/textView10"         //文本视图 id 号
128.           android:layout_width="wrap_content"
                                                    //文本视图宽度：与显示内容相互适应
129.           android:layout_height="wrap_content"
                                                    //文本视图高度：与显示内容相互适应
130.           android:layout_alignParentBottom="true"  //与父控件底部对齐
131.           android:layout_centerHorizontal="true"   //水平居中
132.           android:drawableTop="@drawable/ic_launcher"//设置图片的锚点位置
133.           android:text=""@string/strdrawableTop"/>     //文本视图显示内容
134.       <TextView                                //文本视图标签
135.           android:id="@+id/textView11"         //文本视图 id 号
```

```
136.            android:layout_width="wrap_content"
                                                    //文本视图宽度：与显示内容相互适应
137.            android:layout_height="wrap_content"
                                                    //文本视图高度：与显示内容相互适应
138.            android:layout_alignParentRight="true"    //与父控件右边对齐
139.            android:layout_centerVertical="true"      //垂直居中
140.            android:drawableRight="@drawable/ic_launcher"//设置图片的锚点位置
141.            android:text="@string/strdrawableRight"/>  //文本视图显示内容
142.    </LinearLayout>                                    //线性布局标签
143.</RelativeLayout>                                      //相对布局标签
```

源代码清单 3.2 \Ch03_Ex01_TextView\res\values\strings.xml

```
1. <?xml version="1.0" encoding="utf-8"?>
2. <resources>
3.     <string name="app_name">TextView控件应用案例</string>
4.     <string name="action_settings">设置</string>
5.     <string name="strPhone">拨电话：13760382993</string>
6.     <string name="strWebAddr">百度搜索：http://www.baidu.com</string>
7.     <string name="strEmailAddr">发邮件：lihz@sziit.com.cn</string>
8.     <string name="strShadowColor">设置字符蓝色阴影颜色</string>
9.     <string name="strTextStyle">设置字符蓝色阴影颜色</string>
10.    <string name="strEllipsize">韩愈·《进学解》，业精于勤，荒于嬉，行成于思，毁于随</string>
11.    <string name="strdrawableLeft">在图片左边</string>
12.    <string name="strdrawableBottom">在图片下方</string>
13.    <string name="strdrawableTop">在图片上方</string>
14.    <string name="strdrawableRight">在图片右边</string>    </resources>
```

- 修改 MainActivity.java，增加动态创建 TextView，如源代码清单 3.3 所示。

源代码清单 3.3 \Ch03_Ex01_TextView\src\sziit\lihz\ch03_ex01_textview\MainActivity.java

```
1. public class MainActivity extends Activity {
                                     //从Activity基类派生 Main Activity
2.     protected void onCreate(Bundle savedInstanceState) {//重写 onCreate方法
3.         super.onCreate(savedInstanceState);           //基类调用 onCreate方法
4.         setContentView(R.layout.textviewlayout1);     //设置视图界面
5.         RelativeLayout rl= (RelativeLayout)this.findViewById(R.id.relayout);
           //绑定对象
6.         TextView tv=new TextView(this);              //动态创建 TextView对象
7.         tv.setText("动态创建一个TextView控件");//设置 TextView 控件上的文本内容
8.         tv.setBackgroundColor(Color.BLUE);           //设置 TextView 背景颜色
9.         tv.setTextColor(Color.WHITE);                //设置 TextView 控件上文本的颜色
10.        tv.setTextSize(20);                          //设置 TextView 控件上文本的大小
11.        tv.setGravity(Gravity.CENTER_HORIZONTAL);
                                                        //设置文本内容相对于此控件的对齐方式
           //设置 TextView 的布局参数(布局宽和布局高)
12.        LayoutParams params=new LayoutParams(LayoutParams.WRAP_CONTENT,
           LayoutParams.WRAP_CONTENT);
13.        tv.setLayoutParams(params);                  //设置 TextView 的布局参数
14.        rl.addView(tv, params);                      //将 TextView 控件添加到相对布局容器
15.    }}
```

（a）设计轮廓　　　　　　　　（b）仿真界面

图 3.4　TextView 控件应用案例界面

3.2.2　编辑框控件

1. EditText 作用

EditText（EditText）的结构为：

```
java.lang.Object
  android.view.View
    android.widget.TextView
      android.widget.EditText
```

已知直接子类：AutoCompleteTextView, ExtractEditText。

已知间接子类：MultiAutoCompleteTextView。

从此可知 EditText 继承关系：View→TextView→EditText。EditView 类继承自 TextView 类，EditView 与 TextView 最大的不同就是用户可以对 EditView 控件进行编辑，还可以为 EditView 控件设置监听器，用来判断用户的输入是否合法。属性可分为：布局参数（Layout Parameters）、视图（View）、TextView 和已过时（Deprecated）等属性。

2. EditText 的常用属性

EditText 常用属性如表 3.4 所示（XML 属性名均以 android:开头）。

表 3.4　EditText 常用属性及其对应方法

XML 属性	Java 代码设置方法	功　能
android:cursorVisible	setCursorVisible(boolean)	设置光标是否可见，默认可见
android:lines	setlines(int)	设置编辑文本的行数
android:maxLines	setMaxLines(int)	设置编辑文本的最大行数
android:minLines	setMinLines(int)	设置编辑文本的最小行数
android:maxLength	setFilters(InputFilter[])	设置最大显示长度
android:password	setTransformationMethod(TransformationMethod)	设置文本框中的内容是否显示为密码
android:singleLine		设置文本为单行模式
android:phoneNumber	setKeyListener(KeyListener)	设置文本框中的内容只能是电话号码
android:scrollHorizontally	setHorizontallyScrolling(boolean)	是否设置为水平滚动

续表

XML 属性	Java 代码设置方法	功　　能
android:selectAllOnFocus	setSelectAllOnFocus(boolean)	当文本获得焦点时自动选中全部文本内容
android:lineSpacingExtra	setLineSpacing(float, float)	设置行间距
android:lineSpacingMultiplier		设置行间距的倍数
android:inputType	setInputType(int type)	参数类型说明详见表 3.5
android:inputMethod	InputMethodManager	文本指定输入法
android:imeOptions	setImeOptions(int imeOptions)	设置输入法 Enter 键图标
android:digits	限制输入内容，只可取制定的字符，设置允许输入哪些字符，如 "1234567890.+-*/%\n()"	
android:capitalize	设置英文字母大写类型。设置如下值：sentences，仅第一个字母大写；words，每一个单词首字母大小，用空格区分单词；characters，每一个英文字母都大写	

表 3.5　EditText 的 android:inputType 属性取值含义说明

XML 属性	取　　值	含　　义
android:inputType	"none"	输入普通字符
	"text"	输入普通字符
	"textCapCharacters"	输入普通字符
	"textCapWords"	单词首字母大小
	"textCapSentences"	仅第一个字母大写
	"textAutoComplete"	前两个自动完成
	"textMultiLine"	多行输入
	"textImeMultiLine"	输入法多行
	"textNoSuggestions"	不提示
	"textUri"	URI 格式
	"textEmailAddress"	电子邮件地址格式
	"textEmailSubject"	邮件主题格式
	"textShortMessage"	短消息格式
	"textLongMessage"	长消息格式
	"textPersonName"	人名格式
	"textPostalAddress"	邮政格式
	"textPassword"	密码格式
	"textVisiblePassword"	密码可见格式
	"textWebEditText"	作为网页表单的文本格式
	"textFilter"	文本筛选格式
	"textPhonetic"	拼音输入格式
android:inputType	"number"	数字格式
	"numberSigned"	有符号数字格式
	"numberDecimal"	可以带小数点的浮点格式
	"phone"	拨号键盘
	"datetime"	日期时间键盘
	"date"	日期键盘
	"time"	时间键盘

该控件的重要属性和方法详细描述如下：

（1）android:inputType

该属性为父类 TextView 的属性，会影响 EditText 控件输入值时启动的虚拟键盘风格。例如 android:inputType="numberDecimal"，将显示带小数点的浮点格式虚拟键盘。

（2）Editable getText()

功能：获得 EditText 控件中用户输入的信息。

参数：内容字符串。

示例代码如下：

```
1. EditText etHeight=(EditText)findViewById(R.id.height);//引用XML布局中EditText控件
2. double height=Double.parseDouble(etHeight.getText().toString())/100;//获得用户输入信息
```

3. **EditText 控件应用案例**

实现要点：

- 密码隐藏，代码方法：input.setInputType(InputType.TYPE_CLASS_TEXT |Input Type.TYPE_TEXT_VARIATION_PASSWORD); layout 配置方法：android:inputType= "textPassword"。
- 创建 Ch03_Ex02_EditText 项目，修改布局文件 edittextlayout1.xml，如源代码清单 3.4 所示，其字符串资源如源代码清单 3.5 所示，Java 源代码清单见 3.6。其布局效果如图 3.5 所示。

源代码清单 3.4　\Ch03_Ex02_EditText\res\layout\edittextlayout1.xml

```
1.  <?xml version="1.0" encoding="utf-8"?>        //xml头文件,定义版本和编码方式
2.  <LinearLayout xmlns:android="http://schemas.android.com/apk/res/android"
3.      android:layout_width="fill_parent"        //线性布局宽度：填充父控件
4.      android:layout_height="fill_parent"       //线性布局高度：填充父控件
5.      android:orientation="vertical">           //线性布局方向：垂直方向
6.      <TextView                                  //文本视图标签
7.          android:layout_width="fill_parent"    //文本视图宽度：填充父控件
8.          android:layout_height="wrap_content"  //文本视图高度：与内容高度相适应
9.          android:text="@string/strName"/>      //文本视图显示信息
10.     <EditText                                  //编辑框标签
11.         android:id="@+id/etName"              //编辑框id号
12.         android:layout_width="fill_parent"    //编辑框宽度：填充父控件
13.         android:layout_height="wrap_content"  //编辑框高度：与内容高度相适应
14.         android:hint="@string/strPlsInputName" //编辑框提示信息
15.         android:textColorHint="#0f0"/>        //编辑框提示信息颜色
16.     <TextView                                  //文本视图标签
17.         android:layout_width="fill_parent"    //文本视图宽度：填充父控件
18.         android:layout_height="wrap_content"  //文本视图高度：与内容高度相适应
19.         android:text="@string/strPassword" /> //文本视图显示信息
20.     <EditText                                  //编辑框标签
21.         android:id="@+id/etPassword"          //编辑框id号
22.         android:layout_width="fill_parent"    //编辑框宽度：填充父控件
23.         android:layout_height="wrap_content"  //编辑框高度：与内容高度相适应
24.         android:hint="@string/strPromtPasswd" //编辑框提示信息
25.         android:password="true"/>             //编辑框隐藏密码
```

```
26.     <TextView                                       //文本视图标签
27.         android:id="@+id/textView"                  //文本视图id号
28.         android:layout_width="fill_parent"          //文本视图宽度:填充父控件
29.         android:layout_height="wrap_content"        //文本视图高度:与内容高度相适应
30.         android:textSize="20sp"/>                   //文本视图显示字体大小
31.     <TextView                                       //文本视图标签
32.         android:layout_width="wrap_content"         //文本视图宽度:与内容宽度相适应
33.         android:layout_height="wrap_content"        //文本视图高度:与内容高度相适应
34.         android:text="@string/strPrompt"/>          //文本视图显示信息
35.     <EditText                                       //编辑框标签
36.         android:id="@+id/edittext"                  //编辑框id号
37.         android:layout_width="match_parent"         //编辑框宽度:与父控件相匹配
38.         android:layout_height="wrap_content"/>      //编辑框高度:与内容高度相适应
39.     <Button                                         //按钮标签
40.         android:id="@+id/button"                    //按钮id号
41.         android:layout_width="wrap_content"         //按钮宽度:与内容宽度相适应
42.         android:layout_height="wrap_content"        //按钮高度:与内容高度相适应
43.         android:text="@string/strBtnTitle"/>        //按钮显示信息
44.     <TextView                                       //文本视图标签
45.         android:id="@+id/tvShow"                    //文本视图id号
46.         android:layout_width="wrap_content"         //文本视图宽度:与内容宽度相适应
47.         android:layout_height="wrap_content" />     //文本视图高度:与内容高度相适应
48. </LinearLayout>
```

源代码清单 3.5   \Ch03_Ex02_EditText\res\values\strings.xml

```
1. <?xml version="1.0" encoding="utf-8"?>
2. <resources>        //资源标签
3.     <string name="app_name">EditText 控件应用案例</string> //定义字符串键-值对
4.     <string name="strName">用户名: </string>
5.     <string name="strPlsInputName">请输入用户名</string>
6.     <string name="strPassword">密码: </string>
7.     <string name="strPromtPasswd">请输入密码</string>
8.     <string name="strPrompt">请在下方输入:</string>
9.     <string name="strBtnTitle">单击显示信息</string> </resources>
```

源代码清单 3.6   \Ch03_Ex02_EditText\src\sziit\lihz\ch03_ex02_edittext\MainActivity.java

```
1. public class MainActivity extends Activity implements OnClickListener {
                                        //实现监听器
2.     private EditText metName;        //定义 EditText 对象 etName
3.     private EditText metPassword;    //定义 EditText 对象 etPassword
4.     private TextView mtvMsg;         //定义 TextView 对象 textView
5.     private EditText mEditText;      //定义 EditText 对象
6.     private Button mButton;          //定义 Button 对象
7.     private TextView mTextView;      //定义 TextView 对象
8.     protected void onCreate(Bundle savedInstanceState) {//重写 onCreate 方法
9.         super.onCreate(savedInstanceState);         //基类调用 onCreate 方法
10.        setContentView(R.layout.edittextlayout1);   //设置屏幕布局
```

```
11.      //获取控件对象实例化
12.      metName = (EditText) findViewById(R.id.etName);
                                    //绑定 EditText 控件 etName
13.      metPassword = (EditText) findViewById(R.id.etPassword);
                                    //绑定 EditText 控件 etPassword
14.      mtvMsg = (TextView) findViewById(R.id.textView);
                                    //绑定 TextView 控件 textView
15.      metPassword.setOnKeyListener(new OnKeyListener() {
                                    //为 EditText 监听回车键
16.        public boolean onKey(View v, int keyCode, KeyEvent event) {
17.          if (keyCode == KeyEvent.KEYCODE_ENTER) {//按回车键即显示文本
18.            mtvMsg.setText("您的用户名为:"+metName.getText().toString() + "\n"
19.            + "您的密码为:" + metPassword.getText().toString());}
20.            return false; }});
21.      //获取控件对象实例化
22.      mEditText = (EditText) findViewById(R.id.edittext);
23.      mTextView = (TextView) findViewById(R.id.tvShow);
24.      mButton = (Button) findViewById(R.id.button);
25.      mButton.setId(10);                      //设置按钮 id 号
26.      mButton.setOnClickListener(this);}     //设置按钮单击监听器
27.   public void onClick(View v) {             //单击事件回调函数
28.      if (v.getId() == 10) {                 //根据控件 id 号确定控件对象
29.        String tmp = mEditText.getText().toString();
30.        mTextView.setText(tmp);} }}          //显示输入信息
```

（a）设计轮廓　　　　　　　　（b）仿真界面

图 3.5　EditText 控件应用案例界面

3.2.3　按钮和图片按钮

1. 按钮的作用

按钮的结构为：

```
java.lang.Object
  android.view.View
    android.widget.TextView
      android.widget.Button
```

已知直接子类：CompoundButton。

已知间接子类：CheckBox、RadioButton、ToggleButton。

按钮继承于 TextView，其主要作用在于响应用户的单击事件，当用户单击、按下或长按等操作，会触发 onClick、onLongClick 或 onKey 事件。

2. 按钮比较常用的属性和方法

按钮的属性和方法类似 TextView，比较常用的属性如下：

- android:text：设置按钮的显示文本。
- android:background：设置按钮的颜色或背景图片。

图片按钮（ImageButton）继承于按钮，两者的区别在于图片按钮不可显示文字，只能显示图片，其常用属性是 android:src，用于设置 ImageButton 上显示的图片。

3. 按钮的主要用法

按钮的用法主要是为按钮设置 View.OnClickListener、View. View.OnClickListener 或 View.setOnKeyListener 监听器，并在监听器的实现代码中开发按钮 onClick(View v)、onLongClick(View v)或 onKey(View v, int keyCode, KeyEvent event)事件处理代码。

（1）在 XML 布局文件中可通过按钮的 android:onClick 属性设置单击事件处理方法，例如：

```
android:onClick="DoClick"
```

该方法 DoClick 必须声明为 public 且仅接收一个 View 类型参数，例如：

```
public void DoClick(View view ) { //按钮单击事件处理过程 }
```

（2）在 Java 代码中，设置 View.OnClickListener 监听器和事件处理示例代码如下：

```
button.setOnClickListener(new Button.OnClickListener() {
    public void onClick(View v) {//TODO Auto-generated method stub  }  });
```

提示：类似上面的处理代码，均可在输入以下代码后，选择"Add umimplented method"自动生成：

```
button.setOnClickListener(new Button.OnClickListener() { });
```

（3）可通过 setText (CharSequence text)或 setText(int resid)方法或 android:text 属性设置按钮显示的字符串。

（4）可通过 setBackgroundDrawable(Drawable background)或 setBackground(Drawable background)方法或 android:background 属性设置背景图片。

4. Button 和 ImageButton 典型案例

实现要点：

- setOnClickListener：单击事件响应。
- 事件监听器：一般使用匿名内部类，若包含很多按钮，代码过于相似，则会另外实现一个命名的类。
- setOnLongClickListener：长按事件响应。
- 视图的切换：通过意图（inetnt）。
- 多个事件监听器分开处理，通过它们的 id 识别各视图进行区分。
- 创建 Ch03_Ex03_Button 项目，修改布局文件 activity_main.xml，如源代码清单 3.7 所示，Java 源代码清单见 3.8，其布局效果如图 3.6 所示。

（a）设计轮廓　　　　　　　　　（b）仿真界面

图 3.6　Button 和 ImageButton 典型案例界面

源代码清单 3.7　\Ch03_Ex03_Button\res\layout\activity_main.xml

```
1.  <?xml version="1.0" encoding="utf-8"?>        //xml 文件头,定义版本和编码方式
2.  <LinearLayout xmlns:android="http://schemas.android.com/apk/res/android"
3.      android:id="@+id/llayout"                 //设置线性布局 id 号
4.      android:layout_width="fill_parent"        //线性布局宽度：填充父控件
5.      android:layout_height="fill_parent"       //线性布局高度：填充父控件
6.      android:orientation="vertical" >          //线性布局方向：垂直方向
7.      <TextView                                 //文本视图标签
8.          android:id="@+id/textview1"           //文本视图 id 号
9.          android:layout_width="fill_parent"    //文本视图宽度：填充父控件
10.         android:layout_height="wrap_content"  //文本视图高度：与内容高度相适应
11.         android:text="@string/strMsg" />      //文本视图显示文本
12.     <Button                                   //按钮标签
13.         android:id="@+id/button1"             //按钮 id 号
14.         android:layout_width="wrap_content"   //按钮宽度：与内容宽度相适应
15.         android:layout_height="wrap_content"  //按钮高度：与内容高度相适应
16.         android:layout_gravity="center"       //按钮位置：居中
17.         android:text="单击或长按按钮" />      //按钮显示文本信息
18.     <Button                                   //按钮标签
19.         android:id="@+id/button2"             //按钮 id 号
20.         android:layout_width="wrap_content"   //按钮宽度：与内容宽度相适应
21.         android:layout_height="wrap_content"  //按钮高度：与内容高度相适应
22.         android:layout_gravity="center"       //按钮位置：居中
23.         android:text="设置字体颜色" />        //按钮显示文本信息
24.     <Button                                   //按钮标签
25.         android:id="@+id/button3"             //按钮 id 号
26.         android:layout_width="wrap_content"   //按钮宽度：与内容宽度相适应
27.         android:layout_height="wrap_content"  //按钮高度：与内容高度相适应
28.         android:layout_gravity="center"       //按钮位置：居中
29.         android:text="设置字体背景" />        //按钮显示文本信息
30.     <Button                                   //按钮标签
```

```
31.        android:id="@+id/button4"              //按钮id号
32.        android:layout_width="wrap_content"    //按钮宽度：与内容宽度相适应
33.        android:layout_height="wrap_content"   //按钮高度：与内容高度相适应
34.        android:layout_gravity="center"        //按钮位置：居中
35.        android:text="设置按钮背景" />         //按钮显示文本信息
36.    <Button                                    //按钮标签
37.        android:id="@+id/button5"              //按钮id号
38.        android:layout_width="wrap_content"    //按钮宽度：与内容宽度相适应
39.        android:layout_height="wrap_content"   //按钮高度：与内容高度相适应
40.        android:layout_gravity="center"        //按钮位置：居中
41.        android:onClick="MyClick"              //按钮onClick属性
42.        android:text="设置onClick属性" />      //按钮显示文本信息
43.    <Button                                    //按钮标签
44.        android:id="@+id/button6"              //按钮id号
45.        android:layout_width="wrap_content"    //按钮宽度：与内容宽度相适应
46.        android:layout_height="wrap_content"   //按钮高度：与内容高度相适应
47.        android:layout_gravity="center"        //按钮位：居中
48.        android:text="设置屏幕背景颜色" />     //按钮显示文本信息
49.    <ImageButton                               //图像按钮标签
50.        android:id="@+id/button7"              //图像按钮id号
51.        android:layout_width="wrap_content"    //图像按钮宽度：与内容宽度相适应
52.        android:layout_height="wrap_content"   //图像按钮高度：与内容高度相适应
53.        android:layout_gravity="center"        //图像按钮位置：居中
54.        android:src="@drawable/ic_launcher"/>  //图像按钮显示图片
55.</LinearLayout>
```

源代码清单 3.8   \Ch03_Ex03_Button\src\sziit\lihz\ch03_ex03_button\MainActivity.java

```
1. public class MainActivity extends Activity {
                                    //从活动Activity基类派生子类MainActivity
2.     private TextView tvMsg;      //定义TextView对象tvMsg
3.     private Button btnDemo1;     //定义Button对象btnDemo1
4.     private Button btnDemo2;     //定义Button对象btnDemo2
5.     private Button btnDemo3;     //定义Button对象btnDemo3
6.     private Button btnDemo4;     //定义Button对象btnDemo4
7.     private Button btnDemo5;     //定义Button对象btnDemo5
8.     private Button btnDemo6;     //定义Button对象btnDemo6
9.     private ImageButton btnDemo7;//定义Button对象btnDemo7
10.    private LinearLayout llayout;//定义LinearLayout对象llayout
11.    protected void onCreate(Bundle savedInstanceState) {
12.        super.onCreate(savedInstanceState);setContentView(R.layout.activity_main);
13.        //通过ID找到定义在main.xml里的TextView和Button控件
14.        tvMsg = (TextView) findViewById(R.id.textview1);//得到TextView的引用
15.        btnDemo1 = (Button) findViewById(R.id.button1);//得到Button的引用
16.        btnDemo2 = (Button) findViewById(R.id.button2);//得到Button的引用
17.        btnDemo3 = (Button) findViewById(R.id.button3);//得到Button的引用
18.        btnDemo4 = (Button) findViewById(R.id.button4);//得到Button的引用
19.        btnDemo5 = (Button) findViewById(R.id.button5);//得到Button的引用
20.        btnDemo6 = (Button) findViewById(R.id.button6);//得到Button的引用
21.        btnDemo7 = (ImageButton) findViewById(R.id.button7);
                                                       //得到Button的引用
22.        llayout = (LinearLayout) findViewById(R.id.llayout);
                                                       //得到Linear Layout的引用
```

```
23.    btnDemo1.setOnClickListener(new Button.OnClickListener() {//增加事件响应
24.        public void onClick(View v) {           //单击事件响应
25.            Toast.makeText(MainActivity.this,"注意观察TextView里的文字发生了改变!",
26.            Toast.LENGTH_LONG).show();           //Toast 提示控件
27.            tvMsg.setText("演示按钮单击事件响应!");}});//TextView 的文字发生改变
28.    btnDemo1.setOnLongClickListener(new Button.OnLongClickListener()
       {//增加事件响应
29.        public boolean onLongClick(View v) {    //长按事件响应
30.            Toast.makeText(MainActivity.this,"注意观察TextView里的文字发生
               了改变!",Toast.LENGTH_LONG).show(); //Toast 提示控件
31.            tvMsg.setText("演示按钮长按事件响应!");//TextView 的文字发生改变
32.            return false; }});
33.    btnDemo1.setOnKeyListener(new Button.OnKeyListener() {//增加事件响应
34.        public boolean onKey(View v, int keyCode, KeyEvent event) {
                                                    //键盘事件响应
35.            Toast.makeText(MainActivity.this,"注意观察TextView里的文字发生
               了改变!", Toast.LENGTH_LONG).show();//Toast 提示控件
36.            tvMsg.setText("演示按钮按键事件响应!");//TextView 的文字发生改变
37.            return false; }});
38.    btnDemo2.setOnClickListener(new Button.OnClickListener() {
                                                    //增加事件响应
39.        public void onClick(View v) {           //单击事件响应
40.            tvMsg.setText("设置字体颜色为蓝色"); //设置显示文本
41.            tvMsg.setTextSize(20);              //设置字体大小
42.            tvMsg.setTextColor(Color.BLUE);}});//设置字体颜色
43.    btnDemo3.setOnClickListener(new Button.OnClickListener() {//增加事件响应
44.        public void onClick(View v) {           //单击事件响应
45.            llayout.setBackgroundColor(Color.TRANSPARENT);//设置手机背景色
46.            tvMsg.setText("设置字体背景为红色"); //设置显示文本
47.            tvMsg.setTextSize(20);              //设置字体大小
48.            tvMsg.setBackgroundColor(Color.RED);}});//设置字体颜色
49.    btnDemo4.setOnClickListener(new Button.OnClickListener() {//增加事件响应
50.        public void onClick(View v) {           //单击事件响应
51.            llayout.setBackgroundColor(Color.DKGRAY);   //设置手机背景色
52.            btnDemo1.setBackgroundResource(R.drawable.ic_launcher);
                                                    //设置背景图片
53.            btnDemo1.setText("设置按钮背景图片"); //设置显示文本
54.            btnDemo2.setBackgroundColor(Color.GREEN);   //设置背景色
55.            btnDemo3.setBackgroundColor(Color.YELLOW);  //设置背景色
56.            btnDemo4.setBackgroundColor(Color.RED); }});//设置背景色
57.    btnDemo6.setOnClickListener(new Button.OnClickListener() {
                                                    //增加事件响应
58.        public void onClick(View v) {           //单击事件响应
59.            llayout.setBackgroundColor(Color.LTGRAY);}});//设置手机背景色
60.    btnDemo7.setOnClickListener(new Button.OnClickListener() {
                                                    //增加事件响应
61.        public void onClick(View v) {           //单击事件响应
62.            llayout.setBackgroundColor(Color.YELLOW);   //设置手机背景色
63.            Toast.makeText(MainActivity.this,"注意观察TextView里的文字发生
               了改变!", Toast.LENGTH_LONG).show();  //Toast 提示控件
```

```
64.            tvMsg.setText("演示图像按钮按键事件响应!");  //TextView的文字发生改变
65.            tvMsg.setTextSize(30);                    //设置字体大小
66.            tvMsg.setTextColor(Color.WHITE);}});}     //设置字体颜色
67.    public void MyClick(View view) {                  //实现onClick属性
68.        tvMsg.setText("消息来自于button5的onClick属性");  //设置显示文本
69.        tvMsg.setTextSize(30);                        //设置字体大小
70.        tvMsg.setTextColor(Color.DKGRAY);}}           //设置字体颜色
```

以上按钮控件应用主要涉及按钮单击事件监听器：void setOnClickListener(OnClickListener l)。

功能：该方法用于设置按钮被单击的事件监听器。在 Android 中是通过监听器来完成对控件事件的监视处理的，一旦控件发生了某个事件，监听器就会做出反应，触发相应的方法。

参数：l OnClickListener 是一个接口，其抽象方法为 void onClick(View v)，当按钮被单击时会触发 onClick()方法。

3.2.4 状态开关按钮和开关控件

1. 状态开关按钮和开关简介

状态开关按钮（ToggleButton）和开关（Switch）均派生于 Button，它们的本质也是按钮，从而它们支持 Button 支持的各种属性和方法。从功能上来看，它们与复选框（CheckBox）非常相似，都可以提供两个状态，其区别主要体现在功能上，ToggleButton、Switch 通常用于切换程序中的某种状态。

2. ToggleButton 和 Switch 所支持的 XML 属性及相关方法

ToggleButton 和 Switch 所支持的 XML 属性及相关方法分别如表 3.6 和表 3.7 所示。

表 3.6 ToggleButton 支持的 XML 属性及相关方法说明

XML 属性	Java 代码设置方法	功能
android:checked	setChecked(boolean)	设置该按钮是否被选中
android:textOff	setTextOff(CharSequence)	设置当该按钮处于关闭状态时显示的文本
android:textOn	setTextOn(CharSequence)	设置当该按钮处于打开状态时显示的文本

表 3.7 Switch 控件支持的 XML 属性及相关方法说明

XML 属性	Java 代码设置方法	功能
android:checked	setChecked(boolean)	设置该开关是否被选中
android:switchMinWidth	setSwitchMinWidth(int)	设置该开关的最小宽度
android:switchPadding	setSwitchPadding(int)	设置开关与标题文本之间的空白
android:switchTextAppearance	setSwitchTextAppearance(Context,int)	设置该开关图标上的文本样式
android:textOff	setTextOff(CharSequence)	设置该开关的状态为关闭时显示的文本
android:textOn	setTextOn(CharSequence)	设置该开关的状态为打开时显示的文本
android:textStyle	setSwitchTypeface(Typeface)	设置该开关的文本的风格
android:thumb	setThumbResource(int)	设置采用自定义 Drawable 绘制开关按钮
android:track	setTrackResource(int)	设置采用自定义 Drawable 绘制开关轨道
android:typeface	setSwitchTypeface(Typeface)	设置该开关的文本的字体风格

3. ToggleButton 和 Switch 典型案例

实现要点：

- android:checked：表示按钮 ToggleButton 或开关 Switch 是否被选中。
- android:textoff：表示按钮 ToggleButton 或开关 Switch 关闭状态显示的文本。
- android:texton：表示按钮 ToggleButton 或开关 Switch 开启显示的文本。
- OnCheckedChangeListener：按钮 ToggleButton 或开关 Switch 状态变化监听器。
- 利用 ToggleButton 和 Switch 动态控制布局，创建 Ch03_Ex04_ToggleButton 项目，定义资源文件 strings.xml，修改布局文件 togglebutton_switch.xml，实现 SwitchDemo.java 功能，如源代码清单 3.9、3.10 和 3.11 所示，其布局效果如图 3.7 所示。

源代码清单 3.9　\Ch03_Ex04_ToggleButton\res\values\strings.xml

```
1.  <?xml version="1.0" encoding="utf-8"?>
2.  <resources>         //资源标签
3.    <string name="app_name">ToggleButton 典型案例 </string>
4.    <string name="action_settings">设置</string>    //字符串键-值对
5.    <string name="strTextOff">横向排列</string>
6.    <string name="strTextOn">纵向排列</string>
7.    <string name="strBtn1">第一个按钮</string>
8.    <string name="strBtn2">第二个按钮</string>
9.    <string name="strBtn3">第三个按钮</string>
10.   <string name="strSoundOn">声音已开启</string>
11.   <string name="strSoundOff">声音已关闭</string>
12. </resources>
```

源代码清单 3.10　\Ch03_Ex04_ToggleButton\res\layout\togglebutton_switch.xml

```
1.  <LinearLayout xmlns:android="http://schemas.android.com/apk/res/android"
2.      android:layout_width="fill_parent"           //线性布局宽度
3.      android:layout_height="fill_parent"          //线性布局高度
4.      android:orientation="vertical">              //线性布局方向：垂直方向
5.      <!-- 定义一个 ToggleButton 按钮和 Switch -->
6.      <ToggleButton                                //状态开关按钮 ToggleButton 标签
7.          android:id="@+id/toggleButton"           //ToggleButton 的 id 号
8.          android:layout_width="wrap_content"      //ToggleButton 的宽度
9.          android:layout_height="wrap_content"     //ToggleButton 的高度
10.         android:layout_gravity="center_horizontal"//ToggleButton 位置：水平居中
11.         android:checked="true"                   //ToggleButton 是否选中：是
12.         android:textOff="@string/strTextOff"     //ToggleButton 未选中时显示信息
13.         android:textOn="@string/strTextOn"/>     //ToggleButton 选中时显示信息
14.     <Switch                                      //开关 Switch 标签
15.         android:id="@+id/switcher"               //开关 Switch 的 id 号
16.         android:layout_width="wrap_content"      //开关 Switch 的宽度
17.         android:layout_height="wrap_content"     //开关 Switch 的高度
18.         android:layout_gravity="center_horizontal"//开关 Switch 的位置
19.         android:checked="true"                   //开关 Switch 是否选中：是
20.         android:textOff="@string/strTextOff"     //开关 Switch 未选中时显示信息
21.         android:textOn="@string/strTextOn" />    //开关 Switch 选中时显示信息
22.     <TextView                                    //文本视图标签
23.         android:id="@+id/tvSound"                //文本视图 id 号
```

```
24.         android:layout_width="wrap_content"           //文本视图的宽度
25.         android:layout_height="wrap_content"          //文本视图的高度
26.         android:layout_gravity="center_horizontal"//文本视图的位置
27.         android:text="@string/strSoundOn"             //文本视图显示内容
28.         android:textColor="@android:color/black"    //文本视图的字体颜色
29.         android:textSize="20sp"/>                      //文本视图的字体大小
30.     <ToggleButton                                      //状态开关按钮ToggleButton标签
31.         android:id="@+id/tbSound"                      //ToggleButton的id号
32.         android:layout_width="wrap_content"           //ToggleButton的宽度
33.         android:layout_height="wrap_content"          //ToggleButton的高度
34.         android:layout_gravity="center_horizontal"//ToggleButton位置:水平居中
35.         android:checked="true"/>                       //ToggleButton是否选中:是
36.     <!-- 下面定义了三个按钮 -->
37.     <LinearLayout                                      //线性布局标签
38.         android:id="@+id/llayout"                      //线性布局id号
39.         android:layout_width="fill_parent"            //线性布局的宽度
40.         android:layout_height="fill_parent"           //线性布局的宽度
41.         android:orientation="vertical">                //线性布局的方向:垂直方向
42.         <Button                                        //按钮标签
43.             android:layout_width="wrap_content"       //按钮的宽度
44.             android:layout_height="wrap_content"      //按钮的宽度
45.             android:layout_gravity="center_horizontal"//按钮位置:水平居中
46.             android:text="@string/strBtn1" />         //按钮显示文本
47.         <Button                                        //按钮标签
48.             android:layout_width="wrap_content"       //按钮的宽度
49.             android:layout_height="wrap_content"      //按钮的宽度
50.             android:layout_gravity="center_horizontal"//按钮位置:水平居中
51.             android:text="@string/strBtn2"/>          //按钮显示文本
52.         <Button                                        //按钮标签
53.             android:layout_width="wrap_content"       //按钮的宽度
54.             android:layout_height="wrap_content"      //按钮的宽度
55.             android:layout_gravity="center_horizontal"//按钮位置:水平居中
56.             android:text="@string/strBtn3"/>          //按钮显示文本
57.     </LinearLayout>                                    //线性布局标签
58.</LinearLayout>                                        //线性布局标签
```

源代码清单 3.11   \Ch03_Ex04_ToggleButton\src\sziit\lihz\
ch03_ex04_togglebutton\SwitchDemo.java

```java
1. public class SwitchDemo extends Activity {//从Activity基类派生子类SwitchDemo
2.     private ToggleButton toggleButton;       //定义ToggleButton对象
3.     private Switch switcher;                  //定义Switch对象
4.     private ToggleButton mToggleButton;      //定义ToggleButton对象
5.     private TextView tvSound;                 //定义TextView对象
6.     protected void onCreate(Bundle savedInstanceState) {  //重写onCreate方法
7.         super.onCreate(savedInstanceState);               //调用基类方法
8.         //requestWindowFeature(Window.FEATURE_NO_TITLE); //隐藏标题栏
9.         setContentView(R.layout.togglebutton_switch);    //设置手机界面布局
10.        toggleButton = (ToggleButton) findViewById(R.id.toggleButton);
11.        switcher = (Switch) findViewById(R.id.switcher); //绑定Switch对象
12.        final LinearLayout llayout = (LinearLayout) findViewById(R.id.llayout);
```

```
13.    mToggleButton = (ToggleButton) findViewById(R.id.tglSound);
                                                                    //获取到控件
14.    mToggleButton.setOnCheckedChangeListener(new OnCheckedChangeListener() {
            public void onCheckedChanged(CompoundButton buttonView,
15.           boolean isChecked) {                      //添加监听事件
16.             if (isChecked) {                        //检测开关状态
17.               tvSound.setText(R.string.strSoundOn);}//声音已开启
18.             else {tvSound.setText(R.string.strSoundOff);}}});//声音已关闭
19.    tvSound = (TextView) findViewById(R.id.tvSound);
20.    OnCheckedChangeListener listener = new OnCheckedChangeListener() {
                                  //构造 OnCheckedChangeListener 对象
21.       public void onCheckedChanged(CompoundButton arg0, boolean isChecked)
       {//处理状态变化事件响应
22.         if (isChecked) {                            //检测开关状态
23.           llayout.setOrientation(1);}               //设置 LinearLayout 垂直布局
24.         else {   llayout.setOrientation(0);}} };
                                                  //设置 LinearLayout 水平布局
25.    toggleButton.setOnCheckedChangeListener(listener);
                                                  //设置状态变化监听器
26.    switcher.setOnCheckedChangeListener(listener);}}//设置状态变化监听器
```

（a）设计轮廓

（b）仿真界面

图 3.7 ToggleButton 和 Switch 典型案例界面

3.2.5 单选按钮和复选框

1. 单选按钮和复选框简介

单选按钮（RadioButton）和复选框（CheckBox）都是用户界面中最普通的 UI 组件，它们都继承了 Button 类，因此都可直接使用 Button 支持的各种属性和方法。

RadioButton 和 CheckBox 与普通按钮不同的是，它们多了一个可选中的功能，因此它们都可额外指定一个 android:checked 属性，该属性用于指定 RadioButton 和 CheckBox 初始时是否被选中。RadioButton 与 CheckBox 的不同之处在于，一组 RadioButton 只能选中其中一个，因此 RadioButton 通常要与 RadioGroup 一起使用，用于定义一组单选按钮。

RadioButton 表示单个圆形单选框，而 RadioGroup 是可容纳多个 RadioButton 的容器，它们之间的关系为：

- 每个 RadioGroup 中的 RadioButton 同时只能选中一个。
- 每个 RadioGroup 中的 RadioButton 默认会有一个被选中。
- RadioButton 只有选中和未选中两种状态。

CheckBox 派生于 CompoundButton 类，也只有选中和未选中两种状态。但与 RadioButton 不同的是，同一时刻可有多个 CheckBox 处于选中状态。可先在布局文件中定义多个 CheckBox，然后对每个 CheckBox 进行事件监听（setOnCheckedChangeListener），通过 isChecked 来判断选项是否被选中。

CheckBox 的类结构为：

```
java.lang.Object
  android.view.View
    android.widget.TextView
      android.widget.Button
        android.widget.CompoundButton
          android.widget.CheckBox
```

2. 单选按钮和复选框常用方法

RadioButton 从父类 CompoundButton 继承了一些成员函数，如表 3.8 所示。

表 3.8 RadioButton 和 CheckBox 类成员函数

Java 方法	功 能 描 述
isChecked()	判断是否被选中，若被选中，则返回 true，否则返回 false
performClick()	调用 OnClickListener 监听器，模拟一次单击
setChecked(boolean)	通过传入的参数设置控件是否被选中
toggle()	置反控件当前状态
setOnCheckedChangeListener (OnCheckedChangeListener)	为控件设置 OnCheckedChangeListener 监听器

3. 单选按钮和复选框典型案例

实现要点：

- onCheckedChanged(RadioGroup group,int checkedId)事件监听器：可监听单选按钮、复选框选中状态的改变。
- 掌握在 Android 中建立 CheckBox 的方法。
- 掌握 CheckBox 的常用属性。
- 掌握 CheckBox 选中状态变换的事件(监听器):先设置 CheckBox 的 setOnCheckedChange Listener，再通过 isChecked 来判断选项是否被选中。
- 创建 Ch03_Ex05_RadioButton 项目,修改布局文件 radiobutton_checkbox.xml,实现 Raido ButtonCheckBox.java 功能，如源代码清单 3.12 和 3.13 所示，其运行效果如图 3.8 所示。

源代码清单 3.12　\Ch03_Ex05_RadioButton\res\layout\radiobutton_checkbox.xml

```
1. <TableLayout xmlns:android="http://schemas.android.com/apk/res/android"
2.     android:id="@+id/tableLayout"          //表格布局 TableLayout 的 id 号
3.     android:layout_width="fill_parent"     //表格布局 TableLayout 的宽度
4.     android:layout_height="fill_parent"    //表格布局 TableLayout 的高度
5.     android:orientation="vertical">        //表格布局 TableLayout 的方向
```

```
6.    <TableRow>                                          //表格行 TableRow 标签
7.        <TextView                                       //文本视图标签
8.            android:layout_width="wrap_content"         //文本视图的宽度
9.            android:layout_height="wrap_content"        //文本视图的高度
10.           android:text="操作系统: "                    //文本视图的显示信息
11.           android:textSize="20dp"/>                   //文本视图的字体
12.       <!-- 定义一组单选按钮 -->
13.       <RadioGroup                                     //单选组标签
14.           android:id="@+id/radioGroup"                //单选组的id号
15.           android:layout_gravity="center_horizontal"  //单选组的位置: 水平居中
16.           android:orientation="horizontal">           //单选组的方向
17.           <!-- 定义两个单选按钮 -->
18.           <RadioButton                                //单选按钮标签
19.               android:id="@+id/rbAndroid"             //单选按钮的id号
20.               android:layout_width="wrap_content"     //单选按钮的宽度
21.               android:layout_height="wrap_content"    //单选按钮的高度
22.               android:checked="true"                  //单选按钮是否被选中: 是
23.               android:text="Adroid"/>                 //单选按钮的显示信息
24.           <RadioButton                                //单选按钮标签
25.               android:id="@+id/rbIos"                 //单选按钮的id号
26.               android:layout_width="wrap_content"     //单选按钮的宽度
27.               android:layout_height="wrap_content"    //单选按钮的高度
28.               android:text="IOS"/>                    //单选按钮的显示信息
29.       </RadioGroup>
30.   </TableRow>
31.   <TableRow>                                          //表格行 TableRow 标签
32.       <TextView                                       //文本视图标签
33.           android:layout_width="wrap_content"         //文本视图的宽度
34.           android:layout_height="wrap_content"        //文本视图的高度
35.           android:text="喜欢的音乐: "                  //文本视图的显示信息
36.           android:textSize="20dp"/>                   //文本视图的字体
37.       <!-- 定义一个垂直的线性布局 -->
38.       <LinearLayout                                   //线性布局标签
39.           android:layout_width="wrap_content"         //线性布局的宽度
40.           android:layout_height="wrap_content"        //线性布局的高度
41.           android:layout_gravity="center_horizontal"  //线性布局的位置: 水平居中
42.           android:orientation="vertical">             //线性布局的方向: 垂直
43.           <!-- 定义三个复选框 -->
44.           <CheckBox                                   //复选框 CheckBox 标签
45.               android:layout_width="wrap_content"     //复选框 CheckBox 的宽度
46.               android:layout_height="wrap_content"    //复选框 CheckBox 的高度
47.               android:checked="true"                  //复选框 CheckBox 选中状态: 是
48.               android:text="古典"/>                    //复选框 CheckBox 的显示信息
49.           <CheckBox                                   //复选框 CheckBox 标签
50.               android:layout_width="wrap_content"     //复选框 CheckBox 的宽度
51.               android:layout_height="wrap_content"    //复选框 CheckBox 的高度
52.               android:text="摇滚"/>                    //复选框 CheckBox 的显示信息
53.           <CheckBox                                   //复选框 CheckBox 标签
54.               android:layout_width="wrap_content"     //复选框 CheckBox 的宽度
55.               android:layout_height="wrap_content"    //复选框 CheckBox 的高度
```

```
56.            android:text="流行"/>              //复选框CheckBox的显示信息
57.        </LinearLayout>
58.    </TableRow>
59.    <TableRow>                                  //表格行TableRow标签
60.        <TextView                               //文本视图标签
61.            android:layout_width="wrap_content"   //文本视图的宽度
62.            android:layout_height="wrap_content"  //文本视图的高度
63.            android:text="背景色: "               //文本视图的显示信息
64.            android:textSize="20dp" />           //文本视图的字体
65.        <!-- 定义一组单选按钮 -->
66.        <RadioGroup                              //单选组标签
67.            android:id="@+id/rgBackcolor"        //单选组的id号
68.            android:layout_gravity="center_horizontal"//单选组的位置:水平居中
69.            android:orientation="horizontal">    //单选组的方向
70.            <!-- 定义三个单选按钮 -->
71.            <RadioButton                         //单选按钮标签
72.                android:id="@+id/rbRed"          //单选按钮的id号
73.                android:layout_width="wrap_content"   //单选按钮的宽度
74.                android:layout_height="wrap_content"  //单选按钮的高度
75.                android:text="红色"/>             //单选按钮的显示信息
76.            <RadioButton                         //单选按钮标签
77.                android:id="@+id/rbGreen"        //单选按钮的id号
78.                android:layout_width="wrap_content"   //单选按钮的宽度
79.                android:layout_height="wrap_content"  //单选按钮的高度
80.                android:checked="true"           //单选按钮是否被选中:是
81.                android:text="绿色"/>             //单选按钮的显示信息
82.            <RadioButton                         //单选按钮标签
83.                android:id="@+id/rbBlue"         //单选按钮的id号
84.                android:layout_width="wrap_content"   //单选按钮的宽度
85.                android:layout_height="wrap_content"  //单选按钮的高度
86.                android:text="蓝色"/>             //单选按钮的显示信息
87.        </RadioGroup>
88.    </TableRow>
89.    <TextView                                    //文本视图标签
90.        android:id="@+id/tvShow"                 //文本视图的id号
91.        android:layout_width="wrap_content"       //文本视图的宽度
92.        android:layout_height="wrap_content" />   //文本视图的高度
93.</TableLayout>
```

源代码清单3.13   \Ch03_Ex05_RadioButton\src\sziit\lihz\ch03_ex05_radiobutton\
RaidoButtonCheckBox.java

```
1. public class RaidoButtonCheckBox extends Activity { //从Activity派生子类
2.     private RadioGroup mRadioGroup;       //定义私有单选组RadioGroup对象
3.     private TextView mtvShow;             //定义私有文本视图TextView对象
4.     private RadioGroup mRGBackColor;      //定义私有单选组RadioGroup对象
5.     private String hint;                  //定义私有String对象
6.     private TableLayout tableLayout;      //定义私有TableLayout对象
7.     protected void onCreate(Bundle savedInstanceState) {//重写onCreate方法
8.         super.onCreate(savedInstanceState);           //调用基类方法
9.         setContentView(R.layout.radiobutton_checkbox);   //设置手机屏幕布局
```

```
10.     //获取界面上相关组件
11.     mRadioGroup = (RadioGroup) findViewById(R.id.radioGroup);
12.     mtvShow = (TextView) findViewById(R.id.tvShow);
13.     mRGBackColor = (RadioGroup) findViewById(R.id.rgBackcolor);
14.     tableLayout = (TableLayout) findViewById(R.id.tableLayout);
15.     hint = "背景色为绿色";
16.     tableLayout.setBackgroundColor(Color.GREEN);
17.     //为RadioGroup组件的OnCheck事件绑定事件监听器
18.     mRadioGroup.setOnCheckedChangeListener(new OnCheckedChangeListener() {
19.         public void onCheckedChanged(RadioGroup group, int checkedId) {
20.             //根据用户选中的单选按钮来动态改变hint字符串的值
21.             String hint = checkedId == R.id.rbAndroid ? "您手机的操作系统是
                    Android": "您手机的操作系统是IOS";
22.             mtvShow.setText(hint);}});    //修改show组件中的文本
23.     mRGBackColor.setOnCheckedChangeListener(new OnCheckedChangeListener() {
24.         public void onCheckedChanged(RadioGroup group, int checkedId) {
25.             //根据用户选中的单选按钮来动态改变hint字符串的值
26.             switch (checkedId) {              //根据id号区分不同的RadioButton
27.             case R.id.rbRed: hint = "背景色为红色";
28.                 tableLayout.setBackgroundColor(Color.RED);
29.                 Toast.makeText(RaidoButtonCheckBox.this, hint, Toast.LENGTH_
                    LONG).show(); break;
30.             case R.id.rbGreen:hint = "背景色为绿色";
31.                 tableLayout.setBackgroundColor(Color.GREEN);
32.                 Toast.makeText(RaidoButtonCheckBox.this, hint, Toast.LENGTH_
                    SHORT).show();break;
33.             case R.id.rbBlue:hint = "背景色为蓝色";
34.                 tableLayout.setBackgroundColor(Color.BLUE);
35.                 Toast.makeText(RaidoButtonCheckBox.this, hint, Toast.LENGTH_
                    SHORT).show();break;
36.             default: hint = "背景色为绿色";
37.                 Toast.makeText(RaidoButtonCheckBox.this, hint, Toast.LENGTH_
                    SHORT).show();
38.                 tableLayout.setBackgroundColor(Color.GREEN);  break;   }
39.             mtvShow.setText(hint);}});}} //修改show组件中的文本
```

（a）设计轮廓

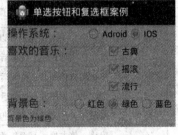
（b）运行界面

图 3.8　RadioButton 和 CheckBox 典型案例

3.2.6 图像视图控件

1. 图像视图简介

图像视图(ImageView)的结构为:
```
java.lang.Object
    android.view.View
        android.widget.ImageView
```

已知直接子类:ImageButton,QuickContactBadge。

已知间接子类:ZoomButton。

ImageView 直接继承自 View 类,它的主要功能是显示图片,实际上它不仅可用来显示图片,还可用来显示任何 Drawable 对象。ImageView 适用于任何布局,且 Android 为其提供了放大、缩小、旋转和着色等操作。

2. ImageView 的常用属性和方法

ImageView 常用的属性及其对应方法如表 3.9 所示(XML 属性名均以 android:开头)。

表 3.9 ImageView 常用的属性及其对应方法

XML 属性名	Java 代码设置方法	功 能
android:adjustViewBounds	setAdjustViewBounds(boolean adjustViewBounds)	设置该属性为 true 时,可在图像视图调整边界时保持图片的纵横比例
android:baseline	setBaseline(int baseline)	设置图像视图内基线的偏移量
android:baselineAlignBottom	setBaselineAlignBottom(boolean aligned)	如果为 true,图像视图将基线与父控件底部边缘对齐
android:cropToPadding	setCropToPadding(boolean cropToPadding)	如果为 true,会剪切图片以适应图像视图内边距的大小
android:maxHeight	setMaxWidth(int maxWidth)	为图像视图提供最大高度的可选参数
android:maxWidth	setMaxWidth(int maxWidth)	为图像视图提供最大宽度的可选参数
android:scaleType	setScaleType(ScaleType scaleType)	为了使图片适合图像视图的大小,控制如何变更图片大小或移动图片。取值常量如表 3.10 所示
android:src	setImageResource(int resId)	设置可绘制对象作为图像视图显示的内容
android:tint	setColorFilter(int color)	将图片渲染成指定的颜色

表 3.10 ScaleType 枚举常量及其含义

常 量	值	含 义 描 述
MATRIX	0	用矩阵来绘图
FIT_XY	1	拉伸图片(不按比例)以填充 ImageView 的宽高
FIT_START	2	按比例拉伸图片,拉伸后图片的高度为 ImageView 的高度,且显示在 ImageView 的左边
FIT_CENTER	3	按比例拉伸图片,拉伸后图片的高度为 ImageView 的高度,且显示在 ImageView 的中间
FIT_END	4	按比例拉伸图片,拉伸后图片的高度为 ImageView 的高度,且显示在 ImageView 的右边
CENTER	5	按原图大小显示图片,但图片宽高大于 ImageView 的宽高时,截取图片中间部分显示
CENTER_CROP	6	按比例放大原图直至等于某边 ImageView 的宽高显示。
CENTER_INSIDE	7	当原图大小等于 ImageView 的大小时,按原图大小居中显示;反之将原图缩放至 ImageView 的大小并居中显示

ImageView 常用接口方法如表 3.11 所示。

表 3.11　ImageView 常用接口方法及功能描述

公　共　方　法	功　能　描　述
void clearColorFilter()	清除颜色过滤
int getBaseline()	返回部件顶端到文本基线的偏移量，若不支持基线对齐，返回 –1
boolean getBaselineAlignBottom()	返回当前视图基线是否将考虑视图的底部
Drawable getDrawable()	返回视图的可绘制对象；如果没有关联可绘制对象，返回空
Matrix getImageMatrix()	返回视图的选项矩阵
ImageView.ScaleType getScaleType()	返回当前 ImageView 使用的缩放类型
void invalidateDrawable(Drawable dr)	使指定的可绘制对象失效
void jumpDrawablesToCurrentState ()	调用与视图相关的所有可绘制对象的 Drawable.jumpToCurrentState()方法
int[] onCreateDrawableState(int extraSpace)	为当前视图生成新的 Drawable 状态时发生
void setAdjustViewBounds(boolean adjust ViewBounds)	当需要在 ImageView 调整边框时保持可绘制对象的比例时，将该值设为真
void setAlpha(int alpha)	设置透明度
void setBaseline(int baseline)	设置部件顶部边界文本基线的偏移量
void setBaselineAlignBottom(boolean aligned)	设置是否设置视图底部的视图基线
void setColorFilter(int color)	为图片设置着色选项，采用 SRC_ATOP 合成模式
void setColorFilter(ColorFilter cf)	为图片应用任意颜色滤镜
void setColorFilter(int color, PorterDuff. Mode mode)	为图片设置着色选项。color：应用的着色颜色；mode：如何着色；标准模式为 SRC_ATOP
void setImageBitmap(Bitmap bm)	设置位图作为该 ImageView 的内容
void setImageDrawable(Drawable drawable)	设置可绘制对象为该 ImageView 显示的内容
void setImageResource(int resId)	通过资源 ID 设置可绘制对象为该 ImageView 显示的内容
void setImageURI(Uri uri)	设置指定的 URI 为该 ImageView 显示的内容
void setImageMatrix(Matrix matrix)	矩阵变换
void setImageState(int[] state, boolean merge)	设置视图可见和不可见
void setSelected(boolean selected)	改变视图的选中状态；视图有选中和未选中两种状态

ImageView 控件的主要属性和公用接口方法详细描述如下：

（1）android:src

该属性设置 ImageView 要显示的图片。例如，android:src="@drawable/ic_launcher"表示该控件要显示 res\drawable 目录下的图片 ic_launcher。

（2）void setImageResource (int resId)

功能：设置 ImageView 将要显示的图片资源。

参数：resId 为图片资源 ID，例如 R.drawable.ic_launcher。

（3）void setImageBitmap(Bitmap bm)

（4）void setImageURI(Uri uri)

功能：设置 ImageView 将要显示的 URI 图片。

参数：uri 为图片的统一资源标识符（uniform resource identifier）。

（5）void setImageDrawable(Drawable drawable)

功能：设置 ImageView 将要显示的 Drawable 对象。

参数：drawable 是 Drawable 类图片。Drawable 是一个抽象类（abstract class），它有好多子类（SubClass）来操作具体类型的资源，如 BitmapDrawable 用来操作位图。

3. 图像视图典型案例

实现要点：

- void setOnClickListener(View.OnClickListener l)：设置 ImageView 控件单击事件监听器。
- void addView(View child)：向 ImageView 控件添加子视图对象。
- void setImageResource(int resId)：将一个 Drawable 资源对象设置为 ImageView 控件的显示内容。
- View.OnClickListener()：当视图被单击时激活的回调接口定义。
- void onClick (View v)：当视图被单击时调用的事件回调函数。
- void setContentView(int layoutResID)：根据布局资源 id 设置活动内容。
- View findViewById(int id)：根据 XML 布局文件中定义的 id 属性查找一个视图对象。
- ImageView(Context context)：图像视图构造器，在当前上下文环境构造图像视图对象。
- void setScaleType (ImageView.ScaleType scaleType)：为了使图片适合图像视图的大小，控制如何变更图片大小或移动图片。
- 创建 Ch03_Ex06_ImageView 项目，修改布局文件 imageview_layout.xml，实现 ImageViewActivity.java 功能，如源代码清单 3.14 和 3.15 所示，其运行效果如图 3.9 所示。

图 3.9　图像视图典型案例

源代码清单 3.14　\Ch03_Ex06_ImageView\res\layout\imageview_layout.xml

```
1. <?xml version="1.0" encoding="utf-8"?>
2. <LinearLayout xmlns:android="http://schemas.android.com/apk/res/android"
3.     android:id="@+id/linearLayout"       android:layout_width="fill_parent"
4.     android:layout_height="fill_parent"  android:orientation="vertical">
5. </LinearLayout>
```

源代码清单 3.15　\Ch03_Ex06_ImageView\src\sziit\lihz\ch03_ex06_imageview

```
1. package sziit.lihz.ch03_ex06_imageview;          //用户自定义包
2. //引入项目所需的功能包
3. import android.os.Bundle;                         //引入绑定包
4. import android.app.Activity;                     //引入活动包
5. import android.view.Menu;                         //引入菜单包
6. import android.view.View;                         //引入视图包
7. import android.view.View.OnClickListener;         //引入单击监听器包
8. import android.widget.ImageView;                  //引入图像视图包
9. import android.widget.LinearLayout;               //引入线性布局包
10.//定义公有派生活动子类，从 Activity 基类派生子类 ImageViewActivity
11.public class ImageViewActivity extends Activity {
12.    //定义一个访问图片的数组,这些图片保存在 drawable-mdpi 文件夹下
13.    int[] images = new int[] { R.drawable.flower1, R.drawable.flower2,
```

```
14.            R.drawable.flower3, R.drawable.flower4, R.drawable.flower5,
15.            R.drawable.flower5 };
16.    int curImageIndex = 0;                              //当前图片索引
17.    protected void onCreate(Bundle savedInstanceState) {//重写 onCreate 方法
18.        super.onCreate(savedInstanceState);             //基类调用 onCreate 方法
19.        setContentView(R.layout.imageview_layout);      //设置手机屏幕布局
20.        //获取 LinearLayout 布局容器
21.        LinearLayout root = (LinearLayout) findViewById(R.id.linearLayout);
22.        final ImageView imageView= new ImageView(this);//创建 ImageView 组件
23.        root.addView(imageView);//将 ImageView 组件添加到 LinearLayout 布局容器
24.        imageView.setImageResource(images[0]);          //初始化时显示第一张图片
25.        imageView.setScaleType(ScaleType.FIT_XY);//设置图片变化的比例类型
26.        imageView.setOnClickListener(new OnClickListener() {
                                                    //设置单击监听器事件接口
27.            public void onClick(View v) {       //单击事件处理函数
28.                if (curImageIndex >= 5) {       //图像数组边界控制
29.                    curImageIndex = -1;}
30.        imageView.setImageResource(images[++curImageIndex]);}
                                                //改变 ImageView 中显示的图片
31.        });}}
```

3.2.7 日期与时间控件

1. 日期与时间控件简介

DatePicker 和 TimePicker 是两个比较易用的控件,它们都从帧布局派生而来,其中 DatePicker 供用户选择日期,而 TimePicker 则供用户选择时间。

DatePicker 是一个选择年月日的日历布局视图, 对于对话框视图,参见 DatePickerDialog。日期选择器(DatePicker)的结构为:

```
java.lang.Object
    android.view.View
        android.view.ViewGroup
            android.widget.FrameLayout
                android.widget.DatePicker
```

TimePicker 是用于选择一天中时间的视图,支持 24 时制及上午/下午模式。时、分及上午/下午(如果可用)都可以用垂直滚动条来控制。可用键盘来输入小时。双数的小时数可以通过输入两个数字来实现,例如在一定时间内输入 1 和 2 即选择了 12 点。分钟能显示输入的单个数字。在 AM/PM 模式下,用户可以输入 'a', 'A'或 'p', 'P'来选取。对于对话框视图,参见 TimePickerDialog。

时间选择器(TimePicker)的结构为:

```
java.lang.Object
    android.view.View
        android.view.ViewGroup
            android.widget.FrameLayout
                android.widget.TimePicker
```

DatePicker 和 TimePicker 在帧布局的基础上提供了一些方法来获取当前用户所选择的日期、时间;如果程序需要获取用户选择的日期、时间,则可通过为 DatePicker 添加 OnDateChanged

Listener、为 TimePicker 添加 OnTimerChangedListener 进行监听来实现。

2. 日期与时间控件支持的 XML 属性及其常用方法

日期选择器和时间选择器支持的 XML 属性及其常用方法分别如表 3.12、表 3.13 和表 3.14 所示。

表 3.12 日期选择器的 XML 属性

XML 属性	属性含义描述
android:calendarViewShown	设置该 DatePicker 是否显示 ClendarView 组件
android:endYear	设置该 DatePicker 允许选择的最后一年
android:startYear	设置该 DatePicker 允许选择的第一年
android:maxDate	设置该 DatePicker 允许选择的最大日期，格式为 mm/dd/yyyy
android:minDate	设置该 DatePicker 允许选择的最小日期，格式为 mm/dd/yyyy
android:spinnersShown	设置该 DatePicker 是否显示 Spinner 日期选择组件

表 3.13 日期选择器的公共方法

公 共 方 法	功 能 描 述
int getDayOfMonth()	获取选择的天数
int getYear()	获取选择的年份
int getMonth()	获取选择的月份
void init(int year, int monthOfYear, int dayOfMonth, DatePicker.OnDateChanged Listener onDateChangedListener)	初始化年月日
void setEnabled(boolean enabled)	设置视图的启用状态
void updateDate(int year, int monthOfYear, int dayOfMonth)	更新日期

表 3.14 时间选择器的公共方法

公 共 方 法	功 能 描 述
int getBaseline()	返回窗口空间的文本基准线到其顶边界的偏移量
Integer getCurrentHour()	获取当前时间的小时部分；返回值当前小时（0～23）
void setCurrentHour(Integer currentHour)	设置当前小时
Integer getCurrentMinute()	获取当前时间的分钟部分
void setCurrentMinute(Integer currentMinute)	设置当前分钟（0～59）
boolean is24HourView()	获取当前系统设置是否为 24 时制。如果是 24 时制则返回 true，否则返回 false
void setEnabled(boolean enabled)	设置可用的视图状态
void setIs24HourView(Boolean is24HourView)	设置是 24 时制还是上午/下午制
void setOnTimeChangedListener (TimePicker.OnTimeChangedListener onTimeChangedListener)	设置时间调整事件的回调函数

● Calendar getInstance()：获得 Calendar 的实例。

● Calendar getInstance(Locale locale)：根据本地 Locale 对象获得 Calendar 的实例。

● void set(int year, int month, int day)：设置 Calendar 对象的年、月和日。

- void set(int year, int month, int day, inthourOfDay,int minute): 设置 Calendar 对象的年、月、日、时和分。
- void onDateChanged(DatePicker view, int year, intmonthOfYear, intdayOfMonth): Calendar 对象的日期发生改变时回调函数。
- void onTimeChanged(TimePicker view, inthourOfDay, int minute): Calendar 对象的时间发生改变时回调函数。

3. 日期与时间控件典型案例

实现要点:
- Calendar getInstance(): 构造当前时区的日历 Calendar 对象实例。
- DatePicker.init(int year, int monthOfYear, int dayOfMonth, OnDateChangedListener onDateChangedListener): 初始化日期选择器对象状态。
- DatePicker.OnDateChangedListener: 日期变化监听器。
- void onDateChanged(DatePicker view, int year, int monthOfYear, int dayOfMonth): 一旦日期发生变化即触发此事件响应函数。
- setOnTimeChangedListener(TimePicker.OnTimeChangedListener onTimeChangedListener): 设置监听用户调整时间的回调函数。
- void onDateChanged(DatePicker view, int year,int month, int day): 日期变化事件响应函数。
- void onTimeChanged(TimePicker view, int hour, int minute): 时间变化事件响应函数。
- void setText (CharSequence text): 设置文本视图对象显示字符串值。
- String getString (int resId): 根据字符串资源 id 获得其字符串信息。
- 创建 Ch03_Ex07_DataPicker 项目,修改字符串文件 strings.xml,修改布局文件 data_time_pickerlayout.xml,实现 DataTimerPickerActivity.java 功能,如源代码清单 3.16、3.17 和 3.18 所示,其运行效果如图 3.10 所示。

源代码清单 3.16 \Ch03_Ex07_DataPicker\res\values\strings.xml

```
1.  <?xml version="1.0" encoding="utf-8"?>
2.  <resources>
3.      <string name="app_name">日期与时间控件典型案例</string>
4.      <string name="action_settings">设置</string>
5.      <string name="choiceBorrowBookTime">请选择借本书的时间</string>
6.      <string name="yourBorrowBookTime">您的借书日期为: </string>
7.      <string name="syear">年</string>     <string name="smonth">月</string>
8.      <string name="sday">日</string>      <string name="shour">时</string>
9.      <string name="sminute">分</string>    </resources>
```

源代码清单 3.17 \Ch03_Ex07_DataPicker\res\layout\data_time_pickerlayout.xml

```
1.  <LinearLayout xmlns:android="http://schemas.android.com/apk/res/android"
2.      android:layout_width="match_parent"  android:layout_height="match_parent"
3.      android:gravity="center_horizontal"  android:orientation="vertical" >
4.      <TextView           android:layout_width="fill_parent"
5.          android:layout_height="wrap_content"    android:background="#00f"
6.          android:text="@string/choiceBorrowBookTime"  android:textColor="#0f0"
7.          android:textSize="20sp" />
8.      <!-- 定义一个DatePicker组件 -->
```

```
9.    <DatePicker
10.        android:id="@+id/datePicker"      android:layout_width="match_parent"
11.        android:layout_height="220dp"     android:layout_gravity="center_horizontal"
12.        android:calendarViewShown="true"  android:endYear="2015"
13.        android:spinnersShown="true"      android:startYear="2000" />
14.    <!-- 定义一个 TimerPicker 组件 -->
15.    <TimePicker
16.        android:id="@+id/timePicker"      android:layout_width="wrap_content"
17.        android:layout_height="100dp"
18.        android:layout_gravity="center_ horizontal"/>
19.    <!-- 显示用户输入日期、时间的控件 -->
20.    <EditText     android:id="@+id/etShow"
21.        android:layout_width="fill_parent"
22.        android:layout_height="wrap_ content"
22.        android:cursorVisible="false"     android:editable="false"
24.        android:text="@string/yourBorrowBookTime"   android:textSize="20sp"
25.        android:textStyle="bold"  android:typeface="sans" />  </LinearLayout>
```

源代码清单 3.18　　\Ch03_Ex07_DataPicker\src\sziit\lihz\
Ch03_ex07_datapicker\DataTimerPickerActivity.java

```
1. package sziit.lihz.ch03_ex07_datapicker;       //用户自定义包
2. //引入本项目所需全部功能包
3. import java.util.Calendar;                     //引入日历包
4. import android.os.Bundle;                      //引入操作系统绑定包
5. import android.app.Activity;                   //引入活动包
6. import android.view.Menu;                      //引入菜单包
7. import android.widget.DatePicker;              //引入日期选择包
8. import android.widget.DatePicker.OnDateChangedListener;
                                                  //引入日期数据改变监听器包
9. import android.widget.EditText;                //引入编辑框包
10.import android.widget.TimePicker;              //引入时间选择包
11.import android.widget.TimePicker.OnTimeChangedListener;//引入时间改变监听器包
12.public class DataTimerPickerActivity extends Activity {
                                                  //从基类 Activity 派生子类
13.   //定义 5 个记录当前时间的私有正形变量
14.   private int mYear;                          //当前年
15.   private int mMonth;                         //当前月
16.   private int mDay;                           //当前日
17.   private int mHour;                          //当前时
18.   private int mMinute;                        //当前分
19.   protected void onCreate(Bundle savedInstanceState) {//子类重写 onCreate 方法
20.      super.onCreate(savedInstanceState);      //调用基类 onCreate 方法
21.      setContentView(R.layout.data_time_pickerlayout);//设置手机屏幕布局
22.      DatePicker datePicker =(DatePicker)findViewById(R.id.datePicker);
                                                  //绑定日期选择对象
23.      TimePicker timePicker =(TimePicker)findViewById(R.id.timePicker);
                                                  //绑定时间选择对象
24.      // 获取当前的年、月、日、小时、分钟
25.      Calendar c = Calendar.getInstance();     //构造 Calendar 对象
26.      mYear = c.get(Calendar.YEAR);            //获取当前年
```

```
27.     mMonth = c.get(Calendar.MONTH);          //获取当前月
28.     mDay = c.get(Calendar.DAY_OF_MONTH);     //获取当前日
29.     mHour = c.get(Calendar.HOUR);            //获取当前小时
30.     mMinute = c.get(Calendar.MINUTE);        //获取当前分钟
31.     //初始化DatePicker组件，初始化时指定监听器
32.     datePicker.init(mYear, mMonth, mDay, new OnDateChangedListener() {
                                                 //初始化datePicker
33.         public void onDateChanged(DatePicker view, int year,
34.                     int month, int day) {    //日期变化事件响应函数
35.             DataTimerPickerActivity.this.mYear = year;//记录当前设置年
36.             DataTimerPickerActivity.this.mMonth = month;//记录当前设置月
37.             DataTimerPickerActivity.this.mDay = day;//记录当前设置日
38.             showDateTime(year, month, day, mHour, mMinute);//显示当期日期、时间
39.             Toast.makeText(DataTimerPickerActivity.this,"您选择的日期: "+mYear+
40.             "年 "+mMonth+"月 "+mDay+"日",Toast.LENGTH_SHORT).show();  }});
41.     timePicker.setOnTimeChangedListener(new OnTimeChangedListener() {
                                                 //设置监听器
42.         public void onTimeChanged(TimePicker view, int hour, int minute)
            {//事件响应函数
43.             DataTimerPickerActivity.this.mHour = hour; //记录当前设置小时
44.             DataTimerPickerActivity.this.mMinute = minute;//记录当前设置分
45.             showDateTime(mYear, mMonth, mDay, hour, minute);//显示当前日期、时间
46.             Toast.makeText(DataTimerPickerActivity.this,"您选择的时间: "+ mHour +
47.             "时  "+ mMinute + "分",Toast.LENGTH_SHORT).show();}});}
48.     //定义在EditText中显示当前日期、时间的方法
49.     private void showDateTime(int year, int month, int day, int hour, int minute) {
50.         EditText show = (EditText) findViewById(R.id.etShow);//绑定编辑框对象
51.         show.setText(getString(R.string.yourBorrowBookTime) + year
52.             + getString(R.string.syear) + (month + 1)
53.             + getString(R.string.smonth) + day + getString(R.string.sday)
54.             + hour + getString(R.string.shour) + minute
55.             + getString(R.string.sminute));}}  //设置编辑框对象显示文本信息
```

（a）设计轮廓　　　　　　　（b）布局界面

图3.10　日期与时间控件典型案例

3.2.8 模拟时钟和数字时钟

1. 模拟时钟和数字时钟简介

时钟控件包括AnalogClock（模拟时钟）和DigitalClock（数字时钟），它们都负责显示时

钟，所不同的是 AnalogClock 控件显示模拟时钟，且只显示时针和分针，而 DigitalClock 显示数字时钟，可精确到秒。

DigitalClock 继承的是 TextView 类，它本身是一个文本框。AnalogClock 继承的是 View 类，可重写 OnDraw()方法。

2. 模拟时钟和数字时钟的 XML 属性及其常用方法

模拟时钟的典型属性如表 3.15 所示。AnalogClock 和 DigitalClock 的其他基本属性均继承于 View 控件。

表 3.15　模拟时钟的典型属性

XML 属性	属性含义描述
android:dial	设置该模拟时钟的表盘使用的图片
android:hand_hour	设置该模拟时钟的时针使用的图片
android:hand_minute	设置该模拟时钟的分针使用的图片

3. AnalogClock 和 DigitalClock 典型案例

实现要点：

- AnalogClock：模拟时钟控件。
- DigitalClock：数字时钟控件。
- boolean android.os.Handler.sendMessage(Message msg)：发送消息
- void handleMessage(Message msg)：处理消息。
- Thread：线程类。
- Handler：句柄类。
- void java.lang.Thread.run()。
- System.currentTimeMillis()：获得系统当前时间。
- Calendar.getInstance()：获得日历对象。
- void Calendar.setTimeInMillis(long milliseconds)：设置日历当前时间。
- 创建 Ch03_Ex08_AnalogClock 项目，修改布局文件 analog_digital_clock.xml，实现 AnalogClockActivity.java 功能，如源代码清单 3.19 和 3.20 所示，其运行效果如图 3.11 所示。

源代码清单 3.19　\Ch03_Ex08_AnalogClock\res\layout\analog_digital_clock.xml

```
1.  <RelativeLayout xmlns:android="http://schemas.android.com/apk/res/android"
2.      xmlns:tools="http://schemas.android.com/tools"
3.      android:layout_width="match_parent"
4.      android:layout_height="match_parent"
5.      android:paddingBottom="@dimen/activity_vertical_margin"
6.      android:paddingLeft="@dimen/activity_horizontal_margin"
7.      android:paddingRight="@dimen/activity_horizontal_margin"
8.      android:paddingTop="@dimen/activity_vertical_margin"
9.      tools:context=".AnalogClockActivity">
10.     <!-- AnalogClock 与 DigitalClock这两个时钟控件都不需要Java 代码,只要在layout
            的 xml 中插入以下代码即可自动显示时间 -->
11.     <!-- 模拟时钟控件 -->
12.     <AnalogClock          android:id="@+id/analogClock"
13.         android:layout_width="wrap_content"
```

```
14.          android:layout_height="wrap_content"
15.          android:layout_centerHorizontal="true"
16.          android:layout_gravity="center_horizontal" />
17.   <!-- 数字时钟控件 -->
18.   <DigitalClock    android:id="@+id/digitalClock"
19.          android:layout_width="wrap_content"
20.          android:layout_height="wrap_content"
21.          android:layout_below="@id/analogClock" android:layout_centerHorizontal="true"
22.          android:layout_gravity="center_horizontal"/>
23.   <TextView    android:id="@+id/textView"
24.          android:layout_width="wrap_content"
25.          android:layout_height="wrap_content"
26.          android:layout_below="@id/digitalClock"
27.          android:layout_centerHorizontal="true"
28.          android:background="#00f"   android:text="时间" android:textColor="#0f0"
29.          android:textSize="20sp" >  </TextView> </RelativeLayout>
```

源代码清单 3.20   \Ch03_Ex08_AnalogClock\src\sziit\lihz\
Ch03_ex08_analogclock\AnalogClockActivity.java

```
1. public class AnalogClockActivity extends Activity {
                                     //从 Activity 基类公有派生子类
2.    protected static final int SHOW_MESSAGE = 0x1234;//定义消息常量
3.    private  Calendar mCalendar;        //定义私有日历 Calendar 对象
4.    private  AnalogClock mAnalogClock;  //定义私有模拟时钟 AnalogClock 对象
5.    private  int mMinutes;              //定义私有分钟变量
6.    private  int mHour;                 //定义私有小时变量
7.    private  TextView mTextView;        //定义私有文本视图 TextView 对象
8.    private  Handler mHandler;          //定义私有句柄 Handler 对象
9.    private  Thread mClockThread;       //定义私有线程 Thread 对象
10.   protected void onCreate(Bundle savedInstanceState) {//子类重写 onCreate 方法
11.       super.onCreate(savedInstanceState);      //基类调用 onCreate 方法
12.       setContentView(R.layout.analog_digital_clock);//设置手机屏幕布局界面
13.       mTextView = (TextView) findViewById(R.id.textView);
                                                   //绑定 TextView 对象
14.       mAnalogClock = (AnalogClock) findViewById(R.id.analogClock);
                                                   //绑定 AnalogClock 对象
15.       mHandler = new Handler() {               //构造 Handler 对象
16.          public void handleMessage(Message msg) {//处理消息循环
17.             switch (msg.what) {                //判断什么消息
18.                case SHOW_MESSAGE:              //处理接收到的消息
19.                   mTextView.setText(mHour + " : " + mMinutes);//显示时间
20.                   break;  }
21.             super.handleMessage(msg);} };      //处理缺省消息(系统消息)
22.       mClockThread = new LooperThread();       //构造用户自定义线程
23.       mClockThread.start();                    //启动用户线程 }
24.   class LooperThread extends Thread {          //从基类 Thread 派生子类
25.       public void run() {                      //线程执行接口
26.          super.run();try {do {long time = System.currentTimeMillis();
                                                   //获得系统当前时间
27.             Calendar mCalendar = Calendar.getInstance();  //获得日历对象
```

```
28.         mCalendar.setTimeInMillis(time);           //设置日历时间
29.         mHour = mCalendar.get(Calendar.HOUR);      //获得当前小时
30.         mMinutes = mCalendar.get(Calendar.MINUTE); //获得当前分钟
31.         Thread.sleep(1000);                        //休眠1秒
32.         Message msg = new Message();               //构造消息Message对象
33.         msg.what = SHOW_MESSAGE;                   //设置消息功能
34.         AnalogClockActivity.this.mHandler.sendMessage(msg);
                                                       //发送空消息
35.       } while (AnalogClockActivity.LooperThread.interrupted() ==
          false);  /* 当系统发出中断讯息时停止本循环 */
36.     } catch (Exception e)
        {//捕获异常
37.       e.printStackTrace();}}}}                     //打印异常栈消息
```

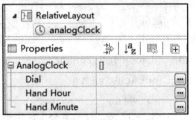

（a）设计轮廓　　　　　　　　　　（b）运行界面

图 3.11　AnalogClock 和 DigitalClock 典型案例

3.2.9　计时器控件

1. 计时器简介

计时器（Chronometer）的结构为：

```
public class Chronometer extends TextView
    java.lang.Object
        android.view.View
            android.widget.TextView
                android.widget.Chronometer
```

Android Chronometer 实现了一个简单的计时器，继承自 TextView 类，其可用 1 秒的时间间隔计时，并显示出计时结果。Chronometer 可用 TextView 类的 text 属性来控制时间显示的颜色、字体大小等；允许使用 setFormat 定义时间显示格式，默认为"MM：SS"，即"分：秒"。Chronometer 可用来显示时间的推移，当需要记录用户进行某项操作所消耗的时间或者在游戏中需要限制时间时，非常实用。

2. 计时器的重要属性和常用方法

计时器包括表 3.16 所示的重要 XML 属性

表 3.16　Chronometer 的重要 XML 属性

XML 属性	功 能 描 述
android:format	定义时间的格式，如 hh:mm:ss。若指定，计时器将根据这个字符串来显示，替换字符串中第一个 "%s" 为当前"MM:SS"或"H:MM:SS"格式的时间显示。若不指定，计时器将简单地显示"MM:SS"或"H:MM:SS"格式的时间

计时器包括以下三个公有构造函数：

- Chronometer(Context context)：初始化计时器对象，设置当前时间为基准时间。
- Chronometer(Context context, AttributeSet attrs)：初始化标准视图布局信息，设置当前时间为基准时间。
- Chronometer(Context context, AttributeSet attrs, int defStyle)：初始化标准视图布局信息和风格，设置当前时间为基准时间。

计时器的主要方法如表 3.17 所示。

表 3.17　计时器的主要方法及功能描述

方　　法	功能描述
long getBase()	返回基准时间
String getFormat()	返回当前字符串格式，此格式是通过 setFormat()实现的
void setBase(long base)	设置计时器开始的时间基线，如为 SystemClock.elapsedRealtime()
void setFormat(String format)	设置显示时间的格式信息，计时器将会显示这个参数所对应的值，如果字符串的值为 null，那么返回的值为 MM:SS 格式
void start()	开始计时
void stop()	停止计时
setOnChronometerTickListener	当计时器改变时调用

Chronometer 类有 3 个重要的方法：start()、stop()和 setBase()，其中 start()表示开始计时；stop()表示停止计时；setBase()表示重新计时。start()和 stop()没有任何参数，setBase()有参数，表示开始计时的基准时间，若要从当前时刻重新计时，可将该参数值设为 SystemClock.elapsedRealtime()。在默认情况下，Chronometer 只输出 MM:SS 格式的时间。例如，当计时到 2 分 30 秒时，Chronometer 会显示 02:30。若要改变显示的时间格式，可使用其 setFormat()方法。该方法需要一个 String 使用 "%s" 表示计时信息。例如，使用 setFormat("计时时间：%s")设置显示信息，Chronometer 会显示计时时间：(计时时间：02:30)。

Chronometer 可通过 onChronometerTick 事件方法来捕捉计时动作。该方法 1 秒调用一次。若要使用 onChronometerTick 事件方法，必须实现接口 android.widget.Chronometer.OnChronometerTickListener。

3. 计时器典型案例

实现要点：

- start()：开始计时。
- stop()：停止计时。
- setBase(SystemClock.elapsedRealtime())：设置初始时间。
- setFormat("计时信息：　(%s)")：设置时间显示格式。
- setFormat(null)：使用非格式显示字符。
- 注册权限：<uses-permission android:name="android.permission.VIBRATE" />
- 创建 Ch03_Ex09_Chronometer 项目，修改字符串资源文件 string.xml，修改布局文件 chronometer.xml，实现 ChronometerActivity.java 功能，如源代码清单 3.21、3.22 和 3.23 所示，其运行效果如图 3.12 所示。

主要步骤：

首先在布局文件中对应的位置声明定义计时器的位置与属性：

```
1. <Chronometer  android:id="@+id/chronometer"
2.   android:layout_width="wrap_content"  android:layout_height="wrap_content"/>
```

其次，在程序中使用定义的计时器，并设置显示时间格式，可使用计时器三个常用方法：

```
1. mChronometer = (Chronometer) findViewById(R.id.chronometer);
2. mChronometer.setFormat("计时时间: (%s)"); mChronometer.start();//计时开始
3. mChronometer.stop();//暂停计时
4. mChronometer.setBase(SystemClock.elapsedRealtime());//复位计时器，停止计时
```

最后，在 AndroidManifest.xml 文件注册权限：

```
<uses-permission android:name="android.permission.VIBRATE" />
```

源代码清单 3.21   \Ch03_Ex09_Chronometer\res\values\strings.xml

```
1. <?xml version="1.0" encoding="utf-8"?>
2. <resources xmlns:xliff="urn:oasis:names:tc:xliff:document:1.2">
3.  <string name="app_name">计时器典型案例</string>
4.  <string name="action_settings">设置</string>
5.  <string name="cm_start">启动</string>
6.  <string name="cm_stop">停止</string>
7.  <string name="cm_reset">复位</string>
8.  <string name="cm_set_format">设置时间格式</string>
9.  <string name="cm_clear_format">清除时间格式</string>
10.  <string name="cm_initial_format">初始格式: <xliff:g id="定时器初始时间格式">%s</xliff:g></string>   </resources>
```

源代码清单 3.22   \Ch03_Ex09_Chronometer\res\layout\chronometer.xml

```
1. <?xml version="1.0" encoding="utf-8"?>
2. <LinearLayout xmlns:android="http://schemas.android.com/apk/res/android"
3.   android:layout_width="fill_parent"  android:layout_height="fill_parent"
4.   android:gravity="center_horizontal"  android:orientation="vertical"
5.   android:padding="4dip" >
6.   <Chronometer   android:id="@+id/chronometer"
7.     android:layout_width="wrap_content"  android:layout_height="wrap_content"
8.     android:layout_weight="0"            android:background="#0f0"
9.     android:format="@string/cm_initial_format"  android:paddingBottom="30dip"
10.    android:paddingTop="30dip"           android:textColor="#00f"
11.    android:textSize="20sp"              android:typeface="sans" />
12.   <LinearLayout    android:layout_width="fill_parent"
13.    android:layout_height="wrap_content"  android:gravity="center_horizontal"
14.    android:orientation="horizontal"      android:padding="4dip" >
15.    <Button     android:id="@+id/btnStart"
16.      android:layout_width="wrap_content"  android:layout_height="wrap_content"
17.      android:text="@string/cm_start" > <requestFocus/> </Button>
18.    <Button     android:id="@+id/btnStop"
19.      android:layout_width="wrap_content"  android:layout_height="wrap_content"
20.      android:text="@string/cm_stop" >    </Button>
21.    <Button     android:id="@+id/btnReset"
22.      android:layout_width="wrap_content"  android:layout_height="wrap_content"
23.      android:text="@string/cm_reset" >   </Button>   </LinearLayout>
24.   <LinearLayout
25.    android:layout_width="fill_parent"   android:layout_height="wrap_content"
```

```
26.         android:gravity="center_horizontal"    android:orientation="horizontal"
27.         android:padding="4dip" >
28.     <Button            android:id="@+id/btnSetFormat"
29.         android:layout_width="wrap_content"    android:layout_height="wrap_content"
30.         android:text="@string/cm_set_format" >        </Button>
31.     <Button            android:id="@+id/btnClearFormat"
32.         android:layout_width="wrap_content"    android:layout_height="wrap_content"
33.         android:text="@string/cm_clear_format" >      </Button>
34.     </LinearLayout>     </LinearLayout>
```

源代码清单 3.23 \Ch03_Ex09_Chronometer\src\sziit\lihz\
Ch03_ex09_chronometer\ChronometerActivity.java

```
1.  public class ChronometerActivity extends Activity {//从Activity基类派生子类
2.      private Chronometer mChronometer;        //创建一个计时器对象
3.      private Button btnStart;                 //创建一个按钮对象
4.      private Button btnStop;                  //创建一个按钮对象
5.      private Button btnReset;                 //创建一个按钮对象
6.      private Button btnSetFormat;             //创建一个按钮对象
7.      private Button btnClearFormat;           //创建一个按钮对象
8.      public void onCreate(Bundle savedInstanceState) {//子类重写onCreate方法
9.          super.onCreate(savedInstanceState); //调用基类onCreate方法
10.         setContentView(R.layout.chronometer);//设置手机屏幕布局界面
11.         mChronometer = (Chronometer) findViewById(R.id.chronometer);
                                                  //将计时器对象实体化
12.         btnStart = (Button) findViewById(R.id.btnStart); //初始化按钮Button组件
13.         btnStart.setEnabled(true);           //使能该按钮
14.         btnStart.setOnClickListener(mChronometerListener);
                                                  //设置按钮单击事件监听器
15.         btnStop = (Button) findViewById(R.id.btnStop);//初始化按钮Button组件
16.         btnStop.setEnabled(false);           //禁用该按钮
17.         btnStop.setOnClickListener(mChronometerListener);
                                                  // 设置按钮单击事件监听器
18.         btnReset = (Button) findViewById(R.id.btnReset);// 初始化按钮Button组件
19.         btnReset.setEnabled(true);           //使能该按钮
20.         btnReset.setOnClickListener(mChronometerListener);
                                                  //设置按钮单击事件监听器
21.         btnSetFormat = (Button) findViewById(R.id.btnSetFormat);
                                                  //初始化按钮Button组件
22.         btnSetFormat.setOnClickListener(mChronometerListener);
                                                  //设置按钮单击事件监听器
23.         btnClearFormat = (Button) findViewById(R.id.btnClearFormat);
                                                  //初始化按钮组件
24.         btnClearFormat.setOnClickListener(mChronometerListener);
                                                  //设置单击事件监听器}
25.     //实现按钮单击监听器事件处理
26.     View.OnClickListener mChronometerListener = new OnClickListener() {
27.         public void onClick(View v) {         //单击事件处理
28.             switch (v.getId()) {              //根据id号识别相应的按钮事件
29.             case R.id.btnStart:               //开始按钮
30.                 mChronometer.start();         //计时器开始
31.                 btnStart.setEnabled(false);   //禁用该按钮
```

```
32.            btnStop.setEnabled(true);       //使能该按钮
33.            btnReset.setEnabled(false);     //禁用该按钮
34.            break;
35.        case R.id.btnStop:                  //停止按钮
36.            mChronometer.stop();            //计时器暂停
37.            btnStart.setEnabled(true);      //使能该按钮
38.            btnStop.setEnabled(false);      //禁用该按钮
39.            btnReset.setEnabled(true);      //使能该按钮
40.            break;
41.        case R.id.btnReset:                 //重置按钮
42.            mChronometer.setBase(SystemClock.elapsedRealtime());// 计时器重置为0
43.            btnStart.setEnabled(true);      //使能该按钮
44.            btnStop.setEnabled(false);      //禁用该按钮
45.            btnReset.setEnabled(true);      //使能该按钮
46.            break;
47.        case R.id.btnSetFormat:             //设置格式按钮
48.            mChronometer.setFormat("定时器时间格式 (%s)");//改变计时器显示内容方式
49.            break;
50.        case R.id.btnClearFormat:           //清除格式按钮
51.            mChronometer.setFormat(null);//恢复原来计时器显示方式
52.            break;
53.        default: break;      }};};)
```

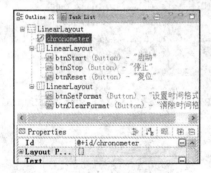

（a）设计轮廓

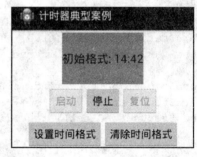

（b）运行界面

图 3.12　计时器典型案例

3.2.10　自动完成文本框

1. 自动完成文本框简介

自动完成文本框（AutoCompleteTextView）继承自 EditText，其实质上也是一个文本框。与普通的文本框不同之处在于，AutoCompleteTextView 多了一个功能：当用户在文本编辑框中输入一定文本之后，AutoCompleteTextView 会显示出一个包含用户输入内容的下拉列表供用户选择，当用户选择其中的某个选项后，AutoCompleteTextView 会将用户选择的选项填写到该文本框，功能类似于在百度或者谷歌的搜索栏输入信息时，弹出的与输入信息接近的提示信息。另外，AutonCompleteTextView 派生了一个子类：MultiAutoCompleteTextView，该子类允许输入多个提示项，每个提示项之间用分隔符分隔。MultiAutoCompleteTextView 提供 setTokenizer()方法用来设置分隔符。

2. 自动完成文本框常用属性及其对应方法

AutoCompleteTextView 继承于 android.widget.EditText，其常用属性及其对应方法如表 3.18 所示。

表 3.18 AutoCompleteTextView 常用属性及其对应方法

XML 属性	Java 方法	功 能 描 述
android:completionThreshold	setThreshold(int)	设置用户至少输入几个字符才会显示提示
android:completionHintView		设置下拉列表中提示标题的视图
android:dropDownHeight	setDropDownHeight(int)	设置下拉列表的高度
android:dropDownWidth	setDropDownWidth(int)	设置下拉列表的宽度
android:completionHint	setCompletionHint(CharSequence)	设置下拉列表中的提示标题
android:dropDownVerticalOffset	setDropDownVerticalOffset(int)	设置下拉列表与文本框之间的垂直偏移
android:dropDownHorizontalOffset	setDropDownHorizontalOffset(int)	设置下拉列表与文本框之间的水平偏移
android:popupBackground	setDropDownBackgroundResource(int)	设置下拉列表的背景

3. 自动完成文本框典型案例

实现要点：

- 使用 AutoCompleteTextView 的主要步骤包括：首先，定义一个字符串数组；然后，将该字符串数组添加到数组适配器（ArrayAdapter）；最后，利用 AutoCompleteTextView 的 setAdapter() 方法，将字符串数组加入到 AutoCompleteTextView 对象中，设置其适配器。
- getAdapter：获取一个可过滤的列表适配器。
- setDropDownHeight() 方法：用来设置提示下拉框的高度。
- setThreshold() 方法：设置从输入第几个字符起出现提示。
- setCompletionHint() 方法：设置提示框最下面显示的文字。
- setOnFocusChangeListener() 方法：设置焦点改变事件，包含监听器 OnFocusChangeListener。
- getDropDownAnchor：获取下拉列表锚记的 View 的 id。
- getDropDownBackground：获取下拉列表的背景色。
- setDropDownBackgroundDrawable：设置下拉列表的背景色。
- setDropDownBackgroundResource：设置下拉列表的背景资源。
- setDropDownAnimationStyle：设置下拉列表的弹出动画。
- setDropDownVerticalOffset：设置下拉列表垂直偏移量。
- getDropDownVerticalOffset：获取下拉列表垂直偏移量。
- setDropDownHorizontalOffset：设置下拉列表水平偏移量。
- setOnItemClickListener：设置下拉列表单击事件。
- getListSelection：获取下拉列表选中为位置。
- getOnItemClickListener：获取选项单击事件。

创建 Ch03_Ex10_AutoCompleteTextView 项目，修改字符串资源文件 string.xml，修改布局文件 auto.xml，实现 AutoActivity.java 功能，如源代码清单 3.24、3.25 和 3.26 所示，其运行效果如图 3.13 所示。

代码清单 3.24　\Ch03_Ex10_AutoCompleteTextView\res\values\strings.xml

```xml
1. <?xml version="1.0" encoding="utf-8"?>
2. <resources>
3.     <string name="app_name">自动完成文本框典型案例</string>
4.     <string name="action_settings">设置</string>
5.     <string name="search">搜索</string>
6.     <string name="completionHint">请选择您喜欢的音乐：</string>
7.     <string-array name="music_array">   <item>美酒加咖啡</item>
8.        <item>雾里看花</item>
9.        <item>梦里水乡</item>   <item>牧羊曲</item>        <item>小城故事</item>
10.       <item>往事随风</item>   <item>花儿为什么这样红</item> <item>敖包相会</item>
11.    </string-array>   </resources>
```

代码清单 3.25　\Ch03_Ex10_AutoCompleteTextView\res\layout\auto.xml

```xml
1. <RelativeLayout xmlns:android="http://schemas.android.com/apk/res/android"
2.     xmlns:tools="http://schemas.android.com/tools"
3.     android:layout_width="match_parent"   android:layout_height="match_parent"....
4.     tools:context=".AutoActivity" >
5.     <AutoCompleteTextView    android:id="@+id/autoCompleteTextView"
6.        android:layout_width="fill_parent"   android:layout_height="wrap_content"
7.        android:completionHint="@string/completionHint"
8.        android:completionThreshold="1"    android:dropDownHorizontalOffset="80px"
9.        android:dropDownVerticalOffset="80px"  android:dropDownWidth="80px"
10.       android:hint="@string/completionHint"  android:inputType="textAutoComplete"
11.       android:popupBackground="#ff0000" />
12.    <MultiAutoCompleteTextView   android:id="@+id/multiAuto"
13.       android:layout_width="fill_parent"   android:layout_height="wrap_content"
14.       android:layout_below="@id/autoCompleteTextView"
15.       android:completionHint="@string/completionHint"
16.       android:completionThreshold="1"    android:dropDownHorizontalOffset="80px"
17.       android:dropDownVerticalOffset="80px"   android:dropDownWidth="80px"
18.       android:hint="@string/completionHint"   android:inputType="textAutoComplete"
19.       android:popupBackground="#ff0000" />
20.    <Button    android:id="@+id/BtnSearch"   android:layout_width="wrap_content"
21.       android:layout_height="wrap_content" android:layout_below="@id/multiAuto"
22.       android:layout_weight="1"            android:text="@string/search"/>
23.    <TextView  android:id="@+id/tvResult" android:layout_width="wrap_content"
24.       android:layout_height="wrap_content" android:layout_below="@id/BtnSearch"/>
25. </RelativeLayout>
```

源代码清单 3.26　\Ch03_Ex10_AutoCompleteTexTview\AutoActivity.java

```java
1. public class AutoActivity extends Activity {           //从 Activity 基类派生子类
2.    private TextView textView;                          //定义私有 TextView 对象
3.    private MultiAutoCompleteTextView multiAuto;        //定义私有对象
4.    protected void onCreate(Bundle savedInstanceState) {//子类重写 onCreate 方法
5.       super.onCreate(savedInstanceState);              //基类调用 onCreate 方法
6.       setContentView(R.layout.auto);                   //设置手机屏幕布局界面
7.       textView = (TextView) findViewById(R.id.tvResult);
8.       final AutoCompleteTextView auto = (AutoCompleteTextView) findViewById
          (R.id.autoCompleteTextView);
9.       //为 auto 提供 Adapter
```

```
10.         String[] musics = this.getResources()
11.            .getStringArray(R.array.music_array);//定义字符串数组,作为提示的文本
12.         //创建一个ArrayAdapter,封装数组
13.         ArrayAdapter<String> ada = new ArrayAdapter<String>(this,
14.             android.R.layout.simple_dropdown_item_1line, musics);
15.         auto.setAdapter(ada);              //设置适配器
16.         auto.setThreshold(1);              //指明当输入多少个字的时候给出响应的提示
17.         auto.setOnItemClickListener(new OnItemClickListener(){
18.         public void onItemClick(AdapterView<?> parent, View view, int position,
                long id) {Toast.makeText(AutoActivity.this,view.toString(),
                Toast.LENGTH_LONG). show();}});
19.         multiAuto = (MultiAutoCompleteTextView)findViewById(R.id.multiAuto);
20.         multiAuto.setAdapter(ada);         //设置适配器
21.         //为MultiAutoCompleteTextView设置分隔符
22.         multiAuto.setTokenizer(new MultiAutoCompleteTextView.CommaTokenizer());
23.         Button button = (Button) findViewById(R.id.BtnSearch);//绑定Button对象
24.         button.setOnClickListener(new View.OnClickListener() {//设置单击监听器
25.         public void onClick(View v) {   //处理单击回调事件
26.             Toast.makeText(AutoActivity.this,auto.getText().toString(), Toast.
                LENGTH_LONG).show();textView.setText(auto.getText().toString());}});}}
```

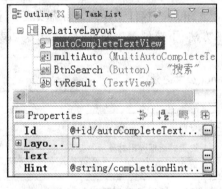
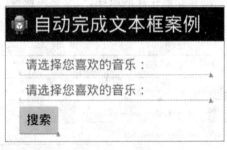

(a) 设计轮廓　　　　　　　　　　　　(b) 运行界面

图 3.13　自动完成文本框典型案例

3.2.11　下拉列表控件

1. 下拉列表控件简介

下拉列表(Spinner)控件的结构为:

```
public final class Spinner extends AbsSpinner
java.lang.Object
    android.view.View
        android.view.ViewGroup
            android.widget.AdapterView<T extends android.widget.Adapter>
                android.widget.AbsSpinner
                    android.widget.Spinner
```

下拉列表控件提供了从一个数据集合中快速选择一项值的方法。默认情况下 Spinner 显示的是当前选择的值,单击 Spinner 会弹出一个包含所有可选值的下拉列表,从该列表中可以为 Spinner 选择一个新值。

2. 下拉列表控件常用属性和方法

下拉列表控件类位于 android.widget 包，其常用属性和方法如表 3.19 所示。

表 3.19　Spinner 常用属性及其对应方法

XML 属性	Java 方法	功 能 描 述
android:prompt	setPrompt(CharSequence prompt) 或 setPromptId(int promptId)	设置弹出选择对话框时的对话框的标题
android:dropDownHorizontalOffset	setDropDownHorizontalOffset(int)	设置下拉列表与文本框之间的水平偏移
android:dropDownVerticalOffset	setDropDownVerticalOffset(int)	设置下拉列表与文本框之间的垂直偏移
android:dropDownSelector		设置列表选择器的显示效果
android:dropDownWidth	setDropDownWidth(int)	设置下拉列表的宽度
android:gravity	setGravity(int)	设置当前选择的项目的对齐方式
android:popupBackground	setPopupBackgroundResource(int)	设置下拉列表的背景
android:spinnerMode		设置下拉列表的显示形式，可以为 dialog（弹出列表）或 dropdown（下拉列表）
android:entries		直接在 XML 布局文件中绑定数据源

通过修改 XML 属性能够在设计阶段快速地设置 Spinner 控件的样式，若需要在程序运行期间动态地修改该控件的属性,则要调用该控件的方法。下拉列表控件类的公共方法如表 3.20 所示。

表 3.20　Spinner 公共方法

公 共 方 法	功 能 描 述
int getBaseline()	返回该控件文本基线的偏移量。
void setPromptId(CharSequence prompt)	设置对话框弹出时显示的提示
void setPromptId(int promptId)	设置对话框弹出时显示的提示
CharSequence getPrompt()	返回对话框弹出时显示的提示
void onClick(DialogInterface dialog, int which)	当单击对话框中的项时该方法将被调用
Boolean performClick()	如果它被定义就调用此视图的 OnClickListener
void setAdapter(SpinnerAdapter adapter)	设置控件所使用的适配器
void setOnItemSelectedListener(OnItemSelectedListener listener)	设置下拉列表被选择监听器
void onItemSelected(AdapterView<?> parent, View view, int position, long id)	选项被选中时触发

在下拉列表控件应用过程中经常使用的重要方法包括：

（1）public void setAdapter(SpinnerAdapter adapter)

功能：设置下拉列表控件所用的适配器。

参数：SpinnerAdapter 类型适配器 adapter。

SpinnerAdapter 派生于 Adapter 适配器。Adapter 适配器是一种连接数据和控件的桥梁，通过 Adapter 可以通知控件显示的数据来源和显示方式。Android 提供了很多 Adapter，表 3.21 列出了常用的几个。

表 3.21 常用适配器

Adapter	含 义
ArrayAdapter<T>	用来绑定一个数组，支持泛型操作
SimpleAdapter	用来绑定在 XML 文件中定义的控件对应的数据
SimpleCursorAdapter	用来绑定游标得到的数据
BaseAdapter	通用的基础适配器

如果连接的数据是数组类型，则需要使用数组适配器 ArrayAdapter<T>。ArrayAdapter<T> 是一种模板类型，<T>中的 T 实际上代表数组元素的类型，如连接字符串数组的适配器应该是 ArrayAdapter<String>，连接整型数组的适配器应该是 ArrayAdapter<Integer>。

（2）public void setOnItemSelectedListener(OnItemSelectedListener listener)

功能：设置下拉列表被选择监听器。该方法是 Spinner 从其父控类 AdapterView 继承而来。

参数：AdapterView.OnItemSelectedListener 是选项监听器接口。该接口需要实现两个方法：一个是 onItemSelected()，选项被选中时触发；另一个是 onNothingSelected()，没有任何选项被选中时触发。

（3）public void onItemSelected(AdapterView<?> parent, View view, int position, long id)

功能：选项被选中时触发。

参数：parent 为单击的 Spinner 控件；view 为单击的那一项视图；position 为单击的那一项的位置；id 为被选中项的行号。

（4）public ArrayAdapter(Context context, int textViewResourceId, T[] objects)

功能：数组适配器 ArrayAdapter 的构造器。

参数：context 为当前活动的环境；Spinner 中的每个选项将会作为 TextView 显示，textViewResourceId 表示所选 TextView 控件的 ID 号；objects 表示该适配器需要连接的数组数据。

（5）void setDropDownViewResource(int resource)

功能：设置 Spinner 下拉列表的显示样式。

参数：resource 为 Spinner 下拉列表的显示样式的 XML 资源文件，可使用 Android 系统自带样式，如 android.R.layout.simple_spinner_dropdown_item 或 android.R.layout.simple_spinner_item。注意：以 "android." 开头的 ID 代表 Android 系统自带资源。

3. 下拉列表控件典型案例

实现要点：

- 通过 OnItemSelectedListener 的回调方法实现 Spinner 选择事件响应。
- 下拉列表有两种显示形式，一种是下拉菜单，另一种是弹出框，由 spinnerMode 属性决定：

android:spinnerMode="dropdown"   android:spinnerMode="dialog"

- android:prompt 属性：必须要引用 strings.xml 中资源 ID，而不能直接用 raw text。
- 使用 Spinner 的主要步骤：

步骤1：添加一个下拉列表项的 list，所添加的项就是下拉列表的选项。

```
List<String> list = new ArrayList<String>();
list.add("中国");list.add("美国");list.add("德国");list.add("英国");
```

步骤 2：为下拉列表定义一个适配器。

```
ArrayAdapter<String> adapter;
adapter= new ArrayAdapter<String>(this,android.R.layout.simple_spinner_item,list);
```

步骤 3：为适配器设置下拉列表下拉时的菜单样式：

```
adapter.setDropDownViewResource(android.R.layout.
                                simple_spinner_dropdown_item);
```

步骤 4：将适配器添加到下拉列表上。

```
Spinner mSpinner; mSpinner = (Spinner)findViewById(R.id.Spinner_guoja);
mSpinner.setAdapter(adapter);
```

步骤 5：为下拉列表设置各种事件的响应，事响应菜单被选中。

```
mSpinner.setOnItemSelectedListener(new Spinner.OnItemSelectedListener(){
   public void onItemSelected(AdapterView<?> arg0, View arg1, int arg2, long 
   arg3) {
   /* 将所选 mSpinner 的值带入 mTextView 中*/
   mTextView.setText("您选择的是: "+ adapter.getItem(arg2));
   arg0.setVisibility(View.VISIBLE);} //mSpinner 显示
   public void onNothingSelected(AdapterView<?> arg0){
     mTextView.setText("NONE");  arg0.setVisibility(View.VISIBLE);}});
/*下拉菜单弹出的内容选项触屏事件处理*/
mSpinner.setOnTouchListener(new Spinner.OnTouchListener(){
  public boolean onTouch(View v, MotionEvent event) {return false;}});
/*下拉菜单弹出的内容选项焦点改变事件处理*/
mSpinner.setOnFocusChangeListener(new Spinner.OnFocusChangeListener(){
   public void onFocusChange(View v, boolean hasFocus) {}});
```

- android:entries = "@array/coutines"：直接在 xml 布局文件中绑定数据源。
- android:prompt = "@string/spin_prompt"：设置对话框弹出的时候显示的提示。
- mSpinner.setAdapter(adapter)：设置适配器。
- mSpinner.setPrompt("请选择城市")：设置对话框弹出的时候显示的提示。

（1）【Spinner 典型案例 1】利用 Spinner 选择旅游城市。

该案例主要实现步骤如下：

步骤 1：创建 Ch03_Ex11_Spinner1 项目，在布局文件 spinner1.xml 中添加 Spinner 控件，如源代码清单 3.27 所示。

代码清单 3.27　\Ch03_Ex11_Spinner1\res\layout\spinner1.xml

```
1. <RelativeLayout xmlns:android="http://schemas.android.com/apk/res/android"
2.   xmlns:tools="http://schemas.android.com/tools"
3.   android:layout_width="match_parent"    android:layout_height="match_parent"
4.   ...tools:context=".MainActivity" >
5.   <TextView  android:id="@+id/tvGoal"   android:layout_width="wrap_content"
6.     android:layout_height="wrap_content"  android:text="@string/strGoal" />
7.   <Spinner    android:id="@+id/spinner_tour_cities"
8.     android:layout_width="wrap_content"   android:layout_height="wrap_content"
9.     android:layout_below="@id/tvGoal"
10.    android:entries="@array/tour_cities" //设置一个数组资源作为该控件显示的内容
11.    android:gravity="center_horizontal"
12.    android:prompt="@string/strPrompt" />//设置对话框弹出时显示的提示
```

```
13.    <TextView android:id="@+id/tvResult" android:layout_width="wrap_content"
14.      android:layout_height="wrap_content"
15.      android:layout_below="@id/spinner_tour_cities"
16.      android:text="@string/strResult"/>
17.    <TextView android:id="@+id/tvSelected" android:layout_width="wrap_content"
18.      android:layout_height="wrap_content" android:layout_alignTop="@id/tvResult"
19.      android:layout_toRightOf="@id/tvResult"/>    </RelativeLayout>
```

步骤2：布局文件中的 android:entries="@array/tour_cities"表示 Spinner 的数据集合是从资源数组 tour_cities 中获取的，tour_cities 数组资源定义在 values/ strings.xml 中，如源代码清单 3.28 所示。

源代码清单 3.28   \Ch03_Ex11_Spinner1\res\values\strings.xml

```
1. <?xml version="1.0" encoding="utf-8"?>
2. <resources>  <string name="app_name">Spinner 典型案例 1</string>
3.   <string name="action_settings">设置</string>
4.   <string name="strGoal">利用 Spinner 选择旅游城市:</string>
5.   <string-array name="tour_cities">
6.       <item>北京市</item>   <item>上海市</item>   <item>广州市</item>
7.       <item>深圳市</item>  <item>哈尔滨市</item>  <item>大连市</item>  <item>武汉市</item>
8.   </string-array>
9.   <string name="strResult">您计划去旅游的城市为:</string>
10.  <string name="strPrompt">请选择您打算去旅游的城市！</string>
11.  <string name="strClickItem">你选择的城市是:</string>
```

步骤3：通过 OnItemSelectedListener 的回调方法，实现 Spinner 选择事件响应如源代码清单 3.29 所示。

源代码清单 3.29   \Ch03_Ex11_Spinner1\src\sziit\lihz\ch03_ex11_spinner1\Spinner1Activity.java

```
1. public class Spinner1Activity extends Activity {//从 Activity 基类派生子类
2.    private String[] tour_cities ;          //定义私有字符串数组
3.    private String strClick ;               //定义私有字符串
4.    private TextView tvSelected;            //定义私有 TextView 对象
5.    private Spinner spinner ;               //定义私有 Spinner 对象
6.    protected void onCreate(Bundle savedInstanceState) {//子类重写 onCreate 方法
7.        super.onCreate(savedInstanceState);  //基类调用 onCreate 方法
8.        setContentView(R.layout.spinner1);   //设置手机屏幕布局
9.        //从资源文件加载字符串数组
10.       tour_cities= getResources().getStringArray(R.array.tour_cities);
11.       strClick= getString(R.string.strClickItem);//从资源文件加载字符串数
12.       tvSelected = (TextView) findViewById(R.id.tvSelected);//引用 TextView 控件
13.       spinner = (Spinner) findViewById(R.id.spinner_tour_cities);
                                              //引用 Spinner 控件
14.       spinner.setOnItemSelectedListener(new OnItemSelectedListener(){
                                              //设置选择监听器
15.         public void onItemSelected(AdapterView<?> parent, View view,
16.             int position, long id) {      //实现 onItemSelected 事件响应
17.           Toast.makeText(Spinner1Activity.this, strClick+ tour_cities
               [position], Toast.LENGTH_SHORT) .show();
18.           tvSelected.setText(tour_cities[position]);}
19.         public void onNothingSelected(AdapterView<?> parent) {}});}}
```

步骤 4：选择【Run As】→【Android Application】命令，运行结果如图 3.14 所示。

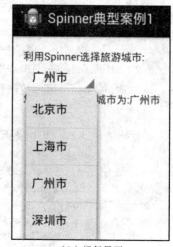

（a）设计轮廓　　　　　　　　　　　（b）运行界面

图 3.14　Spinner 典型案例 1 运行结果

（2）【Spinner 典型案例 2】使用 ArrayAdapter 设置 Spinner 的 Adapter 的案例。

该案例主要实现步骤如下：

步骤 1：创建 Ch03_Ex11_Spinner2 项目，在布局文件 spinner2.xml 中声明两个 TextView 控件和两个 Spinner 控件，如源代码清单 3.30 所示。

源代码清单 3.30　\Ch03_Ex11_Spinner2\res\layout\spinner2.xml

```
1.  <RelativeLayout xmlns:android="http://schemas.android.com/apk/res/android"
2.   ...tools:context=".Spinner2Activity">
3.   <TextView  android:id="@+id/spinnerText1" android:layout_width="fill_parent"
4.     android:layout_height="wrap_content"> </TextView>
5.   <Spinner   android:id="@+id/Spinner01"   android:layout_width="fill_parent"
6.     android:layout_height="wrap_content" android:layout_below="@id/spinnerText1">
7.   </Spinner>
8.   <TextView android:id="@+id/spinnerText2" android:layout_width="fill_parent"
9.     android:layout_height="wrap_content"   android:layout_below="@id/Spinner01">
10.    </TextView>
11.   <Spinner android:id="@+id/Spinner02"  android:layout_width="fill_parent"
12.     android:layout_height="wrap_content" android:layout_below="@id/spinnerText2">
13.   </Spinner>    </RelativeLayout>
```

步骤 2：在 strings.xml 中定义 Spinner 的数据源所使用的字符串数组资源，如源代码清单 3.31 所示。

源代码清单 3.31　\Ch03_Ex11_Spinner2\res\values\strings.xml

```
1. <?xml version="1.0" encoding="utf-8"?>
2. <resources>   <string name="app_name">Spinner 典型案例 2</string>
3.    <string name="action_settings">设置</string>
4.    <string name="best_city">您最喜欢的城市：</string>
5.    <string name="best_sport">您最喜欢的运动：</string>
```

```
6.    <string-array name="cities">   <item>北京</item> <item>上海</item>
7.        <item>深圳</item>   <item>武汉</item>   <item>重庆</item>   <item>福州</item>
8.        <item>厦门</item>   </string-array>   </resources>
```

步骤 3：使用 ArrayAdapter 来设置 Spinner 的数据源，如源代码清单 3.32 所示。

源代码清单 3.32   \Ch03_Ex11_Spinner2\src\sziit\lihz\ch03_ex11_spinner2\Spinner2Activity.java

```java
1.  public class Spinner2Activity extends Activity {//从Activity基类派生子类
2.      private TextView mTextView1;              private Spinner mSpinner1;
3.      private ArrayAdapter<String> mAdapter1; private TextView mTextView2;
4.      private Spinner mSpinner2;    private ArrayAdapter<String> mAdapter2;
5.      private String[] mCities;private String mBestCity; private String mBesSport;
6.      private static final String[] sports = { "游泳", "网球", "篮球", "乒乓球" }
7.      protected void onCreate(Bundle savedInstanceState) {
8.          super.onCreate(savedInstanceState);       setContentView(R.layout.spinner2);
9.          mTextView1 = (TextView) findViewById(R.id.spinnerText1);
10.         mSpinner1 = (Spinner) findViewById(R.id.Spinner01); //初始化Spinner控件
11.         mCities = getResources().getStringArray(R.array.cities);//建立数据源
12.         mBestCity = this.getString(R.string.best_city);
13.         mBesSport = this.getString(R.string.best_sport);
14.         //建立Adapter并且绑定数据源,将可选内容与ArrayAdapter连接起来
15.         mAdapter1 = new ArrayAdapter<String>(this, android.R.layout.simple_spinner_
            item, mCities);    //设置下拉列表的风格
16.         mAdapter1.setDropDownViewResource(android.R.layout.simple_spinner_item);
17.         //绑定Adapter到控件,将adapter添加到spinner中
18.         mSpinner1.setAdapter(mAdapter1);
19.         //添加事件Spinner事件监听
20.         mSpinner1.setOnItemSelectedListener(new SpinnerSelectedListener());
21.         mSpinner1.setVisibility(View.VISIBLE); //设置默认值
22.         mTextView2 = (TextView) findViewById(R.id.spinnerText2);
23.         mSpinner2 = (Spinner) findViewById(R.id.Spinner02); //初始化Spinner控件
24.         //建立Adapter并且绑定数据源,将可选内容与ArrayAdapter连接起来
25.         mAdapter2 = new ArrayAdapter<String>(this,
26.             android.R.layout.simple_spinner_dropdown_item, sports);
27.         //设置下拉列表的风格
28.         mAdapter2.setDropDownViewResource(android.R.layout.simple_spinner_
            dropdown_item);
29.         //绑定Adapter到控件,将adapter添加到spinner中
30.         mSpinner2.setAdapter(mAdapter2);
31.         //添加Spinner事件监听,设置下拉列表被选择监听器
32.         mSpinner2.setOnItemSelectedListener(new OnItemSelectedListener() {
33.             public void onItemSelected(AdapterView<?> parent, View view,
34.                 int position, long id) {     //选项被选中时触发
35.                 mTextView2.setText(mBesSport + sports[position]);
36.                 Toast.makeText(Spinner2Activity.this,
37.                     mBesSport + sports[position], Toast.LENGTH_LONG).show();}
38.             public void onNothingSelected(AdapterView<?> parent){
39.                 //没有任何选项被选中时触发 }});
40.         mSpinner2.setVisibility(View.VISIBLE);} //设置默认值
```

```
41.    //使用数组形式操作
42.    class SpinnerSelectedListener implements OnItemSelectedListener {//设置监听器
43.        public void onItemSelected(AdapterView<?> parent, View view,
44.            int position, long id) {//选项被选中时触发
45.        mTextView1.setText(mBestCity + mCities[position]);
46.        Toast.makeText(Spinner2Activity.this, mBestCity + mCities[position],
47.            Toast.LENGTH_LONG).show();    }
48.    public void onNothingSelected(AdapterView<?> parent){}}}  //没有任何选项被选中时触发
```

步骤4：选择【Run As】→【Android Application】命令，运行结果如图3.15所示。

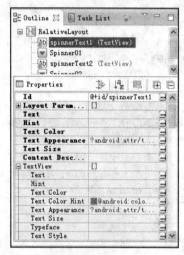
（a）设计轮廓

（b）运行界面

图3.15　Spinner典型案例2运行结果

3.3 案例实施

3.3.1 利用DatePickerDialog和TimePicker设置日期和时间

1. 运行结果

本例运行结果如图3.16所示。单击"设置日期"按钮弹出DatePickerDialog对话框，选择日期后，修改后的日期值显示在绿色背景文本框内。单击"设置时间"按钮弹出TimePicker对话框，选择时间后，修改后的时间值显示在蓝色背景文本框内。

2. 实训目的

（1）掌握DatePickerDialog控件应用技巧。
（2）掌握DatePickerDialog.OnDateSetListener监听器使用方法。
（3）掌握TimePicker控件应用技巧。
（4）掌握TimePickerDialog.OnTimeSetListener监听器使用方法。
（5）掌握Button控件应用技巧。
（6）掌握TextView应用技巧。
（7）了解Calendar使用方法。

图 3.16 设置日期 DatePickerDialog 和时间 TimePicker 控件范例

3. 实训过程

步骤 1：利用 Android Project 项目向导生成名为 Ch03_Ex12_DatePickerDialog 的应用程序。

步骤 2：打开字符串资源文件 strings.xml，利用 Android 可视化资源编辑器，修改并定义相关字符串，如源代码清单 3.33 所示。

源代码清单 3.33　\Ch03_Ex12_DatePickerDialog\res\values\strings.xml

```
1. <?xml version="1.0" encoding="utf-8"?>
2. <resources>    <string name="app_name">设置日期和时间案例</string>
3.    <string name="dateCaption">设置日期</string>
4.    <string name="timeCaption">设置时间</string>   </resources>
```

步骤 3：修改布局文件 datepicker_layout.xml，如源代码清单 3.34 所示。

源代码清单 3.34　\Ch03_Ex12_DatePickerDialog\res\layout\datepicker_layout.xml

```
1.  <?xml version="1.0" encoding="utf-8"?>
2.  <LinearLayout xmlns:android="http://schemas.android.com/apk/res/android"
3.    android:layout_width="wrap_content"    android:layout_height="wrap_content"
4.    android:orientation="vertical" >
5.    <LinearLayout   android:layout_width="wrap_content"
6.       android:layout_height="wrap_content"    android:orientation="horizontal" >
7.       <Button android:id="@+id/pickDate"    android:layout_width="wrap_content"
8.          android:layout_height="wrap_content" android:text="@string/dateCaption" />
9.       <TextView  android:id="@+id/dateDisplay"
10.         android:layout_width= "wrap_content"
11.         android:layout_height="wrap_content"    android:text=""
12.         android:textSize="20sp" />
13.   </LinearLayout>
14.   <LinearLayout
15.      android:layout_width="wrap_content"
16.      android:layout_height="wrap_content"
17.      android:orientation="horizontal" >
18.      <Button android:id="@+id/pickTime"    android:layout_width="wrap_content"
19.         android:layout_height="wrap_content" android:text="@string/timeCaption" />
20.      <TextView android:id="@+id/timeDisplay"  android:layout_width="wrap_content"
21.         android:layout_height="wrap_content"   android:text=""
22.         android:textSize="20sp" />
23.   </LinearLayout>    </LinearLayout>
```

- 创建的第一个 LinearLayout 包含一个用于打开 DatePickerDialog 对话框的按钮和另一个用于显示当前日期的文本控件。
- 创建的第二个 LinearLayout 包含一个用于打开 TimePickerDialog 对话框的按钮和另一个用于显示当前时间的文本控件。

步骤 4：打开 Ch03_Ex12_Activity.java 源代码文件，向类中增加用于保存日期和时间的数据成员，如源代码清单 3.35 所示。

源代码清单 3.35 \Ch03_Ex12_DatePickerDialog\Ch03_Ex12_Activity.java

```java
1. public class Ch03_Ex12_Activity extends Activity {
2.     private TextView mDateDisplay;      //显示日期文本控件
3.     private Button mPickDate;           //打开 DatePickerDialog 对话框的按钮
4.     private int mYear;                  //年
5.     private int mMonth;                 //月
6.     private int mDay;                   //日
7.     static final int DATE_DIALOG_ID = 0;//DatePickerDialog 对话框的唯一标识
8.     private TextView mTimeDisplay;      //显示时间文本控件
9.     private Button mPickTime;           //打开 TimePickerDialog 对话框的按钮
10.    private int mHour;                  //时
11.    private int mMinute;                //分
12.    private int mSecond;                //秒
13.    static final int TIME_DIALOG_ID = 1; //TimePickerDialog 对话框唯一标识
14.    //当用户设置日期对话框时会调用此回调方法，从对话框返回设置的年、月和日
15.    private DatePickerDialog.OnDateSetListener mDateSetListener =
16.      new DatePickerDialog.OnDateSetListener() {
17.        public void onDateSet(DatePicker view, int year,
18.          int monthOfYear, int dayOfMonth) {
19.          mYear = year; mMonth = monthOfYear;
20.          mDay = dayOfMonth; updateDateDisplay(); } };
21.    //当用户设置时间对话框时会调用此回调方法，从对话框返回设置的时和分参数
22.    private TimePickerDialog.OnTimeSetListener mTimeSetListener =
23.      new TimePickerDialog.OnTimeSetListener() {
24.        public void onTimeSet(TimePicker view, int hourOfDay, int minute) {
25.          mHour = hourOfDay;  mMinute = minute; updateTimeDisplay(); } };
26.    /** 当 Activity 第一次创建时被调用. */
27.    public void onCreate(Bundle savedInstanceState) {
28.      super.onCreate(savedInstanceState);
29.      setContentView(R.layout.datepicker_layout);
30.      //从资源中获取日期文本和按钮控件：
31.      mDateDisplay = (TextView) findViewById(R.id.dateDisplay);
                                        //从资源查找文本控件
32.      mPickDate = (Button) findViewById(R.id.pickDate);
                                        //从资源查找设置日期的按钮控件
33.      //增加设置日期按钮的单击侦听者：
34.      mPickDate.setOnClickListener(new View.OnClickListener() {
35.        public void onClick(View v) { showDialog(DATE_DIALOG_ID);}});
36.      //获得当前日期参数值
37.      final Calendar c = Calendar.getInstance(); //获得日历类实例
38.      mYear = c.get(Calendar.YEAR);              //获得当前年
39.      mMonth = c.get(Calendar.MONTH);            //获得当前月
```

```
40.     mDay = c.get(Calendar.DAY_OF_MONTH);        //获得当前日
41.     updateDateDisplay();                          //在日期文本控件显示当前日期
42.   //从资源文件查找设置时间的文本和按钮控件
43.     mTimeDisplay = (TextView) findViewById(R.id.timeDisplay);
                                                    //从资源中查找文本控件
44.     mPickTime = (Button) findViewById(R.id.pickTime);
                                                    //从资源中查找时间按钮控件
45.   //增加设置时间按钮的单击侦听者
46.     mPickTime.setOnClickListener(new View.OnClickListener() {
47.         public void onClick(View v) {showDialog(TIME_DIALOG_ID); } });
48.   //获得当前时间
49.     mHour = c.get(Calendar.HOUR_OF_DAY); mMinute = c.get(Calendar.MINUTE);
50.     updateTimeDisplay();}                        //显示当前时间
51.   private void updateDateDisplay() {             //更新日期TextView中的日期
52.     mDateDisplay.setText( new StringBuilder() //Month 基于0开始,因此要加1
53.         .append(" ").append(mYear).append("年") .append(mMonth + 1).append("月")
54.         .append(mDay).append("日")); }
55.   protected Dialog onCreateDialog(int id) {
56.     switch (id) {
57.       case DATE_DIALOG_ID:
58.         return new DatePickerDialog(this,mDateSetListener,mYear, mMonth, mDay);
59.       case TIME_DIALOG_ID:
60.         return new TimePickerDialog(this,mTimeSetListener, mHour, mMinute, false); }
61.     return null; }
62.   private void updateTimeDisplay() {             //更新时间显示方法
63.     mTimeDisplay.setText( new StringBuilder()
64.         .append(" ").append(pad(mHour)).append("时")
65.         .append(pad(mMinute)).append("分")); }
66.   private Object pad(int c) {
67.     if (c >= 10) return String.valueOf(c);
68.     else   return "0" + String.valueOf(c); } }
```

3.3.2 创建基于多种控件的表单应用案例

1．运行结果

本例运行结果如图3.17所示。本例利用文本框、编辑框、单选按钮、复选框、切换按钮、评分条和图像按钮等控件，创建了一个职工个人信息调查表单。

2．实训目的

（1）掌握图像按钮控件的应用技巧。
（2）掌握开关按钮控件的应用技巧。
（3）掌握编辑框控件的应用技巧。
（4）掌握复选框控件的应用技巧。
（5）掌握单选按钮控件的应用技巧。
（6）掌握评分条控件的应用技巧。

图3.17　基于多种控件的表单应用范例

3. 实训过程

步骤 1：利用 Android Project 项目向导生成名为 Ch03_Ex13_Submit_Form 的应用程序。

步骤 2：修改屏幕布局文件 form_layout.xml，如源代码清单 3.36 所示。

源代码清单 3.36　\Ch03_Ex13_Submit_Form\res\layout\form_layout.xml

```xml
1.  <?xml version="1.0" encoding="utf-8"?>
2.  <TableLayout xmlns:android="http://schemas.android.com/apk/res/android"
3.    android:layout_width="fill_parent"  android:layout_height="fill_parent"
4.    android:stretchColumns="1">
5.    <TextView android:gravity="center_horizontal" android:padding="1dip"
6.      android:text="职工个人信息" android:textSize="32sp" />
7.    <TableRow>
8.      <TextView android:id="@+id/tvName" android:layout_column="1"
9.        android:padding="1dip"android:text="姓名:" android:textSize="24sp" />
10.     <EditText android:id="@+id/edittextName" android:layout_height=
        "wrap_content"
11.       android:layout_gravity="left" android:text="" android:width="200sp" />
12.   </TableRow>
13.   <TableRow>
14.     <TextView android:id="@+id/tvSex" android:layout_column="1"
15.       android:padding="1dip" android:text="性别" android:textSize="24sp" />
16.     <RadioGroup android:layout_width="fill_parent" android:layout_height=
        "wrap_content"
17.       android:orientation="horizontal" >
18.     <RadioButton android:id="@+id/radio_male" android:layout_width=
        "wrap_content"
19.       android:layout_height="wrap_content" android:text="男" />
20.     <RadioButton android:id="@+id/radio_female" android:layout_width=
        "wrap_content"
21.       android:layout_height="wrap_content" android:text="女" />
22.     </RadioGroup>  </TableRow>
23.   <TableRow>
24.     <TextView android:id="@+id/tvHobby" android:layout_column="1"
        android:padding="1dip"
25.       android:text="特长:" android:textSize="24sp" />
26.     <LinearLayout android:layout_width="fill_parent" android:layout_height=
        "wrap_content"
27.       android:layout_weight="1" android:orientation="horizontal" >
28.     <CheckBox android:id="@+id/cbLiteratureArt" android:layout_width=
        "wrap_content"
29.       android:layout_height="wrap_content" android:text="文艺" />
30.     <CheckBox android:id="@+id/cbSports" android:layout_width=
        "wrap_content"
31.       android:layout_height="wrap_content" android:text="体育" />
32.     </LinearLayout>  </TableRow>
33.   <TableRow>
34.     <TextView android:id="@+id/tvMaritalStatus" android:layout_column="1"
35.       android:padding="1dip" android:text="婚否" android:textSize="24sp" />
36.     <LinearLayout android:layout_width="fill_parent" android:layout_height=
        "wrap_content"
```

```
37.         android:layout_weight="1" android:orientation="horizontal" >
38.         <ToggleButton android:id="@+id/togglebutton" android:layout_width=
    "wrap_content"
39.             android:layout_height="wrap_content" android:textOff="未婚"
40.             android:textOn="已婚" />
41.     </LinearLayout>    </TableRow>
42.     <TableRow>
43.         <TextView android:id="@+id/tvHappinessRating" android:layout_column="1"
44.             android:padding="1dip" android:text="幸福等级: " android:textSize=
    "24sp" />
45.     </TableRow>
46.     <LinearLayout>
47.         <RatingBar android:id="@+id/ratingbar" android:layout_width=
    "wrap_content"
48.             android:layout_height="wrap_content" android:numStars="5" android:
    stepSize="0.1" />
49.     </LinearLayout>
50.     <LinearLayout android:gravity="center_horizontal" >
51.       <Button android:id="@+id/submitButton" android:layout_width="wrap_content"
52.             android:layout_height="wrap_content" android:background="@drawable/
    submit_button"
53.             android:text="提交" />
54.     </LinearLayout>    </TableLayout>
```

步骤 3：打开 Ch03_Ex13_Activity.java 源代码文件，编辑功能代码，如源代码清单 3.37 所示。

源代码清单 3.37　\Ch03_Ex13_Submit_Form\src\sziit\lihz\
Ch03_Ex13_SubmitForm\Ch03_Ex13_Activity.java

```
1. package sziit.lihz.Ch03_Ex13_SubmitForm;            //定义用户包
2. import android.app.Activity;                        //引入 Activity 包
3. import android.os.Bundle;                           //引入 Bundle 包
4. import android.view.KeyEvent;                       //引入 KeyEvent 包
5. import android.view.View;                           //引入 View
6. import android.view.View.OnClickListener;           //引入 OnClickListener 包
7. import android.view.View.OnKeyListener;             //引入 OnKeyListener 包
8. import android.widget.Button;                       //引入 Button 包
9. import android.widget.CheckBox;                     //引入 CheckBox 包
10.import android.widget.EditText;                     //引入 EditText 包
11.import android.widget.RadioButton;                  //引入 RadioButton 包
12.import android.widget.RatingBar;                    //引入 RatingBar 包
13.import android.widget.RatingBar.OnRatingBarChangeListener;
                        //引入 OnRatingBarChangeListener
14.import android.widget.Toast;                        //引入 Toast 包
15.import android.widget.ToggleButton;                 //引入 ToggleButton 包
16.public class Ch03_Ex13_Activity extends Activity {
                     //定义父类 Activity 的公有子类 Ch03_02_Activity
17.    String strName="Author";                        //定义保存姓名的字符串数据成员
18.    String strSex;                                  //定义保持性别的字符串数据成员
19.    CheckBox cbLiteratureArt;                       //文艺特长复选框
20.    CheckBox cbSports;                              //体育特长复选框
```

```
21.    boolean bLiteratureArt=false;              //是否爱好文艺?
22.    boolean bSports=false;                     //是否爱好体育?
23.    /** 当Activity第一次创建时调用. */
24.    public void onCreate(Bundle savedInstanceState) {
25.        super.onCreate(savedInstanceState);
26.        setContentView(R.layout.form_layout);
27.        //处理姓名的输入编辑框
28.        final EditText etName=(EditText) findViewById(R.id.edittextName);
                                                   //从资源获取控件
29.        etName.setOnKeyListener(new OnKeyListener() { //设置键盘事件侦听器
30.            public boolean onKey(View v, int keyCode, KeyEvent event) {
                                                   //按键处理方法
31.                //如果事件为编辑框中按"enter"的key-down事件, 则
32.                if ((event.getAction() == KeyEvent.ACTION_DOWN) &&
33.                   (keyCode == KeyEvent.KEYCODE_ENTER)) {//处理键盘按下后响应行为
34.                    strName=etName.getText().toString(); //获取输入的姓名
35.                    Toast.makeText(Ch03_Ex13_Activity.this, etName.getText(),
                       Toast.LENGTH_SHORT).show();
36.                    return true; }
37.                return false; } });
38.        //处理性别的单选按钮单击事件
39.        final RadioButton rb_male = (RadioButton) findViewById(R.id.radio_male);
40.        final RadioButton rb_female = (RadioButton) findViewById(R.id.radio_female);
41.        rb_male.setOnClickListener(radio_listener);  rb_female.setOnClick
           Listener(radio_listener);
42.        //处理特长的复选框单击事件
43.        cbLiteratureArt = (CheckBox) findViewById(R.id.cbLiteratureArt);
44.        cbSports = (CheckBox) findViewById(R.id.cbSports);
45.        cbLiteratureArt.setOnClickListener(checkBox_listener);
46.        cbSports.setOnClickListener(checkBox_listener);
47.        //处理婚姻状态的切换按钮控件的件
48.        final ToggleButton togglebutton = (ToggleButton) findViewById(R.id.
           togglebutton);
49.        togglebutton.setOnClickListener(new OnClickListener() {
50.            public void onClick(View view) {         //执行单击行为
51.                if (togglebutton.isChecked()) { Toast.makeText(Ch03_Ex13_Activity.this,
52.                   "已选择", Toast.LENGTH_SHORT).show(); }
53.                else { Toast.makeText(Ch03_Ex13_Activity.this,
54.                    "未选择", Toast.LENGTH_SHORT).show(); }
55.                String curMaritalStatus = togglebutton.getText().toString();
56.                Toast.makeText(Ch03_Ex13_Activity.this,
57.                    "婚姻状态:" + curMaritalStatus, Toast.LENGTH_SHORT).show(); } });
58.        //处理幸福等级的评分控件状态改变事件:
59.        final RatingBar ratingbar = (RatingBar) findViewById(R.id.ratingbar);
60.        ratingbar.setOnRatingBarChangeListener(new OnRatingBarChangeListener() {
61.            public void onRatingChanged(RatingBar ratingBar, float rating,
               boolean fromUser) {
62.                Toast.makeText(Ch03_Ex13_Activity.this,
```

```
63.              "幸福等级: " + rating, Toast.LENGTH_SHORT).show();} });
64.      //处理带背景图像和文字的提交按钮的单击事件
65.      final Button btnSubmit = (Button) findViewById(R.id.submitButton);
66.      btnSubmit.setOnClickListener(new OnClickListener() {
67.          public void onClick(View view) { //执行按钮单击行为
68.              String strSubmit = ((Button)(view)).getText().toString();
69.              Toast.makeText(Ch03_Ex13_Activity.this,strSubmit, Toast.LENGTH_
                    SHORT).show();} }); }
70.      //单选按钮事件侦听器
71.      private OnClickListener radio_listener = new OnClickListener() {
72.          public void onClick(View view) { //执行单击行为
73.              RadioButton rb = (RadioButton) view;//当前选中的单选按钮
74.              strSex=rb.getText().toString();
75.              Toast.makeText(Ch03_Ex13_Activity.this, rb.getText(),
76.                  Toast.LENGTH_SHORT).show(); } };
77.      //复选框事件侦听器
78.      private OnClickListener checkBox_listener = new OnClickListener() {
79.          public void onClick(View view) { // 根据复选框是否选择,执行不同的行为
80.              CheckBox cb = (CheckBox) view;    String curHobby = cb.getText().toString();
81.              boolean bChecked = cb.isChecked();
82.              switch( cb.getId() )
83.              {
84.              case R.id.cbLiteratureArt:
85.                  if ( bChecked ) { bLiteratureArt = true; Toast.makeText(Ch03_Ex13_
86.                      Activity.this,"已选择"+ curHobby, Toast.LENGTH_SHORT).show();}
87.                  else { bLiteratureArt = false; Toast.makeText(Ch03_Ex13_Activity.this,
88.                      "未选择"+ curHobby, Toast.LENGTH_SHORT).show(); } break;
89.              case R.id.cbSports:
90.                  if ( bChecked ) { bSports = true; Toast.makeText(Ch03_Ex13_
91.                      Activity.this,"已选择"+ curHobby, Toast.LENGTH_SHORT).show();}
92.                  else { bSports = false; Toast.makeText(Ch03_Ex13_Activity.this,
93.                      "未选择"+ curHobby, Toast.LENGTH_SHORT).show(); } break;
94.              default:  Toast.makeText(Ch03_Ex13_Activity.this,
95.                  "未知异常特长!!!", Toast.LENGTH_SHORT).show();   break;}
96.              String strHobby = ( bLiteratureArt) ? (cbLiteratureArt.getText().
97.                  toString()):"" )+ ( (bSports) ? (" " + cbSports.getText().
                    toString() ):"" );
98.              if( (strHobby.length()>0) && ( strName.length()> 0 )){
99.                  Toast.makeText(Ch03_Ex13_Activity.this,
100.                     strName+ "的特长是:" + strHobby, Toast.LENGTH_SHORT).show(); } };}
```

3.3.3 利用下拉列表(Spinner)控件实现歌曲选择功能

1. 运行结果

本例运行的结果如图3.18所示。本例子利用下拉列表控件、文本框控件和数组适配器类实现从给定的中文歌曲下拉列表中选择用户最喜欢的一首歌曲名和歌星。

 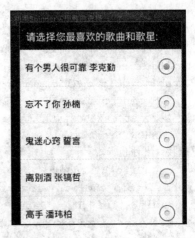

图 3.18　利用下拉列表控件实现歌曲选择功能案例

2. 实训目的

（1）掌握下拉列表控件应用技巧。
（2）掌握数组适配器应用技巧。
（3）掌握 setOnItemSelectedListener 应用技巧。
（4）掌握 void setAdapter(SpinnerAdapter adapter)使用方法。
（5）掌握 string-array 字符串数组资源定义方法。
（6）掌握引用字符串数组资源 getResources().getString(R.string.answer_prompt)方法。
（7）掌握 android:prompt="@string/choice_song"属性使用方法。

3. 实训过程

步骤 1：创建名为 Ch03_Ex14_Spinner 的 Android 应用程序。
步骤 2：编辑字符串资源文件 string.xml，如源代码清单 3.38 所示。

源代码清单 3.38　\Ch03_Ex14_Spinner\res\values\strings.xml

```
1.  <?xml version="1.0" encoding="utf-8"?>
2.  <resources>  <string name="app_name">Ch03_03</string>
3.    <string name="choice_song">请选择您最喜欢的歌曲和歌星:</string>
4.    <string name="answer_prompt">您最喜欢的歌曲和歌星是:</string>
5.    <string-array name="songs_array">
6.      <item>比谁都勇敢　张睿恩</item>      <item>你不对我好我也不对你好　天华</item>
7.      <item>有个男人很可靠　李克勤</item>    <item>忘不了你　孙楠</item>
8.      <item>鬼迷心窍　誓言</item>          <item>离别酒　张镐哲</item>
9.      <item>高手　潘玮柏</item>            <item>神魂颠倒　刘籽彤</item>
10.     <item>网恋新歌　李建科</item>        <item>美人吟　李玲玉</item>
11.   </string-array>  </resources>
```

步骤 3：编辑屏幕布局文件 spinner_layout.xml，如源代码清单 3.39 所示。

源代码清单 3.39　\Ch03_Ex14_Spinner\res\layout\spinner_layout.xml

```
12. <?xml version="1.0" encoding="utf-8"?>
13. <LinearLayout xmlns:android="http://schemas.android.com/apk/res/android"
14.   android:layout_width="fill_parent"  android:layout_height="wrap_content"
15.   android:orientation="vertical"       android:padding="10dip" >
```

```
16. <TextView android:layout_width="fill_parent" android:layout_height="wrap_content"
17.     android:layout_marginTop="10dip" android:text="@string/choice_song"
18.     android:textSize="18sp" />
19. <Spinner android:id="@+id/songSpinner" android:layout_width="wrap_content"
20.     android:layout_height="wrap_content" android:prompt="@string/choice_song" />
21. </LinearLayout>
```

步骤4：编辑 Ch03_Ex13_Activity.java 代码，如源代码清单3.40所示。

源代码清单3.40 \Ch03_Ex13_SubmitForm\Ch03_Ex13_Activity.java

```
1.  public class Ch03_03_Activity extends Activity {
2.    /** 在第一次创建时被调用 */
3.    public void onCreate(Bundle savedInstanceState) {
4.      super.onCreate(savedInstanceState);
5.      setContentView(R.layout.spinner_layout);//设置屏幕布局
6.      Spinner spinner = (Spinner) findViewById(R.id.songSpinner);
7.      ArrayAdapter<CharSequence> adapter = ArrayAdapter.createFromResource(
           this, R.array.songs_array, android.R.layout.simple_spinner_item);
8.      adapter.setDropDownViewResource(android.R.layout.simple_spinner_dropdown_item);
9.      spinner.setAdapter(adapter);  spinner.setSelection(2);
10.     spinner.setOnItemSelectedListener(new Spinner_OnItemSelectedListener()); }
11.   public class Spinner_OnItemSelectedListener implements OnItemSelectedListener {
11.     public void onItemSelected(AdapterView<?> parent, View view, int pos, long id) {
13.       //从资源中获得已定义的字符串
14.       String prompt = getResources().getString(R.string.answer_prompt);
15.       Toast.makeText(parent.getContext(), prompt +
16.         parent.getItemAtPosition(pos).toString(), Toast.LENGTH_LONG).show();
17.     public void onNothingSelected(AdapterView<?> parent) {
18.     }}}//根据需求添加代码，现在什么也不做
```

3.4　知识拓展

3.4.1　滚动视图控件

1. 滚动视图控件简介

滚动视图控件（ScrollView）的继承关系为：

```
public class ScrollView extends FrameLayout
java.lang.Object
    android.view.View
        android.view.ViewGroup
            android.widget.FrameLayout
                android.widget.ScrollView
```

滚动视图控件是一种可供用户滚动的层次结构布局容器，与相同大小的视图相比，能显示更多的内容。ScrollView 是一种 FrameLayout，需要在其中放置有自己滚动内容的子元素。子元素可以是一个复杂的对象布局管理器。通常用的子元素是垂直方向的 LinearLayout，显示在最上层的垂直方向的箭头可以让用户滚动。TextView 类也有自己的滚动功能，所以不需要使用 ScrollView，但是只有两者结合使用，才能保证高效率地显示较多内容。ScrollView 只支持垂直方向的滚动。

2. 滚动视图控件常用属性和方法

滚动视图控件的常用属性和方法如表 3.22、表 3.23 所示。

表 3.22 滚动视图控件常用属性

属 性 名	功 能 描 述
android:background	设置背景色/背景图片,有两种方法设置背景为透明:"@android:color/transparent" 和"@null"
android:clickable	是否响应点击事件
android:contentDescription	设置 View 的备注说明
android:drawingCacheQuality	设置绘图时半透明质量
android:duplicateParentState	若设置此属性,将直接从父容器中获取绘图状态
android:fadingEdge	设置拖动滚动条时,边框渐变的放向:none,边框颜色不变;horizontal,水平方向颜色变淡;vertical,垂直方向颜色变淡
android:fadingEdgeLength	设置边框渐变的长度
android:fitsSystemWindows	设置布局调整时是否考虑系统窗口
android:focusable	设置是否获得焦点。若有 requestFocus()被调用时,后者优先处理
android:focusableInTouchMode	设置在 Touch 模式下 View 是否能获得焦点
android:hapticFeedbackEnabled	设置长按时是否接受其他触摸反馈事件
android:id	给 View 设置一个在当前布局中的唯一编号,可通过调用 View.findViewById()或 Activity.findViewById()根据该编号查找到对应的 View,格式为"@+id/svName"
android:longClickable	设置是否响应长按事件
android:minHeight	设置视图最小高度
android:minWidth	设置视图最小宽度度
android:nextFocusDown	设置下方指定视图获得下一个焦点
android:nextFocusLeft	设置左边指定视图获得下一个焦点
android:nextFocusRight	设置右边指定视图获得下一个焦点
android:nextFocusUp	设置上方指定视图获得下一个焦点
android:onClick	点击时从上下文中调用指定的方法;定义符合如下参数和返回值的函数并将方法名字符串指定为该值即可:public void onClickButton(View view);android:onClick="onClickButton"
android:padding	设置上下左右的边距,以像素为单位填充空白
android:paddingBottom	设置底部的边距,以像素为单位填充空白
android:paddingLeft	设置左边的边距,以像素为单位填充空白
android:paddingRight	设置右边的边距,以像素为单位填充空白
android:paddingTop	设置上方的边距,以像素为单位填充空白
android:saveEnabled	设置是否在窗口冻结时保存 View 的数据,默认为 true
android:scrollX	以像素为单位设置水平方向滚动的偏移值
android:scrollY	以像素为单位设置垂直方向滚动的偏移值
ndroid:scrollbarDefaultDelayBeforeFade	设置 N 毫秒后开始淡化,以毫秒为单位
android:scrollbarSize	设置滚动条的宽度
android:scrollbarStyle	设置滚动条的风格和位置
android:scrollbarThumbHorizontal	设置水平滚动条的 drawable(如颜色)

续表

属 性 名	功 能 描 述
android:scrollbarThumbVertical	设置垂直滚动条的 drawable（如颜色）
android:scrollbarTrackHorizontal	设置水平滚动条背景（轨迹）的 drawable（如颜色）
android:scrollbarTrackVertical	设置垂直滚动条背景（轨迹）的 drawable，注意直接设置颜色值
android:scrollbars	设置滚动条显示状态：none，隐藏；horizontal，水平；vertical，垂直
android:soundEffectsEnabled	设置点击或触摸时是否有声音效果
android:tag	设置一个文本标签
android:visibility	设置是否显示 View

表 3.23 滚动视图控件公共方法

公 共 方 法	功 能 描 述
void addView(View child)	添加子视图
void addView(View child, int index)	添加子视图
void addView(View child, int index, ViewGroup.LayoutParams params)	根据指定的 layout 参数添加子视图
void addView (View child, ViewGroup.LayoutParams params)	根据指定的 layout 参数添加子视图
boolean arrowScroll(int direction)	响应点击上下箭头时对滚动条滚动的处理
void computeScroll()	被父视图调用，用于必要时对其子视图的值（mScrollX 和 mScrollY）进行更新
boolean dispatchKeyEvent(KeyEvent event)	发送一个 key 事件给当前焦点路径的下一个视图
void draw (Canvas canvas)	手动绘制视图（及其子视图）到指定的画布（Canvas）
boolean executeKeyEvent(KeyEvent event)	当接收到 key 事件时，用户可以调用此函数来使滚动视图执行滚动
void fling(int velocityY)	滚动视图的滑动（fling）手势
boolean fullScroll(int direction)	对响应"home/end"短按时响应滚动处理
int getMaxScrollAmount()	当前滚动视图响应箭头事件能够滚动的最大数
boolean isFillViewport ()	指示当前 ScrollView 的内容是否被拉伸以填充视图可视范围
boolean isSmoothScrollingEnabled ()	按箭头方向滚动时，是否显示滚动的平滑效果
boolean onInterceptTouchEvent (MotionEvent ev)	实现此方法是为了拦截所有触摸屏幕时的运动事件
boolean onTouchEvent (MotionEvent ev)	执行此方法为了处理触摸屏幕的运动事件
boolean pageScroll (int direction)	响应短按【Page Up】或【Page Down】时对滚动的处理；direction 滚动方向：FOCUS_UP 表示向上翻一页，FOCUS_DOWN 表示向下翻一页
void requestChildFocus (View child, View focused)	当父视图的一个子视图的要获得焦点时，调用此方法
void setFillViewport (boolean fillViewport)	设置当前滚动视图是否将内容高度拉伸以填充视图可视范围
void setSmoothScrollingEnabled (boolean smoothScrollingEnabled)	设置箭头滚动是否可以引发视图滚动
void smoothScrollBy (int dx, int dy)	平缓滚动到某处
void smoothScrollTo (int x, int y)	平缓滚动到某处

3. **滚动视图控件典型案例**

实现要点：
- View findViewById(int id)：根据在 XML 中定义的视图 ID 查找该视图对象的引用。
- void setContentView(int layoutResID)：根据布局资源 ID，设置活动内容。
- void setOnClickListener(OnClickListener l)：设置单击监听器，捕获鼠标单击事件，并处理单击回调响应 void onClick(View v)，如图 3.19 所示。

图 3.19 按钮单击事件监听器机制

- ScrollView 继承自 FrameLayout，在本案例中其子元素为线性布局容器，TextView 在 LinearLayout 中沿垂直方向线性布局。
- Android Handler 机制：Android 提供的一种异步回调机制，UI 在完成很长时间任务后发出相应的通知。Handle 类的主要作用为：在新启动的线程中发送给消息；在主线程获取、处理消息。其主要包括两个队列：消息队列和线程队列。消息队列使用 sendMessage 和 HandleMessage 的组合来发送和处理消息。线程队列类似一个方法的委托，供用户传递方法。使用 post,postDelayed 添加委托，使用 removeCallbacks 移除委托。在调用 new Handler 实例化句柄对象 mHandler 后，该 mHandler 对象就和主线程的消息队列建立了关系，使用 mHandler.post(Runnable r)方法，发送一个委托的方法 Runnable mScrollToBottom 代理给 mHandler 时，主消息队列会在适当的时候执行这个 mScrollToBottom 中的委托方法，即执行了 runnabler.run 方法，如图 3.20 所示。

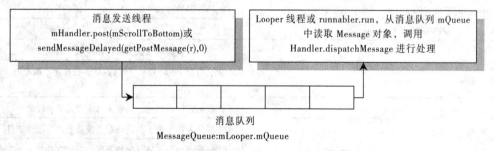

图 3.20 Android Handler 机制

- Handle 发送和处理消息的几个方法：

void handleMessage(Message msg)：处理消息的方法，该方法通常被重写。

boolean hasMessage(int what)：检查消息队列中是否包含有 what 属性为指定值的消息。

boolean hasMessage(int what,Object object)：检查消息队列中是否包含有由 object 属性指定值的消息。

sendEmptyMessage(int what)：发送空消息。

Boolean send EmptyMessageDelayed(int what ,long delayMillis)：设置多少毫秒发送空消息。

boolean sendMessage(Message msg)：立即发送消息。

- boolean sendMessageDelayed(Message msg,long delayMillis)：设置多少秒之后发送消息。
- Editable getText()：获得编辑框 EditText 输入文本信息。
- ScrollView.addView(View child)：将子视图对象添加到滚动视图中。
- ScrollView.scrollTo(int x, int y)：将 ScrollView 滚动到指定坐标(x,y)位置。
- ViewGroup.addView(View child)：将子视图对象添加到视图组（如线性布局）中。
- void setBackgroundColor(int color)：设置视图背景色或背景图片。
- void setText(CharSequence text)：设置视图显示信息。
- String getString(int resId)：获得字符串资源信息。
- Runnable 接口：Android 中实现多线程的一种接口方法。

创建 Ch03_Ex15_ScrollView 项目，修改字符串资源文件 strings.xml，修改布局文件 scrollview_layout.xml，实现 ScrollViewActivity.java 功能，如源代码清单 3.41、3.42 和 3.43 所示，其运行效果如图 3.21 所示。

源代码清单 3.41　\Ch03_Ex15_ScrollView\res\values\strings.xml

```
1. <?xml version="1.0" encoding="utf-8"?>
2. <resources>   <string name="app_name">滚动视图典型案例</string>
3.    <string name="action_settings">设置</string>   <string name="btn_name">发送</string>
4.    <string name="strWho">已发送消息: </string>   </resources>
```

源代码清单 3.42　\Ch03_Ex15_ScrollView\res\layout\scrollview_layout.xml

```
1. <RelativeLayout xmlns:android="http://schemas.android.com/apk/res/android"
2.   xmlns:tools="http://schemas.android.com/tools"
3.   android:layout_width="match_parent"   android:layout_height="match_parent"
4.   ... tools:context=".ScrollViewActivity" >
5.   <ScrollView android:id="@+id/svMessageContent"   android:layout_width=
     "fill_parent"
6.     android:layout_height="160dp"   android:background="#0f0"
7.     android:overScrollMode="always" android:visibility="visible" > </ScrollView>
8.   <EditText android:id="@+id/etMessage"   android:layout_width="220dp"
9.     android:layout_height="wrap_content"
10.     android:layout_below="@+id/svMessageContent" android:inputType="text" />
11.    <Button android:id="@+id/btnSendMessage" android:layout_width="wrap_
       content"
12.     android:layout_height="wrap_content"
13.     android:layout_below="@+id/svMessageContent"
14.     android:layout_marginLeft="220dp" android:text="@string/btn_name"/>
15. </RelativeLayout>
```

源代码清单 3.43　\Ch03_Ex15_ScrollView\src\sziit\lihz\ch03_ex15_scrollview\ScrollViewActivity.java

```
1. public class ScrollViewActivity extends Activity {//从基类派生子类
2.   private static final int[] mBgColor = { Color.GREEN, Color.BLUE };
                             //设置间隔底色
3.   private static int mBgIndex = 0;//ScrollView 间隔底色 mBgColor 的索引值
4.   private LinearLayout mLayout;    //定义线性布局对象
5.   private ScrollView mScroolViewMessage;
                 //定义 ScrollView 对象,记录发送消息滚动列表
```

165

```java
6.    private EditText etMessage;        //定义EditText对象，编辑发送消息编辑框
7.    private String strMessage;         //定义String对象，记录当前编辑框的显示消息
8.    private Button btnSendMessage;     //定义Button对象，发送消息按钮
9.    private final Handler mHandler = new Handler();
                                         //定义Handler对象,控制ScrollView滚动
10.   protected void onCreate(Bundle savedInstanceState) {//子类重写onCreate方法
11.       super.onCreate(savedInstanceState);//基类调用onCreate方法
12.       setContentView(R.layout.scrollview_layout);//设置手机界面布局
13.       mScroolViewMessage=(ScrollView)findViewById(R.id.svMessageContent);
                                         //滚动视图
14.       btnSendMessage = (Button) findViewById(R.id.btnSendMessage);
                                         //引用Button对象
15.       etMessage = (EditText) findViewById(R.id.etMessage);
                                         //引用消息编辑框EditText对象
16.       btnSendMessage.setOnClickListener(new OnClickListener() {
                                         //设置按钮单击监听事件
17.           public void onClick(View v) {//处理单击消息
18.               strMessage=etMessage.getText().toString();//获得当前发送消息
19.               sendMsg(mLayout, ScrollViewActivity.this, getCurrColor(),strMessage);
20.               mHandler.post(mScrollToBottom);}});    //发送滚动消息
21.       mLayout = new LinearLayout(this);             //线性布局方式
22.       mLayout.setOrientation(LinearLayout.VERTICAL); //控件对其方式为垂直排列
23.       mLayout.setBackgroundColor(0xf0ff);            //设置布局背景色
24.       for (int i = 0; i < 8; i++) {  //预先添加部分消息，用于程序调试
25.           sendMsg(mLayout, this, getCurrColor(), i + "掌握ScrollView应用方法!!!!");}
26.       mScroolViewMessage.addView(mLayout);}//将线性布局添加到ScrollView控件中
27.   private int getCurrColor() {        // 功能：获得当前线性布局中文本视图背景色
28.       return mBgColor[(++mBgIndex) % mBgColor.length];}//查询颜色数组
29.   //功能：模拟消息发送，在TextView中设置显示消息，并将该控件添加到线性布局容器
30.   private void sendMsg(LinearLayout layout, Context context, int bgColur, String MSG){
31.       TextView tv = new TextView(context);  //动态构造TextView对象
32.       //获取一个全局的日历实例，用于获取当前系统时间并格式化成小时：分钟形式
33.       tv.setText(this.getString(R.string.strWho)
34.           + DateFormat.format("kk:mm", Calendar.getInstance()) + "]\n" + MSG);
35.       tv.setBackgroundColor(bgColur);       //设置当前TextView控件背景色
36.       layout.addView(tv);}                  //将该控件添加到线性布局容器中
37.   //实现将ScrollView滚动到最底部的Runnable对象实例
38.   private Runnable mScrollToBottom = new Runnable() {
39.       public void run() {
40.           Log.d("ScrollViewActivity", "ScrollY: " + mScroolViewMessage.
                  getScrollY());                //记录日志信息
41.           int off = mLayout.getMeasuredHeight() - mScroolViewMessage.
                  getHeight();
42.           if (off > 0) {mScroolViewMessage.scrollTo(0, off);
                                                //滚动到最底部} }};}
```

 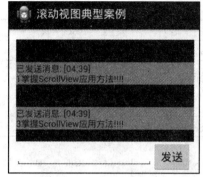

（a）设计轮廓　　　　　　　　　　（b）运行界面

图 3.21　滚动视图控件典型案例

3.4.2　列表视图控件

1. 列表视图控件简介

列表视图控件（ListView）的结构为：

```
class ListView extends AbsListView
    public abstract class AbsListView extends AdapterView<ListAdapter>
```

列表视图控件是 Android 开发中非常常用的组件，其可用来显示一个列表，用户可对这个列表进行操作。ListView 的两个主要职责在于：①将数据填充到布局；②处理用户的选择等操作。要创建并合理应用 ListView 需要三个元素：①ListView 中每一列的 View；②数据：表示将被映射的字符串、图片或者基本组件等数据源；③连接数据与 ListView 的适配器，用来将数据映射到列表视图，根据适配器的种类，可以分成三类：ArrayAdapter、SimpleAdapter 和 SimpleCursorAdapter。其中，ArrayAdapter 最为简单，只能显示一行文字；SimpleAdapter 具有较好的扩充性，可定义各种布局，显示图片视图、按钮和复选框等；SimpleCursorAdapter 主要将数据库内容以列表的形式显示出来。

2. 列表视图控件常用方法

列表视图控件的常用属性和方法如表 3.24、表 3.25 所示。

表 3.24　列表视图控件常用属性及其对应方法

属 性 名	Java 方法	功 能 描 述
android:stackFromBottom		设置是否从底部开始排列列表项，值为 true 和 false；该属性为"true"，则 ListView 中的内容将从底部开始显示
android:transcriptMode		设置该控件的滚动模式（disabled：关闭滚动；normal：当该控件收到数据改变通知且最后一个列表项可见时，该控件会滚动到底部；alwaysScroll：该控件总会自动滚动到底部）
android:cacheColorHint	setCacheColorHint	设置列表将始终以单一、固定的颜色来绘制背景，若想与整体背景色统一，只需将其设为透明（#00000000）即可
android:divider	setDivider(Drawable divider)	设置每项之间间隔线的图片或颜色；也可用#rgb,#argb,#aarrggbb 等格式表示颜色，如 android:divider="@drawable/list_divier"；若不想显示分隔线，则只要设置为 android:divider="@drawable/@null"

续表

属 性 名	Java 方法	功能描述
android:fadingEdge	setFadingEdgeLength	设置上边和下边有黑色的阴影,值为 none 时没有阴影
android:scrollbars		设置一个快速滑动条,只有值为 horizontall vertical 时,才会显示滚动条,并且会自动隐藏和显示
android:fadeScrollbars		设置该属性为"true"可实现滚动条的自动隐藏和显示
android:fastScrollEnabled	setFastScrollEnabled	设置快速滚动效果
android:choiceMode	setChoiceMode(int choiceMode)	设置该 ListView 所使用的选择模式(none:0,无选择模式;singleChoice:1,单选模式;multipleChoice:2,多选模式)
android:dividerHeight	setDividerHeight(intheight)	设置分隔的高度,例如"6px"
android:entries		设置一个数组资源作为该控件显示的内容
android:drawSelectorOnTop		该属性设置为 true 时,单击某项记录,则颜色会显示在最上面,记录内容的文字被遮住;该属性设置为 false 时,按住某项记录不放,颜色会在记录的后面,成为背景色,但记录内容的文字是可见的
android:scrollingCache	setScrollingCacheEnabled	
android:listSelector	setSelector(Drawable)	设置列表项选中时的颜色

表 3.25 列表视图控件公共方法

公 共 方 法	功 能 描 述
void setAdapter(Adapter adapter)	为 ListView 控件设置一个适配器
void setOnItemSelectedListener(AdapterView. OnItemSelectedListener listener)	用于监听该控件某列被单击的事件
getListView()	获得 ListView 的引用
setChoiceMode()	设置 ListView 回应用户单击事件的响应方式
setTextFilterEnabled()	设置是否激活"过滤"功能

列表视图控件的应用主要涉及以下重要方法:

(1) void setAdapter(ListAdapter adapter)

功能:为列表视图控件设置一个 ListAdapter 类型适配器;利用该方法可将创建的适配器与 ListView 控件连接起来。

参数:adapter 为 ListAdapter 类型的适配器。所谓适配器是一个连接数据和 AdapterView (ListView 派生于 AdapterView,它就是一个典型的 AdapterView)的桥梁,通过它能有效地实现数据与 AdapterView 的分离设置,使 AdapterView 与数据的绑定更加简便,修改更加方便。ListAdapter 派生于 Adapter。Android 提供了很多 Adapter,典型的包括:ArrayAdapter、SimpleAdapter 和 SimpleCursorAdapter 等,如图 3.22 所示。

图 3.22 ListView、适配器、数据、监听器和事件响应之间的关系

（2）void setOnItemClickListener(OnItemClickListener listener)

功能：设置列表视图控件选项被单击的监听器。该接口用于监听 ListView 控件某一列表项被单击的事件。

参数：OnItemClickListener 是一个接口，其抽象方法为 onItemClick。

（3）void onItemClick(AdapterView<?> parent, View view,int position, long id)

功能：列表视图控件选项被单击时触发。

参数：parent 代表用户操作的 ListView 控件，view 为当前被单击列表项的视图 view，position 为当前被单击列表项的位置，id 为当前被单击列表项的 id 号。

示例代码如下：

```
1. ListView listView = (ListView) findViewById(R.id.listView1);
                                 //引用XML布局文件ListView对象
2. listView.setAdapter(new MyAdapter());//为列表视图控件设定一个适配器
3. listView.setOnItemClickListener(new OnItemClickListener(){
                                 //设置ListView选项被单击的监听器
4.     @Override
5.     public void onItemClick(AdapterView<?> parent, View view,int position,
       long id) {
6.         //实现选项被单击事件响应方法
7. }} );
```

3. 列表视图控件典型案例

实现要点：

- 列表视图控件在拖动时背景图片消失，变成黑色背景的解决办法：

  XML 中：android:scrollingCache="false" 或者 android:cacheColorHint="#00000000"

  Java 中：setScrollingCacheEnabled(false) 或者 setCacheColorHint(0) 或者 setCacheColorHint(Color.TRANSPARENT);

- 列表视图的上边和下边有黑色阴影的解决办法：

  XML 中：android:fadingEdge="none"
  Java 中：setFadingEdgeLength(0);

- 在列表视图的每一项之间设置一个图片作为间隔的办法：

  XML 中：android:divider="@drawable/list_driver"
  Java 中：Drawable dr = this.getResources().getDrawable(R.colo.list_divider_clour);
          ListView.setDivider(dr);

- 列表视图选中项默认橙黄底色，去掉或改变该种效果的办法：

  XML 中：android:listSelector="@android:color/transparent" 或 "@android:color/blue"
  Java 中：Drawable drawable=this.getResources().getDrawable(R.drawable.select_feedback);
          ListView.setSelector(drawable);

- android:scrollBars="Vertical"与调用 ListView 的如下方法实现的效果相同：
  setVerticalScrollBarEnabled(true)

- android:divider="#00ff00"：设置分隔线颜色。

- android:transcriptMode="alwaysScroll"：该控件总会自动滚动到底部。
- android:dividerHeight="2dp"：设置分隔线边距。
- android:fadingEdge="none"：设置该属性值，去掉上边和下边黑色的阴影。
- android:scrollbars="none"：设置不显示滚动条。
- android:listSelector="@android:color/pink"：设置被选中项的颜色；默认为橙黄底色。
- android:fastScrollEnabled="true"：设置加快滑动速度。
- android:fadeScrollbars="true"：实现滚动条的自动隐藏和显示。
- android:scrollbars="none"：隐藏 ListView 的滚动条。
- android:divider="@drawable/@null"或"#00000000"：设置不显示分隔线。
- android:entries="@array/flower_array"：设置一个数组资源作为该控件显示的内容。
- getResources().getStringArray(R.array.flower_array)：加载资源字符串数组。

（1）列表视图控件典型案例 1：利用 SimpleAdapter 实现 ListView 与数据绑定。

步骤 1：创建 Ch03_Ex16_ListView 项目，修改字符串资源文件 strings.xml，如源代码清单 3.44 所示。

源代码清单 3.44　\Ch03_Ex16_ListView\res\values\strings.xml

```
1.  <?xml version="1.0" encoding="utf-8"?>
2.  <resources>  <string name="app_name">ListView 典型案例 1</string>
3.    <string name="action_settings">设置</string>
4.    <string name="strPrompt">当前选择的花卉名和图片为: </string>
5.    <string-array name="flower_array">
6.      <item>百合花</item>    <item>蝴蝶花</item>    <item>海棠花</item>
7.    </string-array>    </resources>
```

步骤 2：修改布局文件 listview_layout.xml，如源代码清单 3.45 所示。

源代码清单 3.45　\Ch03_Ex16_ListView\res\layout\listview_layout.xml

```
1.  <?xml version="1.0" encoding="UTF-8"?>
2.  <LinearLayout xmlns:android="http://schemas.android.com/apk/res/android"
3.    xmlns:tools="http://schemas.android.com/tools"
4.    android:layout_width="match_parent"  android:layout_height="wrap_content"
5.    android:orientation="vertical"
6.    ... tools:context=".ListViewActivity" >
7.    <ListView android:id="@+id/listView1" android:layout_width="match_parent"
8.      android:layout_height="wrap_content" />
9.    <TextView android:id="@+id/dispTextView" android:layout_width="match_parent"
10.     android:layout_height="wrap_content"    android:text="@string/strPrompt"
11.     android:textColor="#0f0"         android:textSize="20sp" />
12.   <LinearLayout  android:layout_width="match_parent"
13.     android:layout_height="wrap_content"     android:gravity="center"
14.     android:orientation="horizontal" >
15.   <TextView android:id="@+id/curTextView"  android:layout_width="wrap_content"
16.      android:layout_height="wrap_content" android:textColor="#00f"
17.      android:textSize="20sp" />
18.     <ImageView android:id="@+id/curImageView" android:layout_width="40dp"
19.      android:layout_height="40dp" />  </LinearLayout>    </LinearLayout>
```

步骤 3：添加列表项布局 listview_item.xml，如源代码清单 3.46 所示。

源代码清单 3.46　\Ch03_Ex16_ListView\res\layout\listview_item.xml

```xml
1. <?xml version="1.0" encoding="utf-8"?>
2. <LinearLayout xmlns:android="http://schemas.android.com/apk/res/android"
3.     android:layout_width="match_parent"    android:layout_height="wrap_content"
4.     android:gravity="center"               android:orientation="horizontal" >
5.     <ImageView     android:id="@+id/imageView"
6.         android:layout_width="40dp"        android:layout_height="40dp" />
7.     <TextView      android:id="@+id/textView"
8.         android:layout_width="match_parent"  android:layout_height="wrap_content"
9.         android:textSize="20sp" />    </LinearLayout>
```

步骤 4：实现 ListViewActivity.Java 编码，如源代码清单 3.47 所示。

源代码清单 3.47　\Ch03_Ex16_ListView\src\sziit\lihz\ch03_ex16_listview\ListViewActivity.java

```java
1.  public class ListViewActivity extends Activity {//从Activity基类派生子类
2.      private ListView listView1;                  // 定义列表视图对象
3.      private TextView curTextView;                //定义TextView,显示当前选择花名
4.      private ImageView curImageView;              //定义ImageView,显示当前花卉图片
5.      private String[] flower_array;               //定义花卉名字符串数组
6.      private int[] flower_id = { R.drawable.baihe, R.drawable.hudie,
7.          R.drawable.haitang };                    //定义花卉图片资源ID数组
8.      protected void onCreate(Bundle savedInstanceState) {//子类重写onCreate方法
9.          super.onCreate(savedInstanceState);      //基类调用onCreate方法
10.         setContentView(R.layout.listview_layout);    //设置屏幕布局
11.         curTextView = (TextView) findViewById(R.id.curTextView);//绑定TextView对象
12.         curImageView = (ImageView) findViewById(R.id.curImageView);
                                                     //绑定ImageView对象
13.         flower_array = this.getResources().getStringArray(R.array.flower_array);
14.         //步骤1:构建列表视图(ListView)对象
15.         listView1 = (ListView) findViewById(R.id.listView1);
                                                     //引用XML定义的ListView对象
16.         //步骤2: 构造ListView的内容map1、map2和map3
17.         Map<String, Object> map1 = new HashMap<String, Object>();
18.         map1.put("image", flower_id[0]);    //"百合花"图片资源
19.         map1.put("name", flower_array[0]);  //"百合花"
20.         Map<String, Object> map2 = new HashMap<String, Object>();
21.         map2.put("image", flower_id[1]);    //"蝴蝶花"图片资源
22.         map2.put("name", flower_array[1]);  //"蝴蝶花"
23.         Map<String, Object> map3 = new HashMap<String, Object>();
24.         map3.put("image",flower_id[2]);     //"海棠花"图片
25.         map3.put("name", flower_array[2]);  //"海棠花"
26.         //步骤3: 把以上由Map构成的列表视图项放到List中
27.         List<Map<String, Object>> data = new ArrayList<Map<String, Object>>();
28.         data.add(map1);                     //将map1添加到列表容器data中
29.         data.add(map2);                     //将map2添加到列表容器data中
30.         data.add(map3);                     //将map3添加到列表容器data中
31.         //步骤4: 构造SimpleAdapter适配器对象
32.         SimpleAdapter simpleAdapter = new SimpleAdapter(this, data,
33.             R.layout.listview_item, new String[] { "image", "name" },
34.             new int[] { R.id.imageView, R.id.textView });
```

```
35.        //步骤5: 将List中的内容(data)填充到ListView中
36.        listView1.setAdapter(simpleAdapter);//设置适配器,将数据(data)添加到ListView控件
37.        //步骤6: 添加ListView控件事件监听器
38.        listView1.setOnItemClickListener(new OnItemClickListener() {
39.            public void onItemClick(AdapterView<?> parent, View view,
40.            int position, long id) {       //可以通过句柄Handler线程机制优化
41.                curTextView.setText(flower_array[position]);//显示当前选择的花卉名
42.                curImageView.setImageResource(flower_id[position]);}});}}
                                        //显示当前选择的花卉
```

步骤5: 选择【Run As】→【Android Application】命令,运行结果如图3.23所示。

(a)设计轮廓　　　　　　　　　　(b)运行界面

图3.23　列表视图控件典型案例1

(2)列表视图控件典型案例2: 利用ArrayAdaper适配器实现ListView与数据绑定。

实现要点:

- Context getApplicationContext(): 获得当前活动的上下文环境。
- string-array: 在XML中定义字符串数组的标签。
- android:entries="@array/scenic_spots": 设置列表视图数据来源。
- getResources().getStringArray(R.array.scenic_spots): 加载字符串数组资源。
- ArrayAdapter(Context context, int textViewResourceId, T[] objects) {
- init(context, textViewResourceId, 0, Arrays.asList(objects)): 数组适配器构造器。
- void setAdapter(ListAdapter adapter): 设置列表控件适配器。

步骤1: 新建项目Ch03_Ex17_ListViewArrayAdaper,修改字符串文件strings.xml,如源代码清单3.48所示。

源代码清单3.48   \Ch03_Ex17_ListViewArrayAdaper\res\values\strings.xml

```
1. <?xml version="1.0" encoding="utf-8"?>
2. <resources>    <string name="app_name">ListView典型案例2</string>
3.      <string name="tourist_title">中国旅游景点排名:</string>
4.      <string name="strSelectedSpot">您最喜欢的旅游景点:</string>
5.      <string-array name="scenic_spots">
6.          <item>丽江</item>    <item>三亚</item>    <item>黄山</item>
7.          <item>九寨沟</item>  <item>桂林</item>    </string-array>    </resources>
```

步骤2: 修改布局文件listview_layout2.xml,定义两个ListView控件lvScenicSpots和listView1。lvScenicSpots采用android:entries="@array/scenic_spots"属性设置其数据来源,如源代码清单3.49所示。

源代码清单 3.49 \Ch03_Ex17_ListViewArrayAdaper\res\layout\listview_layout2.xml

```xml
1. <RelativeLayout xmlns:android="http://schemas.android.com/apk/res/android"
2.     xmlns:tools="http://schemas.android.com/tools"
3.     android:layout_width="match_parent"    android:layout_height="match_parent"
4.     ... tools:context=".ListViewActivity">
5.     <TextView android:id="@+id/tvTitle"    android:layout_width="wrap_content"
6.       android:layout_height="wrap_content"  android:text="@string/tourist_title"/>
7.     <ListView android:id="@+id/lvScenicSpots" android:layout_width=
8.       "wrap_content" android:layout_height="wrap_content"
9.       android:layout_below="@id/tvTitle"
10.      android:entries="@array/scenic_spots">     </ListView>
11.    <LinearLayout android:id="@+id/linearLayout1"
12.      android:layout_width="fill_parent" android:layout_height="wrap_content"
13.      android:layout_below="@id/lvScenicSpots"
14.      android:orientation="horizontal">
15.    <TextView android:layout_width="wrap_content"
16.      android:layout_height="wrap_content" android:text="@string/strSelectedSpot"/>
17.    <TextView android:id="@+id/tvCurSpots"android:layout_width="wrap_content"
18.      android:layout_height="wrap_content"/>      </LinearLayout>
19.    <ListView android:id="@+id/listView1"android:layout_width="wrap_content"
20.      android:layout_height="wrap_content"android:layout_below="@id/linearLayout1">
21.    </ListView>    </RelativeLayout>
```

步骤 3：修改 ListViewActivity.java 代码，先构造 ArrayAdapter 适配器，并利用 setAdapter 设置其适配器，然后添加 setOnItemClickListener 监听器，如源代码清单 3.50 所示。

步骤 4：选择【Run As】→【Android Application】命令，运行结果如图 3.24 所示。

源代码清单 3.50 \Ch03_Ex17_ListViewArrayAdaper\ListViewActivity.java

```java
1. public class ListViewActivity extends Activity { //从活动基类派生子类
2.    private ListView listView1;                   //定义私有列表视图对象
3.    private ListView lvScenicSpots;               //定义私有列表视图对象
4.    private TextView tvCurSpots;                  //定义私有TextView对象
5.    private String[] scenic_spots;                //定义私有字符串数组
6.    protected void onCreate(Bundle savedInstanceState) {//重写子类onCreate方法
7.       super.onCreate(savedInstanceState);        //调用基类onCreate方法
8.       setContentView(R.layout.listview_layout2); //设置手机屏幕布局
9.       scenic_spots = this.getResources().getStringArray(R.array.scenic_
          spots);                                   //加载字符串数组资源
10.      tvCurSpots = (TextView) this.findViewById(R.id.tvCurSpots);
                                                   //获得TextView对象引用
11.      lvScenicSpots = (ListView) this.findViewById(R.id.lvScenicSpots);
                                                   //获得控件对象
12.      lvScenicSpots.setOnItemClickListener(new OnItemClickListener() {
                                                   //设置监听器
13.         public void onItemClick(AdapterView<?> parent, View view,
14.            int position, long id) {             //列表项被单击回调方法
15.            dispSelectedSpot(view);} });         //显示当前被选项信息
16.      listView1 = (ListView) this.findViewById(R.id.listView1);
                                                   //获得ListView对象引用
```

```
17.      ArrayAdapter<String> adapter = new ArrayAdapter<String>(this,
18.          android.R.layout.simple_list_item_1, scenic_spots);//构造数组适配器
19.      listView1.setAdapter(adapter);        //设置适配器,将列表视图控件与数据绑定
20.      listView1.setOnItemClickListener(new OnItemClickListener() {
21.        public void onItemClick(AdapterView<?> parent, View view,
22.             int position, long id) {       //列表项被单击回调方法
23.             dispSelectedSpot(view);} }); } //显示当前被选项信息
24.  void dispSelectedSpot(View view) {
25.      TextView tv = (TextView) view;        //获取当前选择列表项
26.      tvCurSpots.setText(tv.getText());     //设置当前所选择的旅游景点
27.      Toast.makeText(getApplicationContext(),tv.getText(),Toast.LENGTH_
         LONG).show();}}
```

（a）运行界面　　　　　　　　　　　　（b）设计轮廓

图 3.24　列表视图控件(ListView)典型案例 2

（3）列表视图控件典型案例 3：利用 BaseAdapter 适配器实现 ListView 与数据绑定。

实现要点：

- BaseAdapter：从此基类派生自定义适配器。
- string-array：在 XML 中定义字符串数组的标签。
- getResources().getStringArray(R.array.scenic_spots)：加载字符串数组资源。
- void setAdapter(ListAdapter adapter)：设置列表控件适配器。
- setOnItemClickListener：设置列表项被单击事件监听器。

步骤 1：新建项目 Ch03_Ex18_ListViewBaseAdaper，修改字符串文件 strings.xml。

步骤 2：修改布局文件 listview_layout3.xml，定义三个 TextView 控件、一个 ListView 控件和一个 ImageView 控件，如源代码清单 3.51 所示。

源代码清单 3.51　\Ch03_Ex18_ListViewBaseAdaper\res\layout\listview_layout3.xml

```
1.  <RelativeLayout xmlns:android="http://schemas.android.com/apk/res/android"
2.      xmlns:tools="http://schemas.android.com/tools"
3.      android:layout_width="match_parent"    android:layout_height="wrap_content"
4.      android:paddingBottom="@dimen/activity_vertical_margin"
5.      android:paddingLeft="@dimen/activity_horizontal_margin"
6.      android:paddingRight="@dimen/activity_horizontal_margin"
7.      android:paddingTop="@dimen/activity_vertical_margin"
8.      tools:context=".ListViewMainActivity" >
9.      <TextView android:id="@+id/tvTitle"    android:layout_width="wrap_content"
10.        android:layout_height="wrap_content" android:text="@string/tourist_title" />
11.     <ListView android:id="@+id/listView1"  android:layout_width="wrap_content"
12.        android:layout_height="wrap_content" android:layout_below="@id/tvTitle" >
13.     </ListView>
14.     <TextView android:id="@+id/dispTextView" android:layout_width="match_parent"
15.        android:layout_height="wrap_content" android:layout_below="@id/listView1"
16.        android:text="@string/strSelectedSpot"     android:textColor="#0f0"
17.          android:textSize="20sp" />
18.     <LinearLayout  android:id="@+id/linearLayout"   android:layout_width=
    "match_parent"
19.        android:layout_height="wrap_content"     android:layout_below=
    "@id/dispTextView"
20.        android:gravity="center"           android:orientation="horizontal" >
21.        <TextView  android:id="@+id/curTextView"   android:layout_width=
    "wrap_content"
22.          android:layout_height="wrap_content"     android:textColor="#00f"
23.          android:textSize="20sp" />
24.        <ImageView android:id="@+id/curImageView" android:layout_width=
    "82dp"
25.          android:layout_height="82dp" />  </LinearLayout>  </RelativeLayout>
```

步骤 3：定义列表项布局文件 listview_item.xml，其中包含一个 TextView 控件和一个 ImageView 控件，如源代码清单 3.52 所示。

源代码清单 3.52　\Ch03_Ex18_ListViewBaseAdaper\res\layout\listview_item.xml

```
1. <?xml version="1.0" encoding="utf-8"?>
2. <LinearLayout xmlns:android="http://schemas.android.com/apk/res/android"
3.    android:layout_width="match_parent"    android:layout_height="wrap_content"
4.    android:gravity="center"           android:orientation="horizontal" >
5.    <ImageView   android:id="@+id/imageView1"
6.       android:layout_width="81dp"         android:layout_height="81dp" />
7.    <TextView    android:id="@+id/textView1"
8.       android:layout_width="match_parent" android:layout_height="wrap_content" />
9. </LinearLayout>
```

步骤 4：修改 ListViewActivity.java 代码，先派生 BaseAdapter 适配器子类，然后构造自定义适配器 CustomAdapter 对象，利用 setAdapter 设置其适配器，再添加 setOnItemClickListener 监听器，如源代码清单 3.53 所示。

源代码清单 3.53 \Ch03_Ex18_ListViewBaseAdaper\ListViewMainActivity.java

```java
1.  public class ListViewMainActivity extends Activity {//从活动基类派生子类
2.      private ListView listView1;              // 定义私有列表视图对象
3.      private TextView curTextView;            // 定义私有TextView对象
4.      private ImageView curImageView;          //定义ImageView，显示当前旅游景点图片
5.      private String[] scenic_spots;           // 定义私有字符串数组
6.      private int[] images = { R.drawable.lijiang, R.drawable.sanya,
7.          R.drawable.huangshan, R.drawable.jiuzhaigou };
8.      protected void onCreate(Bundle savedInstanceState) {//重写子类onCreate方法
9.          super.onCreate(savedInstanceState);          //调用基类onCreate方法
10.         setContentView(R.layout.listview_layout3); //设置手机屏幕布局
11.         curTextView = (TextView) findViewById(R.id.curTextView);
                                                    //获得TextView控件对象
12.         curImageView = (ImageView) findViewById(R.id.curImageView);
                                                    //绑定ImageView对象
13.         scenic_spots = this.getResources().getStringArray(R.array.scenic_spots);                                        //加载资源
14.         listView1 = (ListView) findViewById(R.id.listView1);
                                                    //获得ListView控件对象引用
15.         listView1.setAdapter(new CustomAdapter());
                                                    //设置适配器，将列表视图控件与数据绑定
16.         listView1.setOnItemClickListener(new OnItemClickListener() {
17.             public void onItemClick(AdapterView<?> parent, View view,
18.                 int position, long id) {
19.                 curTextView.setText(scenic_spots[position]);//显示当前选择的花卉名
20.                 curImageView.setImageResource(images[position]);
                                                    //显示当前选择的花卉} });}
21.     //自定义适配器
22.     class CustomAdapter extends BaseAdapter {    //从BaseAdapter基类派生子类
23.         public int getCount() {                  //获得列表项总数
24.             return scenic_spots.length;}
25.         public Object getItem(int position) {    //获得指定位置列表项对象
26.             return scenic_spots[position];}
27.         public long getItemId(int position) {    //获得指定位置列表项对象ID
28.             return position;}
29.         public View getView(int position, View convertView, ViewGroup parent) {
30.             ViewHolder vh = new ViewHolder();
31.             //通过以下条件判断语句，来循环利用
32.             if (convertView == null) {
                        //若等于null,则表示屏幕上没有可以被重复利用的对象
33.                 convertView = getLayoutInflater().inflate(R.layout.listview_item, null);    // 创建View
34.                 vh.iv = (ImageView) convertView.findViewById(R.id.imageView1);
35.                 vh.tv = (TextView) convertView.findViewById(R.id.textView1);
36.                 convertView.setTag(vh);
37.             } else { vh = (ViewHolder) convertView.getTag();}
38.             vh.iv.setImageResource(images[position]); vh.tv.setText(scenic_spots[position]);
39.             return convertView;  }}
40.     static class ViewHolder { ImageView iv; TextView tv; }}
```

步骤 5：选择【Run As】→【Android Application】命令，运行结果如图 3.25 所示。

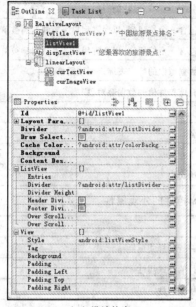
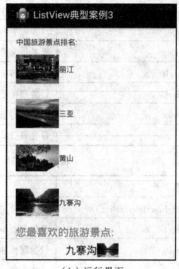

（a）设计轮廓　　　　　　　　　　（b）运行界面

图 3.25　列表视图控件典型案例 3

3.4.3　网格视图控件

1. 网格视图控件简介

网格视图（GridView）的结构为：

```
public final class GridView extends AbsListView
    java.lang.Object
        android.view.View
            android.view.ViewGroup
                android.widget.AdapterView
                    android.widget.AbsListView
                        android.widget.GridView
```

网格视图类位于 android.widget 包下，是显示二维的 ViewGroup、可滚动的网格，一般用来显示多张图片。该组件中的条目通过 ListAdapter 和该组件进行关联。

2. 网格视图控件的属性和常用方法

GridView 控件的 XML 属性分为：行为属性、外观属性、状态属性（分别见表 3.26、表 3.27 和表 3.28），其常见属性和方法如表 3.29 所示。GridView 控件可激发的事件列表如表 3.30 所示。

表 3.26　GridView 控件的行为属性

属性名	功能描述
android:AllowPaging	设置该控件是否支持分页
android:AllowSorting	设置该控件是否支持排序
android:AutoGenerateColumns	设置是否自动为数据源中的每个字段创建列；默认为 true
android:AutoGenerateDeleteButton	设置该控件是否包含一个按钮列，以允许用户删除映射到被单击行的记录

续表

属 性 名	功 能 描 述
android:AutoGenerateEditButton	设置该控件是否包含一个按钮列,以允许用户编辑映射到被单击行的记录
android:AutoGenerateSelectButton	设置该控件是否包含一个按钮列,以允许用户选择映射到被单击行的记录
android:DataMember	设置一个多成员数据源中的特定表绑定到该网格
android:DataSource	设置包含用来填充该控件值的数据源对象
android:DataSourceID	设置所绑定的数据源控件
android:RowHeaderColumn	设置列标题的列名
android:SortDirection	设置列的当前排序方向
android:SortExpression	设置当前排序表达式

表 3.27 GridView 控件的外观属性

属 性 名	功 能 描 述
android:AlternatingRowStyle	设置控件中每隔一行的样式属性
android:EditRowStyle	设置正在编辑的行的样式属性
android:HeaderStyle	设置网格的标题样式属性
android:FooterStyle	设置网格的页脚样式属性
android:EmptyDataRowStyle	设置空行的样式属性
android:PagerStyle	设置网格分页器的样式属性
android:RowStyle	设置表中行的样式属性
android:SelectedRowStyle	设置当前所选行的样式属性

表 3.28 GridView 控件的状态属性

属 性 名	功 能 描 述
android:BackImageUrl	设置要在控件背景中显示的图像的 URL
android:Caption	设置在该控件的标题中显示的文本
android:CaptionAlign	设置标题文本的对齐方式
android:CellPadding	设置单元的内容与边界之间的间隔(px)
android:CellSpacing	设置单元之间的间隔(px)
android:GridLines	设置该控件的网格线样式
android:HorizontalAlign	设置该页面上的控件水平对齐
android:EmptyDataText	设置当该控件绑定到一个空的数据源时生成的文本
android:PagerSettings	引用一个允许设置分页器按钮的属性的对象
android:ShowHeader	设置是否显示标题行
android:ShowFooter	设置是否显示页脚行

表 3.29 GridView 控件的常见属性及其方法

属 性 名	Java 方法	功 能 描 述
android:numColumns	setNumColumns(int)	设置控件列数
android:horizontlSpacing	setHorizontlSpacing(int)	设置列间距

续表

属 性 名	Java 方法	功 能 描 述
android:verticalSpacing	setVerticalSpacing(int)	设置行间距
android:stretchMode	setStretchMode(int)	设置缩放模式,可取值为"spacingWidth"、"columnWidth"或"spacingWidthUniform"
android:columnWidth	setColumnWidth(int)	设置列宽
android:gravity	setGravity (int)	设置该组件中的内容在组件中的位置,可选的值有 top、bottom、left、right、center_vertical、fill_vertical、center_horizontal、fill_horizontal、center、fill、clip_vertical,可以多选,用"|"分开

表 3.30 GridView 控件的事件

事 件 名	功 能 描 述
PageIndexChanging, PageIndexChanged	这两个事件都是在一个分页器按钮被单击时发生。它们分别在网格控件处理分页操作之前和之后激发
RowCancelingEdit	在一个处于编辑模式的行的 Cancel 按钮被单击,且在该行退出编辑模式之前发生
RowCommand	单击一个按钮时发生
RowCreated	创建一行时发生
RowDataBound	一个数据行绑定到数据时发生
RowDeleting, RowDeleted	这两个事件都是在一行的 Delete 按钮被单击时发生。它们分别在该网格控件删除该行之前和之后激发
RowEditing	当一行的 Edit 按钮被单击,且该控件进入编辑模式之前发生
RowUpdating, RowUpdated	这两个事件都是在一行的 Update 按钮被单击时发生。它们分别在该网格控件更新该行之前和之后激发
SelectedIndexChanging, SelectedIndexChanged	这两个事件都是在一行的 Select 按钮被单击时发生。它们分别在该网格控件处理选择操作之前和之后激发
Sorting, Sorted	这两个事件都是在对一个列进行排序的超链接被单击时发生。它们分别在网格控件处理排序操作之前和之后激发

该组件的重要构造函数和方法详细描述:

(1) GridView 的构造函数

- public Grid View(Context context):创建一个默认属性的 GridView 实例。
- public GridView(Context context,AttributeSet attrs):创建一个带有 attrs 属性的 GridView 实例。
- public GridView(Contextcontext,AttributeSetattrs,intdefStyle):创建一个带有 attrs 属性,并且指定其默认样式的 GridView 实例。

(2) GridView 的公共方法

- ListAdapter getAdapter():获得与此组件相关的适配器。
- void setAdapter (ListAdapter adapter):设置 GridView 的数据。adapter 为 GridView 提供数据的适配器。
- void setSelection(int position):设置选中的条目。
- boolean onKeyDown (int keyCode,KeyEvent event):按键按下事件处理。

- boolean onKeyUp(int keyCode,KeyEvent event):按键释放事件处理。
- boolean onKeyMultiple(int keyCode,int repeatCount,KeyEvent event):按键多选事件处理。
- void setOnItemClickListener(OnItemClickListener listener):设置 GridView 项被单击事件监听器。
- void onItemClick(AdapterView<?> parent, View view, int position, long id):GridView 项单击事件响应接口方法。parent 表示 GridView 对象;view 表示被单击视图项;position 表示单击项位置;id 表示单击项视图 ID 号。

3. 网格视图控件典型案例

实现要点:

- private GridView mGridView:声明私有网格视图对象。
- mGridView=(GridView)this.findViewById(R.id.gridView):引用 GridView 对象。
- private int[] mImageIds:定义私有 GridView 菜单图片资源数组。
- private int[] mTitleTexts:定义私有 GridView 菜单字符串资源数组。
- mGridView.setAdapter(new gridViewAdapter(mImageIds, mTitleTexts)):设置适配器。
- mGridView.setOnItemClickListener():设置 GridView 项被单击事件监听器。
- void onItemClick(AdapterView<?> parent, View view,int position, long id):单击事件响应接口方法。
- getSystemService(Context.LAYOUT_INFLATER_SERVICE):获得 LayoutInflater 系统服务接口,利用 View inflate(int resource, ViewGroup root)方法根据布局文件构建视图对象。

步骤 1:新建项目 Ch03_Ex19_GridView,修改字符串文件 strings.xml,如源代码清单 3.54 所示。

源代码清单 3.54　\Ch03_Ex19_GridView\res\values\strings.xml

```
1.  <?xml version="1.0" encoding="utf-8"?>
2.  <resources>  <string name="app_name">网格视图(GridView)典型案例</string>
3.    <string name="action_settings">设置</string>
4.    <string name="strPrompt">请点播您爱听的歌曲:</string>
5.    <string name="SelectPrompt">您点播的歌曲类型为:</string>
6.    <string name="cooldoghotsong">酷狗热歌</string>
7.    <string name="ktvsong">KTV 必点曲</string>
8.    <string name="networksong">网络红歌</string>
9.    <string name="djhotdish">DJ 热碟</string> <string name="classicsong">经典</string>
10.   <string name="oldlovesong">老情歌</string>
11.   <string name="publicsquaredancing">广场舞</string>
12.   <string name="newsong">新歌</string>    <string name="oldsong">老歌</string>
13.   <string name="lonelysong">寂寞</string>
14.   <string name="shop">店铺</string>    <string name="singer">我是歌手</string>
15. </resources>
```

步骤 2:修改布局文件 gridview_layout1.xml,其中包含三个 TextView 控件、一个 GridView 控件和一个 ImageView 控件,如源代码清单 3.55 所示。

源代码清单 3.55　\Ch03_Ex19_GridView\res\layout\gridview_layout1.xml

```
1. <RelativeLayout xmlns:android="http://schemas.android.com/apk/res/android"
2.     xmlns:tools="http://schemas.android.com/tools"
```

```
3.      android:layout_width="match_parent"   android:layout_height="wrap_content"
4.      ...tools:context=".GridViewActivity" >
5.      <TextView
6.          android:id="@+id/textView"         android:layout_width="wrap_content"
7.          android:layout_height="wrap_content"   android:text="@string/strPrompt"
8.          android:textSize="20sp" />
9.      <GridView    android:id="@+id/gridView"
10.         android:layout_width="match_parent"    android:layout_height=
            "wrap_content"
11.         android:layout_below="@id/textView"    android:columnWidth="100dip"
12.         android:gravity="center"           android:horizontalSpacing="10dip"
13.         android:numColumns="auto_fit"          android:paddingRight="5dip"
14.         android:stretchMode="columnWidth"  android:verticalSpacing="10dip" />
15.     <TextView    android:id="@+id/tvSelected"
16.         android:layout_width="wrap_content"    android:layout_height="wrap_content"
17.         android:layout_below="@id/gridView"    android:text="@string/SelectPrompt"
18.         android:textSize="20sp" />
19.     <LinearLayout  android:layout_width="wrap_content"
20.         android:layout_height="wrap_content"       android:layout_below=
            "@id/tvSelected"
21.         android:orientation="horizontal" >
22.         <TextView  android:id="@+id/curTextView"   android:layout_width=
            "wrap_content"
23.             android:layout_height="wrap_content"   android:layout_center
                Horizontal="true" >
24.         </TextView>
25.         <ImageView  android:id="@+id/curImageView"
26.             android:layout_width="wrap_content"   android:layout_height="wrap_
                content"
27.             android:layout_centerHorizontal="true" >
28.         </ImageView>    </LinearLayout>    </RelativeLayout>
```

步骤 3：定义 GridView 单项布局文件 gridview_item.xml，其中包含一个 TextView 控件和一个 ImageView 控件，如源代码清单 3.56 所示。

源代码清单 3.56   \Ch03_Ex19_GridView\res\layout\gridview_item.xml

```
1. <?xml version="1.0" encoding="utf-8"?>
2. <RelativeLayout xmlns:android="http://schemas.android.com/apk/res/android"
3.      android:layout_width="match_parent"    android:layout_height="wrap_content" >
4.      <ImageView    android:id="@+id/imageItem"
5.          android:layout_width="wrap_content"    android:layout_height="wrap_content"
6.          android:layout_centerHorizontal="true" >    </ImageView>
7.      <TextView     android:id="@+id/textItem"
8.          android:layout_width="wrap_content"    android:layout_height="wrap_content"
9.          android:layout_below="@+id/imageItem"  android:layout_centerHorizontal=
            "true" >
10.     </TextView>    </RelativeLayout>
```

步骤 4：修改 GridViewActivity.java 代码，先派生 BaseAdapter 适配器子类，然后构造自定义适配器 GridViewAdapter 对象，利用 setAdapter 设置其适配器，再添加 setOnItemClick Listener 监听器，如源代码清单 3.57 所示。

代码清单 3.57 \Ch03_Ex19_GridView\src\sziit\lihz\ch03_ex19_gridview\GridViewActivity.java

```java
1.  public class GridViewActivity extends Activity {// 从活动基类公有派生子类
2.      private GridView mGridView;                 //声明私有网格视图对象
3.      private ImageView mCurImageView;            //声明私有 ImageView 对象
4.      private TextView mCurTextView;              //声明私有 TextView 对象
5.      private int[] mImageIds = { R.drawable.cooldoghotsong, R.drawable.ktvsong,
6.          R.drawable.networksong, R.drawable.djhotdish,
7.          R.drawable.classicsong, R.drawable.oldlovesong,
8.          R.drawable.publicsquaredancing, R.drawable.newsong,
9.          R.drawable.oldsong, R.drawable.lonelysong, R.drawable.shop,
10.         R.drawable.singer };              //定义私有 GridView 菜单图片资源数组
11.     private int[] mTitleTexts = { R.string.cooldoghotsong, R.string.ktvsong,
12.         R.string.networksong, R.string.djhotdish, R.string.classicsong,
13.         R.string.oldlovesong, R.string.publicsquaredancing,
14.         R.string.newsong, R.string.oldsong, R.string.lonelysong,
15.         R.string.shop, R.string.singer };//定义私有 GridView 菜单字符串资源数组
16.     protected void onCreate(Bundle savedInstanceState) {//子类重写 onCreate 方法
17.         super.onCreate(savedInstanceState);//调用基类 onCreate 方法
18.         setContentView(R.layout.gridview_layout1);       //设置视图界面布局
19.         mCurImageView = (ImageView) this.findViewById(R.id.curImageView);
20.         mCurTextView = (TextView) this.findViewById(R.id.curTextView);
21.         mGridView = (GridView) this.findViewById(R.id.gridView);//引用 GridView 对象
22.         mGridView.setAdapter(new GridViewAdapter(mImageIds, mTitleTexts));
                                                            //设置适配器
23.         mGridView.setOnItemClickListener(new OnItemClickListener() {
                                                            //设置 GridVie 监听器
24.             public void onItemClick(AdapterView<?> parent, View view,
25.                 int position, long id) {    //单击事件响应接口方法
26.                 mCurImageView.setImageResource(mImageIds[position]);//显示图片资源
27.                 mCurTextView.setText(mTitleTexts[position]);//显示字符串资源}});}
28.     public class GridViewAdapter extends BaseAdapter {
                                                    //从 BaseAdapter 派生适配器
29.         private View[] mItemViews;               //定义私有视图对象数组
30.         public GridViewAdapter(int[] imageIds, int[] titleTexts) {
                                                    //自定义适配器类的构造器
31.             mItemViews = new View[imageIds.length];//动态构造视图对象数组
32.             for (int i = 0; i < mItemViews.length; i++) {
33.                 mItemViews[i] = makeItemView(imageIds[i], titleTexts[i]);
                                                    //构造视图对象}}
34.         public int getCount() {          //获得数据总数
35.             return mItemViews.length;}
36.         public View getItem(int position) {   //获得给定位置的视图对象
37.             return mItemViews[position];}
38.         public long getItemId(int position) {   //获得给定位置的视图对象 ID
39.             return position;}
40.         private View makeItemView(int imageResIds, int strTitleTexts) {
41.             LayoutInflater inflater = (LayoutInflater) GridViewActivity.this
42.                 .getSystemService(Context.LAYOUT_INFLATER_SERVICE);
                                                    //获得系统服务接口
43.             View itemView = inflater.inflate(R.layout.gridview_item, null);
                                                    //构造视图对象
```

```
44.         TextView title = (TextView) itemView.findViewById(R.id.textItem);
45.         title.setText(strTitleTexts);            //设置文本视图显示信息
46.         ImageView image = (ImageView) itemView.findViewById(R.id.imageItem);
47.         image.setImageResource(imageResIds);   //设置图像视图显示图片
48.         image.setScaleType(ImageView.ScaleType.FIT_CENTER);
                                                    //设置图片显示方式
49.         return itemView;}
50.      public View getView(int position, View convertView, ViewGroup parent) {
51.         if (convertView == null)
52.            return mItemViews[position];        //获得给定位置的视图对象
53.         return convertView; }}}
```

步骤5：选择【Run As】→【Android Application】命令，运行结果如图3.26所示。

（a）设计轮廓

（b）运行界面

图3.26 网格视图控件典型案例

3.4.4 进度条

1. 进度条简介

ProgressBar 即进度条，Android 系统自带的进度条有两种，分别是圆形进度条和水平进度条。ProgressBar 常常作为某些操作的进度中的可视指示器，为用户呈现操作的进度，它还有一个次要的进度条，用来显示中间进度。

2. 进度条的重要属性和常用方法

进度条的重要属性及常用方法如表3.31、表3.32所示。

表3.31 进度条的重要属性

属 性 名	功 能 描 述
android:style	设置进度条的风格
android:max	设置进度条的范围
android:progress	设置当前第一进度值
android:secondaryProgress	设置当前第二进度值
android:indeterminate	设置是否开启进度条的"不确定模式"，即显示滚动的动画效果而不显示实际进度
android:indeterminateBehavior	设置"不确定模式"的动画效果，可取为"cycle"或"repeat"
android:indeterminateDrawable	设置"不确定模式"进度条上显示的 Drawable 对象

续表

属性名	功能描述
android:indeterminateDuration	设置"不确定模式"的持续时间
android:visibility	设置是否显示，默认显示

android:style 包括 6 种典型的风格：

- @android:style/Widget.ProgressBar.Horizontal：水平进度条（可以显示刻度，常用）。
- @android:style/Widget.ProgressBar.Small：小进度条。
- @android:style/Widget.ProgressBar.Large：大进度条。
- @android:style/Widget.ProgressBar.Inverse：不断跳跃、旋转画面的进度条。
- @android:style/Widget.ProgressBar.Large.Inverse：不断跳跃、旋转动画的大进度条。
- @android:style/Widget.ProgressBar.Small.Inverse：不断跳跃、旋转动画的小进度条。

表 3.32　进度条的重要方法

Java 方法名	功能描述
getMax()	返回这个进度条的范围的上限
setMax(int max)	设置进度条的范围
getProgress()	返回进度值
setProgress(int progress)	设置进度条的当前进度
getSecondaryProgress()	返回次要进度
setSecondaryProgress(int)	设置次要进度值
incrementProgressBy(int diff)	设置增加的进度
incrementSecondaryProgressBy(int)	设置增加的第二进度
isIndeterminate()	指示进度条是否在不确定模式下
setIndeterminate(boolean indeterminate)	设置不确定模式
setVisibility(int v)	设置该进度条是否可视
onSizeChanged(int w, int h, int oldw, int oldh)	当进度值改变时引发此事件

3. 进度条典型案例

实现要点：

- style="@android:style/Widget.ProgressBar.Horizontal"：设置进度条风格。
- requestWindowFeature(Window.FEATURE_INDETERMINATE_PROGRESS)：设置成不明确的进度条风格。
- setProgressBarVisibility(true)：设置进度条可视。
- requestWindowFeature(Window.FEATURE_PROGRESS)：设置窗口风格进度条。
- android:max="100"：最大进度值为 100。
- android:progress="20"：初始化的进度值。
- android:secondaryProgress="30"：初始化的底层第二个进度值。
- 演示了三种实现按钮单击事件监听器 View.OnClickListener 接口的方法。
- Intent intent = new Intent(MainActivity.this, NormalstyleActivity.class)：构建意图对象。
- startActivity(intent)：启动意图，实现活动页面的切换。

● 在 AndroidManifest.xml 文件中注册活动的方法。

步骤 1：新建项目 Ch03_Ex20_ProgressBar，修改字符串文件 strings.xml，如源代码清单 3.58 所示。

源代码清单 3.58 \Ch03_Ex20_ProgressBar\res\values\strings.xml

```
1.  <?xml version="1.0" encoding="utf-8"?>
2.  <resources>    <string name="app_name">进度条(ProgressBar)案例</string>
3.      <string name="mainActivity_title">主活动标题</string>
4.      <string name="NormalstyleActivity_Title">NormalstyleActivity 标题</string>
5.      <string name="normalstyle">常见风格进度条</string>
6.      <string name="ordinary_circular">普通圆形 ProgressBar</string>
7.      <string name="StyleLarge">超大号圆形 ProgressBar</string>
8.      <string name="StyleSmall">小号圆形 ProgressBar</string>
9.      <string name="BarStyleHorizontal">水平进度条</string>
10.     <string name="strExit">返回主界面</string>
11.     <string name="strUpdown">布局中的长形进度条</string>
12.     <string name="strTitlePb">页面标题中的长形进度条</string>
13.     <string name="strPbEx">演示进度条功能</string>
14.     <string name="strPbDialog">Dialog 样式的 ProgressBar</string>
15.     <string name="strShowIndeterminate">带标题的进度条对话框</string>
16.     <string name="strShowIndeterminateNoTitle">不带标题的进度条对话框</string>
17. </resources>
```

步骤 2：修改布局文件 main_layout.xml，其中包含 5 个 Button 控件，单击不同按钮后演示不同风格的进度条，如源代码清单 3.59 所示。

源代码清单 3.59 \Ch03_Ex20_ProgressBar\res\layout\main_layout.xml

```
1.  <LinearLayout xmlns:android="http://schemas.android.com/apk/res/android"
2.      android:layout_width="match_parent"    android:layout_height="wrap_content"
3.      android:orientation="vertical" >
4.      <Button     android:id="@+id/btnNormalstyle"
5.         android:layout_width="wrap_content" android:layout_height="wrap_content"
6.         android:text="@string/normalstyle" >
7.      </Button>
8.      <Button     android:id="@+id/btnUpdown"
9.         android:layout_width="wrap_content"  android:layout_height="wrap_content"
10.        android:text="@string/strUpdown" >
11.     </Button>
12.     <Button     android:id="@+id/btnTitlePb" android:layout_width= "wrap_content"
13.        android:layout_height="wrap_content"    android:text="@string/strTitlePb" >
14.     </Button>
15.     <Button     android:id="@+id/btnPbEx"
16.        android:layout_width="wrap_content"   android:layout_height="wrap_content"
17.        android:text="@string/strPbEx" >
18.     </Button>
19.     <Button     android:id="@+id/btnPbDialog"
20.        android:layout_width="wrap_content"   android:layout_height="wrap_content"
21.        android:text="@string/strPbDialog" >
22.     </Button>    </LinearLayout>
```

步骤 3：修改 MainActivity.java 代码，实现 View.OnClickListener，按钮对象调用 setOnClickListener(this)设置单击监听，再实现 onClick(View v)接口方法，从而多个按钮对应一个监听，根据按钮对象 ID 识别不同对象按钮，然后分别进行处理，如源代码清单 3.60 所示。完整源代码见 Ch03_Ex20_Progress Bar.rar。

源代码清单 3.60   \Ch03_Ex20_ProgressBar\src\sziit\lihz\ch03_ex20_progressbar\MainActivity.java

```java
1.  //从 Activity 基类派生子类，并实现单击监听器接口 View.OnClickListener
2.  public class MainActivity extends Activity implements View.OnClickListener {
3.      private Button btnNormalstyle;          //定义私有按钮对象
4.      private Button btnUpdown;               //定义私有按钮对象
5.      private Button btnTitlePb;              //定义私有按钮对象
6.      private Button btnPbEx;                 //定义私有按钮对象
7.      private Button btnPbDialog;             //定义私有按钮对象
8.      private Intent intent;                  //定义私有意图对象
9.      protected void onCreate(Bundle savedInstanceState) {//子类重写 onCreate 方法
10.         super.onCreate(savedInstanceState);//基类调用 onCreate 方法
11.         //请求窗口特色风格，这里设置成不明确的进度风格
12.         requestWindowFeature(Window.FEATURE_INDETERMINATE_PROGRESS);
13.         setContentView(R.layout.main_layout);// 设置活动视图布局或加载布局界面
14.         btnNormalstyle = (Button) findViewById(R.id.btnNormalstyle);
                                                        //绑定按钮对象
15.         btnNormalstyle.setOnClickListener(this);    //设置按钮单击监听器
16.         btnUpdown = (Button) findViewById(R.id.btnUpdown);//绑定按钮对象
17.         btnUpdown.setOnClickListener(this);         //设置按钮单击监听器
18.         btnTitlePb = (Button) findViewById(R.id.btnTitlePb);//绑定按钮对象
19.         btnTitlePb.setOnClickListener(this);        //设置按钮单击监听器
20.         btnPbEx = (Button) findViewById(R.id.btnPbEx);//绑定按钮对象
21.         btnPbEx.setOnClickListener(this);           //设置按钮单击监听器
22.         btnPbDialog = (Button) findViewById(R.id.btnPbDialog);//绑定按钮对象
23.         btnPbDialog.setOnClickListener(this);       //设置按钮单击监听器
24.         setProgressBarIndeterminateVisibility(true);}
                                                //设置标题栏中的不明确的进度条
25.     public void onClick(View v) {           //多个 button 对应一个监听
26.         switch (v.getId()) {    //根据按钮对象 ID 识别不同对象按钮，然后分别进行处理
27.         case R.id.btnNormalstyle:           //切换到 NormalstyleActivity 活动
28.             intent=new Intent(MainActivity.this, NormalstyleActivity.class);
                                                //构建意图对象
29.             startActivity(intent); break;   //启动意图，实现活动页面的切换
30.         case R.id.btnUpdown:                //切换到 NormalstyleActivity 活动
31.             intent=new Intent(MainActivity.this, UpdownStyleActivity.class);
                //构建意图对象
32.             startActivity(intent); break;   //启动意图，实现活动页面的切换
33.         case R.id.btnTitlePb:               //切换到 TitleStyleActivity 活动
34.             intent = new Intent(MainActivity.this, TitleStyleActivity.class);
                //构建意图对象
35.             startActivity(intent); break;   //启动意图，实现活动页面的切换
```

```
36.         case R.id.btnPbEx:                      //切换到 TitleStyleActivity 活动
37.             intent = new Intent(MainActivity.this, PBDemoActivity.class);
                                                    //构建意图对象
38.             startActivity(intent); break;       //启动意图，实现活动页面的切换
39.         case R.id.btnPbDialog:                  //切换到 PbDialogActivity 活动
40.             intent = new Intent(MainActivity.this, PbDialogActivity.class);
                                                    //构建意图对象
41.             startActivity(intent); break;       //启动意图，实现活动页面的切换
42.         default: break;      }}}
```

步骤 4：选择【Run As】→【Android Application】命令，运行结果如图 3.27 所示。

（a）运行主界面　　　　　　　　　　（b）典型风格进度条

图 3.27　进度条（ProgressBar）典型案例

3.4.5　滑块控件

1. 滑块控件简介

滑块控件（SeekBar）的结构为：

```
public class SeekBar extends AbsSeekBar
    java.lang.Object
      android.view.View
        android.widget.ProgressBar
          android.widget.AbsSeekBar
            android.widget.SeekBar
```

滑块控件继承于进度条，是用来接收用户输入的控件，其应用非常广，比如用来显示音量条、播放进度条等。

2. 滑块控件属性和常用方法

滑块控件的属性和常用方法如表 3.33、表 3.34 所示。

表 3.33　滑块控件的属性

属 性 名 称	功 能 描 述
android:thumb	seekbar 上可绘制的 thumb（可拖动的那个图标）
android:layout_height	设置控件高度，建议使用 wrap_content，否则一定要保证设置的值不小于 seekbar 图片资源中的最高值

续表

属性名称	功能描述
android:maxHeight	设置进度条的最大高度，例如"12px"
android:minHeight	设置进度条的最低高度，例如"12px"
android:paddingLeft	设置左边间距，解决拖动按钮在最左显示不全问题，是 thumb 的一半宽度
android:paddingRight	设置右边间距，解决拖动按钮在最右显示不全问题，是 thumb 的一半宽度
android:progressDrawable	设置自定义的进度条样式，例如"@drawable/seekbar_style"，在其中可以设置进度条背景图，进度条图，缓冲条图

表 3.34 滑块控件常用方法

Java 方法	功能描述
void setMax(int max)	设置 SeekBar 的进度最大值
int getMax()	获得 SeekBar 的进度最大值
int getProgress()	获得 SeekBar 当前进度值
void setProgress(int progress)	设置 SeekBar 当前进度值
SeekBar.OnSeekBarChangeListener	一个回调函数，用来在进度等级发生改变时通知客户端，该函数的格式如下

```
void setOnSeekBarChangeListener(OnSeekBarChangeListener l)
```

功能：设置一个监听器以接收 SeekBar 进度改变时的通知，同时提供用户在 SeekBar 上开始和停止触摸手势的通知。在 SeekBar 中需要监听三个事件：onProgressChanged（数值的改变）、onStartTrackingTouch（开始拖动）、onStopTrackingTouch（停止拖动）。在 onProgressChanged 中用户可得到当前数值的大小。

参数：l 是 SeekBar 的通知监听对象。

3. 滑块控件典型案例

实现要点：

● SeekBar.OnSeekBarChangeListener 接口。

● Handler 对象：用来演示自动调整 SeekBar 进度值。

● ImageView 对象：用来演示手动调整显示图片。

● setOnSeekBarChangeListener：演示两种设置 SeekBar 改变监听器的方法。

步骤 1：新建项目 Ch03_Ex21_SeekBar，修改字符串文件 strings.xml，如源代码清单 3.61 所示。

源代码清单 3.61 \Ch03_Ex21_SeekBar\res\values\strings.xml

```
1.  <?xml version="1.0" encoding="utf-8"?>
2.  <resources>   <string name="app_name">滑块控件(SeekBar)典型案例</string>
3.      <string name="strPrompt">演示滑块控件(SeekBar)功能及其监听器</string>
4.      <string name="strValue">滑块(SeekBar)当前值:</string>
5.      <string name="strSwitchImage">拖动切换图片:</string>
6.      <string name="strChangeScreenLight">改变屏幕亮度:</string>   </resources>
```

步骤 2：修改布局文件 seekbar_layout.xml，其中包含五个 TextView 控件、一个 ImageView 控件和三个 SeekBar 控件，演示 SeekBar 的基本功能和用法，如源代码清单 3.62 所示。

源代码清单 3.62 \Ch03_Ex21_SeekBar\res\layout\seekbar_layout.xml

```xml
1.  <RelativeLayout xmlns:android="http://schemas.android.com/apk/res/android"
2.      xmlns:tools="http://schemas.android.com/tools"
3.      android:layout_width="match_parent"    android:layout_height="match_parent"
4.      ... tools:context=".SeekBarActivity" >
5.      <TextView android:id="@+id/textView1" android:layout_width="wrap_content"
6.          android:layout_height="wrap_content" android:text="@string/strPrompt" />
7.      <TextView  android:id="@+id/textViewProgress"
8.          android:layout_width="wrap_content"  android:layout_height="wrap_content"
9.          android:layout_below="@id/textView1" />
10.     <!-- android:max="100" 设置 seekbar 显示的最大值 -->
11.     <SeekBar    android:id="@+id/seekBar"
12.         android:layout_width="fill_parent"   android:layout_height="wrap_content"
13.         android:layout_below="@id/textViewProgress"    android:max="100"
14.         android:progress="50"            android:secondaryProgress="75" />
15.     <TextView   android:id="@+id/textViewTracking"
16.         android:layout_width="wrap_content"  android:layout_height="wrap_content"
17.         android:layout_below="@id/seekBar" />
18.     <ImageView   android:id="@+id/imageView"  android:layout_width="80dp"
19.         android:layout_height="80dp"  android:layout_below="@id/textViewTracking"
20.         android:layout_centerHorizontal="true"  android:scaleType="fitCenter"
21.         android:src="@drawable/dujuan" />
22.     <LinearLayout    android:id="@+id/linearLayout1"
23.         android:layout_width="fill_parent"       android:layout_height=
            "wrap_content"
24.         android:layout_below="@id/imageView"  android:orientation="horizontal" >
25.         <TextView    android:id="@+id/tvImageView"
26.            android:layout_width="wrap_content" android:layout_height="wrap_content"
27.             android:textSize="20sp"          android:textColor="#00f"
28.             android:text="@string/strSwitchImage" />
29.         <SeekBar     android:id="@+id/seekbarImageView"
30.             android:layout_width="fill_parent" android:layout_height="wrap_content" />
31.     </LinearLayout>
32.     <LinearLayout    android:id="@+id/linearLayout2"
33.         android:layout_width="fill_parent"  android:layout_height="wrap_content"
34.         android:layout_below="@id/linearLayout1"   android:orientation=
            "horizontal" >
35.         <TextView    android:id="@+id/tvChangeSreen"
36.             android:layout_width="wrap_content"   android:layout_height=
                "wrap_content"
37.             android:text="@string/strChangeScreenLight"  android:textSize="20sp"
38.             android:textColor="#0f0" />
39.         <SeekBar    android:id="@+id/seekBarChangeSreen"
40.             android:layout_width="fill_parent"    android:layout_height=
                "wrap_content" />
41.     </LinearLayout>    </RelativeLayout>
```

步骤3：修改 SeekBarActivity.java 代码，设置 setOnSeekBarChangeListener 接口，实现其三个接口方法 onProgressChanged()、onStartTrackingTouch()和 onStopTrackingTouch()，如源代码清单 3.63 所示。完整源代码见 Ch03_Ex21_SeekBar.rar。

源代码清单 3.63　　\Ch03_Ex21_SeekBar\src\sziit\lihz\ch03_ex21_seekbar\SeekBarActivity.java

```java
1.  public class SeekBarActivity extends Activity {         //从 Activity 基类派生子类
2.      private TextView textViewProgress = null;           //声明 TextView 对象
3.      private TextView textViewTracking = null;           //声明 TextView 对象
4.      private SeekBar seekBar = null;                     //声明 SeekBar 对象
5.      private SeekBar seekbarImageView = null;            //声明 SeekBar 对象
6.      private SeekBar seekBarChangeSreen = null;          //声明 SeekBar 对象
7.      private String strValue = null;                     //声明字符串
8.      private ImageView imageView;                        //声明图像视图对象
9.      private boolean flag = true;                        //标记是否需要刷新
10.     private Handler hangler = new Handler();            //声明句柄对象
11.     private int[] tenflowersArray = new int[] { R.drawable.hehua,
12.         R.drawable.juhua, R.drawable.lanhua, R.drawable.meihua,
13.         R.drawable.mudan, R.drawable.shanca, R.drawable.shuixian,
14.         R.drawable.yueji, R.drawable.zhiwei, R.drawable.dujuan, };
15.     protected void onCreate(Bundle savedInstanceState) {//子类重写 onCreate 方法
16.         super.onCreate(savedInstanceState);             //基类调用 onCreate 方法
17.         setContentView(R.layout.seekbar_layout);        //加载布局资源
18.         strValue = this.getString(R.string.strValue);   //加载资源字符串
19.         textViewProgress = (TextView) findViewById(R.id.textViewProgress);
            //绑定 TextView
20.         textViewTracking=(TextView)findViewById(R.id.textViewTracking);
                                                            //绑定 TextView 对象
21.         seekBar = (SeekBar) findViewById(R.id.seekBar);//绑定 SeekBar 对象
22.         seekBar.setMax(100);                           //设置 SeekBar 的进度最大值
23.         textViewProgress.setText(strValue + seekBar.getProgress());
                                                           //设置显示文本信息
24.         seekBar.setOnSeekBarChangeListener(new OnSeekBarChangeListener()
            {//绑定监听器
25.             //实现进度变化接口：当进度条的进度改变时，会调用此方法
26.             public void onProgressChanged(SeekBar seekBar,int progress,boolean
                fromUser) {
27.                 textViewProgress.setText(strValue + progress);
28.                 seekBar.setSecondaryProgress((progress + seekBar.getMax()) / 2);
29.                 Toast.makeText(SeekBarActivity.this, (strValue + progress),
30.                     Toast.LENGTH_LONG).show();
31.                 if (fromUser) { System.out.println("用户改变滑块到:" + progress);
32.                 } else { System.out.println("系统改变滑块到:" + progress);  }}
33.             //当手动开始改变进度滑块的位置时调用此方法
34.             public void onStartTrackingTouch(SeekBar seekBar) {
35.                 System.out.println("开始拖动时:" + seekBar.getProgress());
36.                 textViewTracking.setText("滑块正在调节!");  flag = false;} //停止刷新
```

```java
37.        public void onStopTrackingTouch(SeekBar seekBar){
                                               //当手动改变滑块位置结束时调用
38.            System.out.println("停止拖动时:" + seekBar.getProgress());
39.            textViewTracking.setText("滑块已停止调节!");
40.            flag = true;                     //设置标记为需要刷新
41.            refresh();}});                   //创建时就开始自动更新该拖动条
42.    this.imageView = (ImageView) findViewById(R.id.imageView); refresh();
43.    this.seekbarImageView = (SeekBar) findViewById(R.id.seekbarImageView);
       this.seekbarImageView.setMax(tenflowersArray.length);
44.    this.seekbarImageView.setOnSeekBarChangeListener(aOSBCListener);
45.    this.seekBarChangeSreen=(SeekBar)findViewById(R.id.seekBarChangeSreen);
46.    this.seekBarChangeSreen.setMax(100);
47.    this.seekBarChangeSreen.setOnSeekBarChangeListener(aOSBCListener); }
48. private OnSeekBarChangeListener aOSBCListener = new OnSeekBarChange
    Listener() {
49.    public void onStopTrackingTouch(SeekBar seekBar) { }
50.    public void onStartTrackingTouch(SeekBar seekBar) { }
51.    public void onProgressChanged(SeekBar seekBar, int progress, boolean
       fromUser) {
52.        switch (seekBar.getId()) {
53.        case R.id.seekbarImageView:
54.            imageView.setImageResource(tenflowersArray[seekbarImageView.
               getProgress()]); break;
55.        case R.id.seekBarChangeSreen:
56.            setScreenBrightness((float) seekBarChangeSreen.getProgress() /
               100); break;
57.        default: break; }} };
58. //表示屏幕亮度的方法
59. private void setScreenBrightness(float num) { //表示亮度值
60.    WindowManager.LayoutParams layoutParams = super.getWindow().getAtt
       ributes();                                 //取得屏幕属性
61.    layoutParams.screenBrightness = num;       //设置屏幕亮度
62.    super.getWindow().setAttributes(layoutParams); //重新设置窗口的属性}
63. //该方法自动更新该拖动条
64. private void refresh() { new Thread(new Runnable() {
65.    public void run() {
66.        while (flag && seekBar.getProgress() < 100) {
                                                 //当进度不到100,就更新状态
67.            try {Thread.sleep(1000);          //暂停1秒
68.            } catch (InterruptedException e) {//处理异常
69.                e.printStackTrace();}
70.            //将一个Runnable对象添加到消息队列当中,
71.            //并且当执行到该对象时执行run()方法
72.            hangler.post(new Runnable() {
73.                public void run() {          //重新设置进度条当前的值
74.                    seekBar.setProgress(seekBar.getProgress() + 1);} });}
                   }}) .start(); }}
```

步骤4：选择【Run As】→【Android Application】命令，运行结果如图3.28所示。

（a）设计轮廓

（b）运行主界面

图 3.28 滑块控件典型案例

3.4.6 评分条控件

1. 评分条控件简介

评分条控件（RatingBar）的结构为：

```
public class RatingBar extends AbsSeekBar
java.lang.Object
    android.view.View
        android.widget.ProgressBar
            android.widget.AbsSeekBar
                android.widget.RatingBar
```

评分条控件是基于 SeekBar 和 ProgressBar 的扩展，用星形来显示等级评定；在使用默认 RatingBar 时，用户可通过触摸/拖动/按键（如遥控器）来设置评分；RatingBar 自带有两种模式：小风格 ratingBarStyleSmall 和大风格 ratingBarStyleIndicator，大的只适合做指示，不适合与用户交互。

2. 评分条控件常用属性和方法

评分条控件的常用属性和方法如表 3.35 所示。

表 3.35 评分条控件的常用属性和方法

XML 属性	Java 方法	功 能 描 述
android:numStars	setNumStars(int numStars)	显示的星形数量，必须是整数，如 10
android:rating	setRating(float rating)	默认的评分，必须是浮点类型，如 1.4
android:isIndicator	setIsIndicator(boolean)	RatingBar 是否为一个指示器
android:stepSize	setStepSize(float stepSize)	评分的步长，必须是浮点类型，如 1.5

评分条控件类的公共方法如表 3.36 所示。

表 3.36 评分条控件的公共方法

公 共 方 法	功 能 描 述
void setNumStars(int numStars)	设置显示的星形数量
int getNumStars()	返回显示的星形数量
void setRating(float rating)	设置分数（星型的数量）
float getRating()	获取当前的评分（填充的星形的数量）
void setStepSize(float stepSize)	设置当前评分条的步长

续表

公 共 方 法	功 能 描 述
float getStepSize()	获取评分条的步长
boolean isIndicator()	判断当前的评分条是否仅仅是一个指示器
void setIsIndicator(boolean isIndicator)	设置当前的评分条是否仅仅是一个指示器
void setMax(int max)	设置评分等级的范围，从 0 到 max
void setOnRatingBarChangeListener(RatingBar.OnRatingBarChangeListener(listener)	设置当评分等级发生改变时回调的监听器

3. 评分条控件典型案例

实现要点：

- RatingBar.OnRatingBarChangeListener 接口：一个回调函数，当星级进度改变时修改客户端的星级。
- 为了响应星级评分条的改变，要绑定 OnRatingBarChangeListener 监听器。
- android:numStars="5"：设置总级别（总分或星形个数）。
- android:rating="1.5"：设置当前级别（分数，星形个数）。
- android:stepSize="0.5"：设置调整步长。

步骤 1：新建项目 Ch03_Ex22_RatingBar，修改字符串文件 strings.xml，如源代码清单 3.64 所示。

源代码清单 3.64 \Ch03_Ex22_RatingBar\res\values\strings.xml

```
1. <?xml version="1.0" encoding="utf-8"?>
2. <resources>   <string name="app_name">评分条控件(RatingBar)典型案例</string>
3.     <string name="action_settings">设置</string>
4.     <string name="strPrompt">演示 RatingBar</string>
5.     <string name="strAddRatingBarStar">增加 RatingBar 的星数</string>
6.     <string name="strCurRatingBarStar">当前RatingBar 的星数:</string>   </resources>
```

步骤 2：修改布局文件 ratingbar_layout.xml，其中包含两个 TextView 控件、一个 RatingBar 控件、一个 EditText 控件和一个 Button 控件，演示 RatingBar 的基本功能和用法，如源代码清单 3.65 所示。

源代码清单 3.65 \Ch03_Ex22_RatingBar\res\layout\ratingbar_layout.xml

```
1. <RelativeLayout xmlns:android="http://schemas.android.com/apk/res/android"
2.    xmlns:tools="http://schemas.android.com/tools"
3.    android:layout_width="match_parent"    android:layout_height="match_parent"
4.    ... tools:context=".RatingBarActivity" >
5.    <TextView    android:id="@+id/textView"    android:layout_width="wrap_content"
6.        android:layout_height="wrap_content"    android:text="@string/strPrompt" />
7.    <RatingBar    android:id="@+id/ratingBar"    android:layout_width="wrap_content"
8.        android:layout_height="wrap_content"    android:layout_below="@id/textView"
9.        android:numStars="5"    android:stepSize="0.5" />
10.   <Button    android:id="@+id/button"    android:layout_width="match_parent"
11.       android:layout_height="wrap_content"    android:layout_below=
           "@id/ratingBar"
```

```
12.         android:text="@string/strAddRatingBarStar" />
13.     <LinearLayout    android:layout_width="match_parent"
14.         android:layout_height="wrap_content"    android:layout_below="@id/button"
15.         android:orientation="horizontal" >
16.         <TextView    android:layout_width="wrap_content"
17.           android:layout_height="wrap_content"    android:text="@string/
            strCurRatingBarStar" />
18.         <EditText    android:id="@+id/editText"    android:layout_width=
            "wrap_content"
19.            android:layout_height="wrap_content"
20.            android:layout_gravity="center_horizontal"    android:textSize=
            "20sp" >
21.         </EditText>    </LinearLayout>    </RelativeLayout>
```

步骤 3：修改 RatingBarActivity.java 代码，设置 setOnRatingBarChangeListener 接口，实现其接口方法 onRatingChanged()，如源代码清单 3.66 所示。

源代码清单 3.66 \Ch03_Ex22_RatingBar\src\sziit\lihz\ch03_ex22_ratingbar\RatingBarActivity.java

```
1. public class RatingBarActivity extends Activity {//从Activity基类派生子类
2.    private RatingBar ratingBar;            //声明 RatingBar 对象
3.    private Button button;                  //声明 Button 对象
4.    private EditText editText;              //声明 EditText 对象
5.    protected void onCreate(Bundle savedInstanceState) {//子类重写 onCreate 方法
6.       super.onCreate(savedInstanceState);  //基类调用 onCreate 方法
7.       setContentView(R.layout.ratingbar_layout);//设置视图布局
8.       editText = (EditText) findViewById(R.id.editText);//绑定 EditText 对象
9.       ratingBar = (RatingBar) findViewById(R.id.ratingBar);//绑定 RatingBar 对象
10.      ratingBar.setNumStars(5);            //设置 RatingBar5 颗星
11.      ratingBar.setRating(1.0f);           //设置当前的星数
12.      ratingBar.setStepSize(0.5f);         //设置 RatingBar 评分的步长
13.      ratingBar.setOnRatingBarChangeListener(new OnRatingBarChangeListener(){
                                              //设置监听器
14.   //rattingBar 当前触发的 RatingBar 控件 rating 当前的星数 fromUser 是否是用户触发
15.          public void onRatingChanged(RatingBar ratingBar, float rating,
             boolean fromUser) {
16.            editText.setText(String.valueOf(ratingBar.getRating()));}});
                                              //显示当前评分
17.      editText.setText(String.valueOf(ratingBar.getRating()));//显示当前评分
18.      button = (Button) findViewById(R.id.button);
19.      button.setOnClickListener(new OnClickListener() {
20.          public void onClick(View v) {
21.            ratingBar.setRating(ratingBar.getRating() + 0.5f);
                                              //增加 RatingBar 颗星数
22.            editText.setText(String.valueOf(ratingBar.getRating()));
                                              //显示当前评分
23.      }});  }}
```

步骤 4：选择【Run As】→【Android Application】命令，运行结果如图 3.29 所示。

（a）设计轮廓

（b）运行主界面

图 3.29　评分条控件典型案例

3.4.7　选项卡控件

1. 选项卡控件简介

选项卡控件(TabHost)是整个 Tab 的容器,包含 TabWidget 和 FrameLayout 两部分,TabWidget 是每个 Tab 的表情,该组件就是 TabHost 标签页中上部或者下部的按钮,可以单击按钮切换选项卡；FrameLayout 是 Tab 内容,代表了选项卡界面,添加一个 TabSpec 即可添加到 TabHost 中。选项卡控件(TabHost)实现方式有两种：①继承 TabActivity,从 TabActivity 中用 getTabHost() 方法获取 TabHost,然后设置标签内容；②继承 Activity 类,注意必须调用 setup(),否则会有 NullPointer 异常。两种实现方法的布局要求如表 3.37 所示。

表 3.37　TabHost 两种实现方法的布局要求

方法类型	布局	设置要求
继承 TabActivity	TabHost	必须设置 android:id 为 @android:id/tabhost
	TabWidget	必须设置 android:id 为 @android:id/tabs
	FrameLayout	必须设置 android:id 为 @android:id/tabcontent
继承 Activity 类	TabHost	可自定义 id
	TabWidget	必须设置 android:id 为 @android:id/tabs
	FrameLayout	必须设置 android:id 为 @android:id/tabcontent

2. 选项卡控件常用方法

选项卡控件常用方法如表 3.38 所示。

表 3.38　TabHost 公共方法

公共方法	功能描述
newTabSpec(String tag)	创建一个选项卡。tag 是字符串,即选项卡的唯一标识
addTab(tabSpec)	添加选项卡
getHost()	获取 TabHost。前提是在布局文件中,设置了 android 自带的 id android:id="@android:id/tabhost"
TabSpec setIndicator(View view)	设置按钮名称
TabSpec setContent(int viewId)	设置选项卡内容。可为视图组件、活动或 Fragement
void addTab(TabSpec tabSpec)	添加选项卡。传入的参数是创建选项卡 TabSpec 对象
void setCurrentTab(int index)	设置当前现实哪一个标签
setOnTabChangedListener	设置 TabHost 改变监听器

续表

公 共 方 法	功 能 描 述
void onTabChanged(String tabId)	TabHost 改变监听器接口方法
View getChildTabViewAt(int index)	获得给定位置的按钮视图

3. 选项卡控件典型案例

实现要点：

- 切换按钮下方显示：若想要将按钮放到下面，可将该组件定义在下面，但注意，FrameLayout 要设置 android:layout_widget = "1"。
- 设置 tabhost 位于底部的三种方法：①tabcontent 和 tabs 交换位置；设置 tabcontent 的属性：android:layout_weight="1"。②tabcontent 和 tabs 交换位置；将 tabs 放到一个 relativeLayout 中，然后加上属性 android:layout_alignParentBottom="true"。③将 tabcontent 和 tabs 交换位置（tabs 移动到 LinearLayout 标签以下）；在 tabcontent 中加入属性 android:layout_gravity="top"；在 tabs 中加入属性 android:layout_gravity="bottom"。

TabHost 使用步骤：

（1）定义布局：在 XML 文件中使用 TabHost 组件，并在其中定义一个 FrameLayout 选项卡内容。使用 setContentView()方法显示界面。

（2）继承 TabActivity：显示选项卡组件的 Activity 继承 TabActivity。

（3）获取组件：通过调用 getTabHost()方法，获取 TabHost 对象。

（4）创建添加选项卡：通过 TabHost 创建添加选项卡。

步骤 1：新建项目 Ch03_Ex23_TabHost，修改字符串文件 strings.xml，如源代码清单 3.67 所示。

源代码清单 3.67　\Ch03_Ex23_TabHost\res\values\strings.xml

```
1. <?xml version="1.0" encoding="utf-8"?>
2. <resources>      <string name="app_name">选项卡控件(TabHost)典型案例</string>
3.     <string name="action_settings">设置</string>
4.     <string name="strGoal">演示 TabHost 实现 Tab 切换两种方法：</string>
5.     <string name="strTabActivity">继承 TabActivity 方法</string>
6.     <string name="strActivity">继承 Activity 方法</string>
7.     <string name="strTabhostBelow">设置 tabhost 位于底部</string>   </resources>
```

步骤 2：修改布局文件 activity_main.xml，其中包含一个 TextView 控件、三个 Button 控件，演示 TabHost 的基本功能和用法。

步骤 3：修改布局文件 tabhost_demo1.xml，用来演示选项卡控件。实现方式 1：继承 TabActivity，如源代码清单 3.68 所示。

源代码清单 3.68　\Ch03_Ex23_TabHost\res\layout\tabhost_demo1.xml

```
1. <?xml version="1.0" encoding="utf-8"?>
2. <TabHost xmlns:android="http://schemas.android.com/apk/res/android"
3.     android:id="@android:id/tabhost"
4.     android:layout_width="match_parent" android:layout_height="match_parent" >
5.     <LinearLayout  android:layout_width="match_parent"
6.       android:layout_height="match_parent" android:orientation="vertical" >
7.       <TabWidget android:id="@android:id/tabs"
```

```
8.            android:layout_width="match_parent"   android:layout_height=
              "wrap_content" >
9.         </TabWidget>
10.        <FrameLayout   android:id="@android:id/tabcontent"
11.            android:layout_width="match_parent"   android:layout_height="0dp"
12.            android:layout_weight="1" >
13.            <LinearLayout      android:id="@+id/linearlayout_red"
14.                android:layout_width="match_parent"   android:layout_height=
                   "match_parent"
15.                android:background="#ff0000"    android:orientation="vertical" >
16.                <TextView   android:layout_width="wrap_content"
17.                    android:layout_height="wrap_content"
18.                    android:layout_gravity="center_horizontal"  android:text="
                       红色页面"
19.                    android:textColor="#ffffff"       android:textSize="40sp" >
20.                </TextView>       </LinearLayout>
21.            <LinearLayout
22.                android:id="@+id/linearlayout_green"   android:layout_width=
                   "match_parent"
23.                android:layout_height="match_parent"   android:background=
                   "#00ff00"
24.                android:orientation="vertical" >
25.                <TextView
26.                    android:layout_width="wrap_content"
27.                    android:layout_height="wrap_content"
28.                    android:layout_gravity="center_horizontal" android:text="
                       绿色页面"
29.                    android:textColor="#ffffff"       android:textSize="40sp" >
30.                </TextView>      </LinearLayout>
31.            <LinearLayout
32.                android:id="@+id/linearlayout_blue"    android:layout_width=
                   "match_parent"
33.                android:layout_height="match_parent"   android:background=
                   "#0000ff"
34.                android:orientation="vertical" >
35.                <TextView
36.                  android:layout_width="wrap_content"   android:layout_height=
                     "wrap_content"
37.                    android:layout_gravity="center_horizontal"   android:text=
                       "蓝色页面"
38.                    android:textColor="#ffffff"   android:textSize="40sp" >
39.            </TextView>  </LinearLayout>  </FrameLayout>  </LinearLayout>
                </TabHost>
```

步骤 4：修改 TabActivityDemo1.java 代码，从 TabActivity 基类派生子类，重写 onCreate 方法，调用 getTabHost()方法等实现 TabHost 功能，如源代码清单 3.69 所示。

源代码清单 3.69　\Ch03_Ex23_TabHost\src\sziit\lihz\ch03_ex23_tabhost\TabActivityDemo1.java

```
1. public class TabActivityDemo1 extends TabActivity {    //从TabActivity基类派生子类
2.     private TabHost tabhost;                            //声明私有 TabHost 对象
3.     protected void onCreate(Bundle savedInstanceState) {//子类重写 onCreate 方法
```

```
4.    super.onCreate(savedInstanceState);        //基类调用 onCreate 方法
5.    setContentView(R.layout.tabhost_demo1); //设置 TabActivity 界面布局
6.    tabhost = getTabHost();       //从 TabActivity 上面获取放置 Tab 的 TabHost
7.    TabHost.TabSpec tab1 = tabhost.newTabSpec("one");//构造新的 Tab 标签
8.    tab1.setIndicator("红色");                //设置指示器或按钮名称
9.    tab1.setContent(R.id.linearlayout_red); //设置内容
10.   TabHost.TabSpec tab2 = tabhost.newTabSpec("two");//构造新的 Tab 标签
11.   tab2.setIndicator("绿色");                //设置指示器或按钮名称
12.   tab2.setContent(R.id.linearlayout_green);//设置内容
13.   TabHost.TabSpec tab3 = tabhost.newTabSpec("three");//构造新的 Tab 标签
14.   tab3.setIndicator("蓝色");                //设置指示器或按钮名称
15.   tab3.setContent(R.id.linearlayout_blue);//设置内容
16.   tabhost.addTab(tab1);                     //将 Tabs 添加到 TabHost
17.   tabhost.addTab(tab2);                     //将 Tabs 添加到 TabHost
18.   tabhost.addTab(tab3);}}                   //将 Tabs 添加到 TabHost
```

步骤 5：修改布局文件 tabhost_demo2.xml，用来演示选项卡控件。实现方式 2：继承 Activity，源代码见 Ch03_Ex23_TabHost.zip。

步骤 6：修改 ActivityDemo2.java 代码，从 Activity 基类派生子类，重写 onCreate 方法，从资源文件得到 TabHost 对象实例，调用 TabHost.setup()等实现 TabHost 功能，如源代码清单 3.70 所示。

源代码清单 3.70   \Ch03_Ex23_TabHost\src\sziit\lihz\ch03_ex23_tabhost\ActivityDemo2.java

```
1. public class ActivityDemo2 extends Activity {     //从活动基类派生子类
2.    private TabHost tabhost;                //声明私有 TabHost 对象
3.    private TabWidget mTabWidget;           //声明私有 TabWidget 对象
4.    protected void onCreate(Bundle savedInstanceState) {//子类重写 onCreate 方法
5.       super.onCreate(savedInstanceState);  //基类调用 onCreate 方法
6.       setContentView(R.layout.tabhost_demo2);    //设置活动界面布局
7.       tabhost = (TabHost) findViewById(R.id.tabHost);//得到 TabHost 对象实例
8.       tabhost.setup();                     //调用 TabHost.setup()
9.       tabhost.setBackgroundColor(Color.argb(152,24,80,152));
                                              //设置 TabHost 的背景颜色
10.      //创建 Tab 标签，设置指示器或按钮名称和内容
11.      tabhost.addTab(tabhost.newTabSpec("one").setIndicator("虎刺梅")
12.          .setContent(R.id.euphorbia_milii_layout));
13.      tabhost.addTab(tabhost.newTabSpec("two").setIndicator("观音莲")
14.          .setContent(R.id.guanyin_lotus_layout));
15.      tabhost.addTab(tabhost.newTabSpec("two").setIndicator("桃美人")
16.          .setContent(R.id.peach_beauty_layout));
17.      tabhost.setCurrentTab(1);            //设置当前现实哪一个标签
18.      tabhost.setOnTabChangedListener(new OnTabChangeListener(){
                                              //设置监听器
19.         public void onTabChanged(String tabId) {//重写接口方法
20.            Toast.makeText(ActivityDemo2.this, tabId, Toast.LENGTH_LONG).
               show();  }});
21.      mTabWidget = tabhost.getTabWidget(); //得到 TabWidget 对象
22.      View v = mTabWidget.getChildTabViewAt(1);
23.      v.setOnClickListener(new OnClickListener(){
24.         public void onClick(View v) {
```

25.   Toast.makeText(ActivityDemo2.this,"观音莲:"+String.valueOf(v.
             getTag(), Toast.LENGTH_LONG).show();}});}}

步骤 7：选择【Run As】→【Android Application】命令，运行结果如图 3.30 所示。

图 3.30　选项卡控件典型案例

3.4.8　画廊控件

1. 画廊控件简介

画廊（Gallery）控件的结构为：

```
public class Gallery extends AbsSpinner
    implements GestureDetector.OnGestureListener
        java.lang.Object
            android.view.View
                android.view.ViewGroup
                    android.widget.AdapterView<T extends android.widget.Adapter>
                        android.widget.AbsSpinner
                            android.widget.Gallery
```

Gallery 控件主要用于横向显示图像列表。

2. 画廊控件的常用属性和方法

画廊控件的常用属性和方法如表 3.39、表 3.40 所示。

表 3.39　画廊控件的常用属性

属 性 名	功 能 描 述
android:animationDuration	设置布局变化时动画的转换所需时间
android:gravity	设置在对象的 x 和 y 轴上如何放置内容
android:spacing	设置图片之间的间隔
android:unselectedAlpha	设置未选中条目的透明度
android:entries	直接在 XML 布局文件中绑定数据源

表 3.40　画廊控件的公共方法

公 共 方 法	功 能 描 述
boolean dispatchKeyEvent(KeyEvent event)	在焦点路径上分发按钮事件到下一个视图
void dispatchSetSelected(boolean selected)	分发 setSelected 给视图的子类

续表

公共方法	功能描述
ViewGroup.LayoutParams generateLayoutParams(AttributeSet attrs)	返回一个新的已设置属性集合的布局参数
boolean onDown(MotionEvent e)	当轻击和按下手势事件发生时通知该方法
boolean onFling(MotionEvent e1, MotionEvent e2, float velocityX, float velocityY)	当初始化的按下动作事件和松开动作事件匹配时通知Fling事件
boolean onKeyDown(int keyCode, KeyEvent event)	处理按键按下事件
boolean onKeyUp(int keyCode, KeyEvent event)	处理按键释放事件
void onLongPress(MotionEvent e)	处理长按事件
boolean onScroll(MotionEvent e1, MotionEvent e2, float distanceX, float distanceY)	当初始按下动作事件和当前移动动作事件导致滚动时通知本方法
void onShowPress(MotionEvent e)	用户已经执行按下动作还没有执行移动或者弹起动作通知本方法
boolean onSingleTapUp(MotionEvent e)	在轻击动作和弹起动作事件触发时通知本方法
boolean onTouchEvent(MotionEvent event)	实现该方法来处理触摸屏动作事件
void setAnimationDuration(int animationDurationMillis)	设置当子视图改变位置时动画转换时间
void setCallbackDuringFling(boolean shouldCallback)	设置抛滑的过程中是否回调
void setGravity(int gravity)	设置子视图的对齐方式
void setSpacing(int spacing)	设置 Gallery 中项的间距
void setUnselectedAlpha(float unselectedAlpha)	设置 Gallery 中未选中项的透明度值
boolean showContextMenu()	显示该视图上下文菜单

要实现 Gallery 画廊控件功能，需要一个容器来存放 Gallery 显示的图片，用户可使用一个继承自 BaseAdapter 类的派生类 ImageAdapter 来装这些图片，并实现以下四个抽象方法：

（1）public int getCount()：获取 ImageAdapter 适配器中图片个数。

（2）public Object getItem(int position)：获取图片在 ImageAdapter 适配器中的位置。

（3）public long getItemId(int position)：获取图片在 ImageAdapter 适配器中的位置。

（4）public View getView(int position, View convertView, ViewGroup parent)：获取 ImageAdapter 适配器中指定位置的视图对象。

3. 画廊控件典型案例

实现要点：

- 画廊控件。
- BaseAdapter。
- ImageView。
- AdapterView.OnItemClickListener。
- void setAdapter(SpinnerAdapter adapter)：设置适配器。
- onItemClick(AdapterView, View, int, long)回调方法的参数 AdapterView 表示单击事件发生的地方，参数 View 负责接收单击事件，参数 int 表示单击视图从零开始的位置，参数 long 表示单击项的行 ID。

步骤1：利用 Android 项目向导，创建名为 Ch03_Ex24_Gallery 的应用程序。

步骤 2：将欲浏览的图片复制到 res/drawable/目录。例如，图片 photo_0.JPG～photo_7.JPG。或 photo_0.GIF～photo_7.GIF。

步骤 3：打开 res/values/strings.xml 文件，增加字符串定义，如源代码清单 3.72 所示。

源代码清单 3.71   \Ch03_Ex24_Gallery\res\values\strings.xml

```
1. <?xml version="1.0" encoding="utf-8"?>
2. <resources>     <string name="app_name">画廊控件(Gallery)典型案例</string>
3.     <string name="description">摄影爱好者画廊</string>   </resources>
```

步骤 4：打开 res/layout/ gallery_layout.xml 文件，插入源代码清单 3.72 所示的代码。

源代码清单 3.72   \Ch03_Ex24_Gallery\res\layout\gallery_layout.xml

```
1. <?xml version="1.0" encoding="utf-8"?>
2. <LinearLayout xmlns:android="http://schemas.android.com/apk/res/android"
3.     android:layout_width="fill_parent"    android:layout_height="fill_parent"
4.     android:orientation="vertical" >
5.  <TextView  android:layout_width="fill_parent"
6.     android:layout_height="wrap_content"   android:background="#0f0"
7.     android:gravity="center_horizontal"    android:text="@string/
       description"
8.     android:textSize="20sp" />
9.  <Gallery android:id="@+id/picGallery"  android:layout_width="fill_parent"
10.    android:layout_height="wrap_content" />   </LinearLayout>
```

步骤 5：创建命名为 attrs.xml 的 XML 文件，并保存在 res/values/目录下，如源代码清单 3.73 所示。

源代码清单 3.73   \Ch03_Ex24_Gallery\res\values\attrs.xml

```
1. <?xml version="1.0" encoding="utf-8"?>
2. <resources><declare-styleable name="PhotoGallery"> <attr name="android:
   galleryItemBackground" />
3.    </declare-styleable> </resources>
```

步骤 6：打开 GalleryActivity.java 文件，实现 setOnItemClickListener 接口，如源代码清单 3.74 所示。

源代码清单 3.74   \Ch03_Ex24_Gallery\src\sziit\lihz\ch03_ex24_gallery\GalleryActivity.java

```
1. public class GalleryActivity extends Activity {  //从Activity基类派生子类
2.    protected void onCreate(Bundle savedInstanceState) {//子类重写onCreate方法
3.     super.onCreate(savedInstanceState);       //基类调用onCreate方法
4.     setContentView(R.layout.gallery_layout);  //设置活动布局界面
5.     Gallery g = (Gallery) findViewById(R.id.picGallery);//引用Gallery对象
6.     g.setAdapter(new ImageAdapter(this));     //设置适配器
7.     g.setOnItemClickListener(new OnItemClickListener() {//设置控件单项单击监听器
8.     public void onItemClick(AdapterView parent, View v, int position,long
       id) {                                    //实现接口方法
9.      Toast.makeText(GalleryActivity.this, "" + position,Toast.LENGTH_
        SHORT).show();}}); }
10. public class ImageAdapter extends BaseAdapter {
                                    //ImageAdapter类，继承自BaseAdapter
11.    int mGalleryItemBackground;             //背景图片资源ID
12.    private Context mContext;               //上下文对象
```

```
13.    private Integer[] mImageIds = { R.drawable.photo_0, R.drawable.photo_1,
14.            R.drawable.photo_2, R.drawable.photo_3, R.drawable.photo_4,
15.            R.drawable.photo_5, R.drawable.photo_6, R.drawable.photo_7 };
16.    public ImageAdapter(Context c) {                    //构造方法
17.        mContext = c;
18.        TypedArray a = obtainStyledAttributes(R.styleable.PhotoGallery);
19.        mGalleryItemBackground = a.getResourceId(
20.            R.styleable.PhotoGallery_android_galleryItemBackground, 0);
        a.recycle();}
21.    public int getCount() {return mImageIds.length;}  //获取图片的个数
22.    public Object getItem(int position) {return position;}
                                           //获取图片在库中的位置
23.    public long getItemId(int position) {return position;}
                                           //获取图片在库中的位置
24.    //获取适配器中指定位置的视图对象
25.    public View getView(int position, View convertView, ViewGroup parent) {
26.        ImageView i = new ImageView(mContext);
27.        i.setImageResource(mImageIds[position]);
28.        i.setLayoutParams(new Gallery.LayoutParams(150, 100));
29.        i.setScaleType(ImageView.ScaleType.FIT_XY);
30.        i.setHorizontalScrollBarEnabled(true);
31.        i.setBackgroundResource(mGalleryItemBackground);
32.        return i;    }}}
```

步骤7：选择【Run As】→【Android Application】命令，运行结果如图3.31所示。

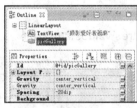
（a）设计轮廓

（b）运行主界面

图3.31 画廊控件典型案例

本章小结

本章首先简要介绍了Android视图类（View）和视图组类（ViewGroup）的基本概念及其公用属性和方法；其次，重点讲解了Android常用控件（TextView、Button、ImageButton、ToggleButton、RadioButton、CheckBox、Switch和Chronometer等）和Android高级控件（AutoCompleteTextView、Spinner、ScrollView、ListView、GridView、ProgressBar和Gallery等）等XML属性设置方法和Java编程技巧；最后，通过典型案例帮助用户掌握这些Android常用控件和高级控件实现用户界面方法和技巧。

强化练习

1. 填空题

（1）在Android编程过程中，可用（   ）方法根据控件的ID号获取对控件的引用。

（2）当 android:indeterminate 取值为（　　　　）时，开启了进度条的"不确定模式"。

（3）Adapter 配置好以后，需要用（　　　　）函数将 ListView 和 Adapter 绑定。

（4）Android（　　　　）类为所有可视化控件的基类，主要提供控件绘制和事件处理的方法。

（5）Android（　　　　）是视图类 View 的子类，也可充当其他控件的容器。

（6）View 的属性 android:id 定义了视图控件的唯一（　　　　）。

（7）TextView 的 android:textColor 属性用来设置 TextView 的（　　　　）。

（8）TextView 的（　　　　）属性用来将指定格式文本转化成可单击的超链接显示。

（9）EditText 的 android:password 可用来设置文本框中的内容是否显示为（　　　　）。

（10）EditText 的（　　　　）会影响 EditText 控件输入值时启动的虚拟键盘风格。

（11）ImageButton 的（　　　　）属性用来设置 ImageButton 上显示的图片。

（12）按钮的 void setOnClickListener(OnClickListener l)方法用于设置按钮被单击的事件（　　　　）。

（13）状态开关按钮（ToggleButton）和开关（Switch）均派生于 Button，它们的本质也是（　　　　）。

（14）android:checked 属性表示按钮 ToggleButton 或开关 Switch 是否被（　　　　）。

（15）ImageView 控件的（　　　　）属性可用来设置 ImageView 要显示的图片。

（16）ImageView 控件的 void setImageResource(int resId)方法设置 ImageView 将要显示的（　　　　）。

（17）Handler.sendMessage(Message msg)用来（发送消息），handleMessage(Message msg)用来（　　　　）。

（18）System.currentTimeMillis()用来获得系统（　　　　）。

（19）Android Chronometer 实现了一个简单的（　　　　），继承自 TextView。

（20）Android 的 Chronometer 类有三个重要的方法：start、stop 和 setBase，其中 start 表示（　　　　）；stop 表示停止计时；setBase 表示重新计时。

（21）Android 的下拉列表（Spinner）控件有两种显示形式，一种是下拉菜单，一种是弹出框，由（　　　　）属性决定。

（22）void setAdapter(SpinnerAdapter adapter)方法用来设置控件所使用的（　　　　）。

（23）（　　　　）是一种连接数据和控件的桥梁，通过其可通知控件显示的数据来源和显示方式。

（24）适配器 ArrayAdapter<T>用来绑定一个（　　　　），支持泛型操作。

（25）Android 的 Spinner 控件在 xml 布局文件中绑定数据源 coutines 的方法为：（　　　　）= "@array/coutines"。

（26）getResources().getStringArray(R.array.tour_cities)实现从资源文件加载（　　　　）功能。

（27）android.R.layout.simple_spinner_item 以"android."开头的，代表（　　　　）自带布局中的控件，本质上是 Android 系统自带的用于显示选项的 TextView 控件 ID。

（28）在 XML 中定义字符串数组的标签名为：（　　　　）。

（29）Android 的 GridView 控件的 android:numColumns 用来设置控件（　　　　）。

（30）进度条（ProgressBar）的（　　　　）属性用来设置进度条的范围。

（31）Android 的 ListView 继承于（　　　　）。
（32）requestWindowFeature(Window.FEATURE_NO_TITLE)的功能为隐藏（　　　　）。

2. 单选题

（1）在 Android 中常用来显示程序执行进度的控件为（　　　）。
　　A. 文本视图（TextView）　　　　　　B. 编辑框（EditText）
　　C. 按钮（Button）　　　　　　　　　D. 进度条（ProgressBar）
（2）获得进度条（ProgressBar）控件当前进度值的方法为（　　　）。
　　A. getProgress()　　　　　　　　　B. getSecondaryProgress()
　　C. setMax(int max)　　　　　　　　D. setProgress(int progress)
（3）在 Android 中用于显示图片的控件为（　　　）。
　　A. 按钮　　　　B. 文本视图　　　　C. ImageView　　　　D. EditText
（4）在 Android 中能简单有效地处理单击事件响应 OnClickListener 的控件为（　　　）。
　　A. 按钮　　　　　　　　　　　　　B. 文本视图
　　C. ImageView　　　　　　　　　　D. EditText
（5）若要捕获某类视图控件的事件，需要为该控制设置（　　　）。
　　A. 方法　　　　B. 监听器　　　　　C. 回调函数　　　　D. 属性
（6）图像视图控件中使用什么属性引用图片资源（　　　）。
　　A. android:src　　B. android:text　　C. android:id　　　D. android:img
（7）控制虚拟键盘输入类型的属性为（　　　）。
　　A. android:text　　B. android:src　　C. android: inputType　　D. android:id
（8）在 Android 中，若要向工程中导入图片资源，应将图片放在什么目录（　　　）。
　　A. res\picture　　B. res\string　　C. res\icon　　　　D. res\drawable
（9）在 Android 中，若要向工程添加字符串资源，应将其添加到（　　　）文件。
　　A. dimens.xml　　B. styles.xml　　C. strings.xml　　　D. value.xml
（10）ListView 是常用的（　　　）类型控件。
　　A. 按钮　　　　B. 图像视图　　　　C. 列表　　　　　　D. 下拉列表
（11）ListView 与数组或 List 集合的多个值进行数据绑定时使用（　　　）。
　　A. ArrayAdapter　　　　　　　　　B. BaseAdapter
　　C. SimpleAdapter　　　　　　　　　D. SimpleCursorAdapter
（12）在 Android 列表视图中以下表示系统自定义的，只显示一行文字的布局文件是（　　　）。
　　A. android.R.layout.simple_list_item_0　　B. android.R.layout.simple_list_item_1
　　C. android.layout.simple_list_item_0　　　D. android.layout.simple_list_item_1
（13）Android 中包含多种基本 UI 控件和高级 UI 控件，这些 UI 控件均派生于（　　　）类。
　　A. 视图组　　　B. 控件　　　　　　C. 文本视图　　　　D. 视图

3. 问答题

（1）什么是 Android 视图类？
（2）什么是 Android 视图组类？

(3)简述 Android 的句柄(Handler)机制。

(4)简述将进度条与句柄结合在一起进行使用的方法。

4. 编程题

(1)编程实现图 3.32 所示的发送短消息功能,其中包括两个 TextView、两个 EditText 和一个 Button 控件。主要核心知识点包括:输入手机号码的 EditText 应设置 android:numeric="integer",只允许输入整数;获得 SmsManager 实例,SmsManager sms = SmsManager.getDefault();拆分长短信,ArrayList<String> list = sms.divideMessage(message);发送短信,sms.sendTextMessage(telNum, null, text, null, null);设置按钮监听器,void setOnClickListener(OnClickListener l);在 AndroidManifest.xml 中添加发送短信权限:<uses-permission android:name="android.permission.SEND_SMS"/>。

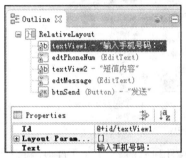
(a)设计轮廓

(b)运行主界面

图 3.32 开发短信发送器

(2)在本章画廊控件典型案例的基础上,在布局文件中增加 ImageSwitcher 控件实现图 3.33 所示的图片浏览器效果。主要核心知识点包括:(ImageSwitcher) findViewById(R.id.switcher),查找布局文件中的 ImageSwitcher 对象;void setInAnimation(Animation inAnimation),设置图片进入屏幕动画;void setOutAnimation(Animation outAnimation),设置图片退出屏幕动画;mGallery.setAdapter(newImageAdapter(this)),设置适配器;mGallery.setOnItemSelectedListener(this),设置某项被选择的监听器;mImageSwitcher.setImageResource(mImageIds[position]),显示图片。

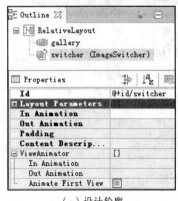
(a)设计轮廓

(b)运行主界面

图 3.33 结合 Gallery 和 ImageSwitcher 的图片浏览器

（3）利用 Android 的 TextView、EditText 和 Button 控件实现三种不同方式设置 TextView 字体颜色的方法，其显示内容根据 EditText 输入的内容变化，设置字体大小和字形功能，如图 3.34 所示。主要核心知识点包括：void setTextColor(ColorStateList colors)，设置字体颜色；void setTypeface(Typeface tf)，设置字形；void setTextSize(float size)，设置字体大小。

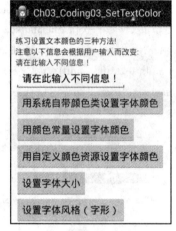

（a）设计轮廓　　　　　　　　　　　　（b）运行主界面

图 3.34　设置字体内容、颜色、大小和字形

（4）已知每月月供额=〔贷款本金×月利率×(1+月利率)^{还款月数}〕÷〔(1+月利率)^{还款月数}−1〕计算公式，请设计一个如图 3.35 所示的贷款计算器，根据输入的贷款额、还款年数和年利率，计算每月月供额。主要核心知识点包括：void setOnClickListener(OnClickListener l)，设置单击监听器，处理单击事件响应；double parseDouble(String string)，将字符串转换成浮点数；String valueOf(double value)，将浮点数转换成字符串；Editable getText()，获得输入信息；android:inputType="numberDecimal"，设置输入类型为浮点数；android:inputType="number"，设置输入类型为整数。

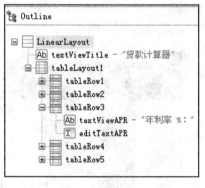
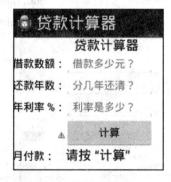

（a）设计轮廓　　　　　　　　　　　　（b）界面布局

图 3.35　贷款计算器

第 4 章 Android 应用编程基本项目实训

【知识目标】
- 了解实现动画功能的通用工具类 AnimationUtils；
- 了解 HTTP 通信编程相关知识；
- 了解 JSON 基本知识。

【能力目标】
- 掌握如何编写 Android 动画资源文件；
- 掌握如何在程序中使用 Android 动画资源文件；
- 掌握 setAnimationListener 的方法及其接口实现技巧；
- 掌握如何在程序中使用 Drawable 资源文件；
- 掌握 Shared Preferences 数据存储方法；
- 掌握如何在程序中使用 GridView 组件。

【重点、难点】
- 掌握淡入淡出/左进右出的程序过渡动画实现方法；
- 掌握九宫格菜单实现及其应用技巧；
- 掌握多线程编程方法及其应用技巧。

4.1 利用动画资源文件实现程序过渡动画

4.1.1 项目实训目标

- 掌握如何编写 Android 动画资源文件。
- 掌握如何在程序中使用 Android 动画资源文件。

4.1.2 知识准备

本节主要介绍如何定义 AnimationDrawable 资源，以及在程序中使用 Android 动画资源文件为程序添加动画效果。

1. Android 动画资源文件知识详解

Android 的动画由四种类型组成，分别是：
（1）alpha：设置透明度的改变。
（2）scale：设置图片进行缩放。
（3）translate：设置图片进行位置变换。

（4）rotate：设置图片进行旋转。

下面先通过几个动画资源文件来了解动画资源文件的结构，示意代码如源代码清单 4.1～4.4 所示。

源代码清单 4.1　alpha.xml

```
1. <?xml version="1.0" encoding="UTF-8"?>
2. <alpha xmlns:android="http://schemas.android.com/apk/res/android"
3.     android:duration="2000"    android:fromAlpha="0.3"
4.     android:toAlpha="1.0" />
```

源代码清单 4.2　scale.xml

```
1. <?xml version="1.0" encoding="UTF-8"?>
2. <scale xmlns:android="http://schemas.android.com/apk/res/android"
3.     android:fromXScale="0.5"    android:toXScale="1.5"
4.     android:pivotX="40%"        android:repeatCount="2"/>
```

源代码清单 4.3　translate.xml

```
1. <?xml version="1.0" encoding="utf-8"?>
2. <set xmlns:android="http://schemas.android.com/apk/res/android">
3.     <translate   android:fromXDelta="100%"   android:toXDelta="0"
4.     android:duration="300"    android:fillBefore="true" />    </set>
```

源代码清单 4.4　rotate.xml

```
1. <?xml version="1.0" encoding="utf-8"?>
2. <rotate xmlns:android="http://schemas.android.com/apk/res/android"
3.     android:fromDegrees="-45"    android:toDegrees="30"
4.     android:pivotX="50%"         android:pivotY="50%"/>
```

通过上面的文件可以大概了解动画资源文件的结构及编写方式，接下来通过两个表格详细了解四种类型动画效果及其各自的属性，如表 4.1 所示。

表 4.1　Android 四种类型动画效果及其属性

动画类型	说明	属性[类型]	作用
alpha	渐变透明度动画效果	fromAlpha[float]	动画起始时的透明度（0.0 表示完全透明，1.0 表示完全不透明）
		toAlpht[float]	动画结束时的透明度
scale	渐变尺寸伸缩动画效果	fromXScale[float] fromYScale[float]	动画起始时 X、Y 坐标上的伸缩尺寸（0.0 表示收缩到没有，1.0 表示正常无伸缩，值小于 1.0 表示收缩，大于 1.0 表示放大）
		toXScale[float] toYScale[float]	动画结束时 X、Y 坐标上的伸缩尺寸
		pivotX[float] pivotY[float]	动画相对于对象的 X 或 Y 方向坐标上的中点位置（取值 0%～100%，50% 为对象的 X 或 Y 方向坐标上的中点位置）
translate	画面转换位置移动动画效果	fromXDelta[float] toXDelta[float]	动画起始、结束时 X 坐标
		fromYDelta[float] toYDelta[float]	动画起始、结束时 Y 坐标

续表

动画类型	说　明	属性[类型]	作　用
rotate	画面转移旋转动画效果	fromDegrees[float]	动画起始时对象的角度（角度为负数表示逆时针旋转，为正数表示顺时针旋转，from 负数 to 正数：顺时针；from 负数 to 负数：逆时针；form 正数 to 正数：顺时针；from 正数 to 负数：逆时针）
		toDegrees[float]	动画结束时物件旋转的角度（可以大于360°）
		pivotX[float] pivotY[float]	为动画相对于物件 X、Y 坐标的开始位置

表 4.2 是以上四种 Android 动画共有的属性。

表 4.2　四种 Android 动画共有的属性

属性[类型]	作　用	说　明
Duration[long]	动画持续时间	时间以毫秒为单位
fillAfter[boolean]	当设置为 true 时，该动画转化在动画结束后被应用	
fillBefore[boolean]	当设置为 true 时，该动画转化在动画开始前被应用	
Interpolator	指定一个动画的插入器	常见的有：accelerate_decelerate_interpolator 加速－减速动画插入器；accelerate_interpolator 加速动画插入器；decelerate_interpolator 减速动画插入器，其他属于特定的动画效果
repeatCount[int]	动画的重复次数	
repeatMode[String]	定义重复的行为	1：重新开始；2：plays backward
startOffset[long]	动画之间的时间间隔	
zAdjustment[int]	定义动画的 Z Order 的改变	0：保持 Z Order 不变；1：保持在最上层；-1：保持在最下层

2. 如何编写 Android 动画资源文件

动画资源文件存放在项目的 res/anim/filename.xml 文件中，其中 filename 作为资源文件的 id 被程序引用。在实现一些较为复杂的动画效果时，需将根标签定义成<set>，即动画集。下面通过在 Eclipse 上新建一个动画资源文件进行详细说明。

（1）在项目的 res 目录下面新建一个文件夹 anim：将鼠标指针移动到 res 目录并右击，选择【New】→【Folder】命令（见图 4.1），在弹出的对话框的【Folder name】文本框中输入 anim，如图 4.2 所示。

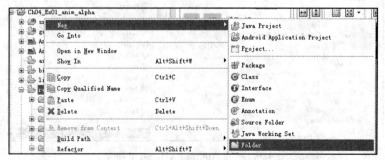

图 4.1　新建文件夹命令

图 4.2 新建文件夹对话框

（2）在 anim 文件夹下新建 xml 文件：将鼠标指针移动到文件夹【anim】上并右击，选择【New】→【Android XML File】命令（见图 4.3），在弹出的新建 Android XML 文件对话框中选择动画类型，在【File】文本框中输入自定义文件名称，单击【Finish】按钮完成创建，如图 4.4 所示。

（3）根据程序需要实现的动画效果编写动画资源文件。

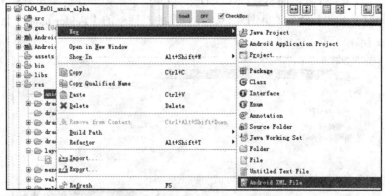

图 4.3 新建 Android XML 菜单命令

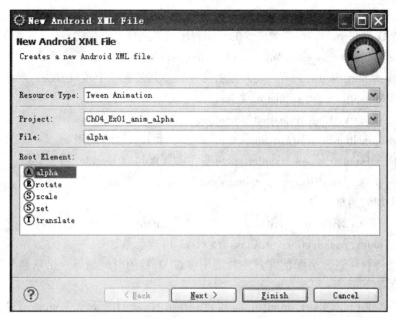

图 4.4 新建 Android XML 对话框

3. 如何在程序中使用动画资源文件

程序中使用动画资源文件有以下两种方式：

（1）在 Java 代码中引用：R.anim.filename，如下所示。

```
Animation animation=AnimationUtils.loadAnimation(this, R.anim.alpha);
```

（2）在 XML 文件中引用：@[package:]anim/filename，如源代码清单 4.5 所示。

源代码清单 4.5

```
<?xml version="1.0" encoding="utf-8"?>
<animation-list xmlns:android="http://schemas.android.com/apk/res/android"
  android:oneshot="true" >
  <item android:drawable="@[package:]anim/alpha" android:duration="2000"/>
</animation-list>
```

4.1.3 项目实践

1. 淡入淡出的程序过渡动画

（1）项目简介

为了避免在打开Android应用程序时出现程序卡顿或者界面卡死的现象，通常会为程序添加一个过渡动画（或叫引导页），用于展现一些与程序相关的介绍信息或者广告信息，同时在后台完成耗时操作，以提高程序的流畅度，改善用户体验。

（2）实现步骤

实现要点：

- class AnimationUtils：实现动画功能的通用工具类。
- Animation loadAnimation(Context context, int id)：从资源文件加载动画；参数 context 为访问资源文件的应用程序上下文环境；参数 id 为动画资源 ID。

- View inflate(Context context, int resource, ViewGroup root)：根据 XML 资源文件构造视图对象；参数 context 为访问资源文件的应用程序上下文环境；参数 resource 为布局资源 ID。
- void setAnimationListener(AnimationListener listener)：设置动画监听器。
- interface AnimationListener：动画监听器接口；其包含三个接口方法。
  - void onAnimationStart(Animation animation)：动画开始时调用；
  - void onAnimationEnd(Animation animation)：动画结束时调用；
  - void onAnimationRepeat(Animation animation)：动画重复时调用。
- Handler mHandler = new Handler()：声明并实例化句柄 Handler 对象。
- boolean postDelayed(Runnable r, long delayMillis)：延迟 delayMillis 毫秒后，将 Runnable 对象添加到消息对象。
- void startActivity(Intent intent)：启动意图对象 intent 所指定的活动。
- Intent(Context packageContext, Class<?> cls)：构造意图对象。

该动画是使程序过渡页面的背景图片透明度在 3 秒的时间内有一个从暗到明的变化过程，具体步骤如下：

步骤 1：创建 Ch04_Ex01_anim_alpha 项目，在项目 res 目录下新建 anim 目录，定义动画资源文件 alpha.xml，如源代码清单 4.6 所示。

源代码清单 4.6　\Ch04_Ex01_anim_alpha\res\anim\alpha.xml

```
1. <?xml version="1.0" encoding="UTF-8"?>
2. <alpha xmlns:android="http://schemas.android.com/apk/res/android"
3.     android:duration="3000"  android:fromAlpha="0.3" android:toAlpha="1.0" />
```

步骤 2：在程序界面 activity_main.xml 添加背景图片，如源代码清单 4.7 所示。

源代码清单 4.7　\Ch04_Ex01_anim_alpha\res\layout\activity_main.xml

```
1. <?xml version="1.0" encoding="utf-8"?>
2. <LinearLayout xmlns:android="http://schemas.android.com/apk/res/android"
3.     android:layout_width="fill_parent"       android:layout_height="fill_parent"
4.     android:layout_gravity="center_vertical"  android:background="@drawable/splash_bg"
5.     android:orientation="vertical"  android:paddingBottom="7.0dip"
6.     android:paddingLeft="7.0dip"    android:paddingRight="7.0dip"
7.     android:paddingTop="7.0dip" >  </LinearLayout>
```

步骤 3：实现 MainActivity.java 代码，设置过渡页面无标题、隐藏状态栏全屏显示，调用自定义的动画文件 alpha.xml，如源代码清单 4.8 所示。

源代码清单 4.8　\Ch04_Ex01_anim_alpha\src\sziit\lihz\ch04_ex01_anim_alpha\MainActivity.java

```
1. public class MainActivity extends Activity {//从Activity基类派生子类
2.     private Handler mHandler = new Handler();//声明并实例化句柄Handler对象
3.     protected final int SPLASH_TIME = 500;   //声明并初始化闪屏时间常量
4.     public void onCreate(Bundle savedInstanceState) {//子类重写onCreate方法
5.         super.onCreate(savedInstanceState);    //基类调用onCreate方法
6.         requestWindowFeature(Window.FEATURE_NO_TITLE); //设置无标题
7.         getWindow().addFlags(WindowManager.LayoutParams.FLAG_FULLSCREEN);
           //全屏显示
8.         View view = View.inflate(this, R.layout.activity_main, null);
```

```
                                                   //根据布局文件构造视图
9.      setContentView(view);                      //设置活动显示视图
10.     Animation animation=AnimationUtils.loadAnimation(this, R.anim.alph
        a);                                        //加载动画效果
11.     animation.setAnimationListener(new AnimationListener() {
                                                   //设置动画监听器
12.        public void onAnimationStart(Animation arg0) {}//动画开始接口
13.        public void onAnimationRepeat(Animation arg0) { }//动画重复接口
14.        public void onAnimationEnd(Animation arg0) {    //动画结束接口
15.           mHandler.postDelayed(new Runnable() {//创建被执行的 Runnable 对象
16.              public void run() {               //启动线程执行体
17.                 startActivity(new Intent( getApplicationContext(),
                    SecondActivity.class));}
18.           }, SPLASH_TIME); }});
19.     view.startAnimation(animation);}}          //启动动画
```

步骤 4：在页面跳转后的 SecondActivity.java 中，为了演示效果，程序只是新建了一个 TextView 对象提示这是 SecondActivity，没有界面布局，如源代码清单 4.9 所示。

源代码清单 4.9   \Ch04_Ex01_anim_alpha\src\sziit\lihz\ch04_ex01_anim_alpha\SecondActivity.java

```
1. public class SecondActivity extends Activity{
2.     protected void onCreate(Bundle savedInstanceState) {super.onCreate(sav
       edInstanceState);
3.        TextView view=new TextView(this);         //创建文本视图对象
4.        view.setText("SecondActivity");           //设置文本视图显示内容
5.        view.setTextColor(Color.GREEN);           //设置文本颜色
6.        view.setBackgroundColor(Color.BLACK);     //设置背景颜色
7.        setContentView(view);}}
```

步骤 5：在 AndroidManifest.xml 文件中，增加 SecondActivity 活动的注册，如源代码清单 4.10 所示。

源代码清单 4.10   \Ch04_Ex01_anim_alpha\AndroidManifest.xml

```
1.  <?xml version="1.0" encoding="utf-8"?>
2.  <manifest xmlns:android="http://schemas.android.com/apk/res/android"
3.      package="sziit.lihz.ch04_ex01_anim_alpha"
4.      android:versionCode="1"   android:versionName="1.0" >
5.      ...
6.      <application
7.      ...
8.          android:theme="@style/AppTheme" >
9.          <activity
10.             android:name="sziit.lihz.ch04_ex01_anim_alpha.MainActivity"
11.             android:label="@string/MainActivity_Title" >
12.             <intent-filter>
13.                 <action android:name="android.intent.action.MAIN" />
14.                 <category android:name="android.intent.category.LAUNCHER" />
15.             </intent-filter>
16.         </activity>
17.         <activity
18.             android:name="sziit.lihz.ch04_ex01_anim_alpha.SecondActivity"
```

```
19.         android:label="@string/SecondActivity_Title"></activity>
20.   </application>   </manifest>
```

（3）运行结果

选择【Run As】→【Android Application】命令，运行结果如图 4.5 所示。

（a）主屏动画延迟淡出　　（b）淡入淡出动画第二界面

图 4.5　淡入淡出的程序过渡动画效果

2. 左进右出的程序过渡动画

（1）项目简介

在 Android 程序中，通常会定义程序中 Activity 切换的动画资源文件，以提高程序的流畅度，改善用户体验。下面实例中定义了左进右出的资源文件。

（2）实现步骤

实现要点：

- void overridePendingTransition(int enterAnim, int exitAnim)：在 startActivity(Intent) 执行后立即被调用，显示设置执行的动画效果；参数 enterAnim 为活动进入时动画资源 ID；参数 exitAnim 为活动退出时动画效果。
- void onBackPressed()：当活动检测到用户按下返回键后被调用。

程序定义了两个动画（左进和右出），分别在进入一个 Activity 和退出该 Activity 时显示，步骤如下：

步骤 1：创建 Ch04_Ex02_anim_translate 项目，在项目 res 目录下新建 anim 目录，该项目动画资源文件有两个：slide_left_in.xml（左进）和 slide_right_out.xml（右出），如源代码清单 4.11 和 4.12 所示。

源代码清单 4.11　\Ch04_Ex02_anim_translate\res\anim\slide_left_in.xml

```
1. <?xml version="1.0" encoding="utf-8"?>
2. <set xmlns:android="http://schemas.android.com/apk/res/android" >
3.    <translate  android:duration="300" android:fromXDelta="-100.0%p"
4.          android:toXDelta="0.0" />    </set>
```

源代码清单 4.12　\Ch04_Ex02_anim_translate\res\anim\slide_right_out.xml

```
1. <?xml version="1.0" encoding="utf-8"?>
```

```
2. <set xmlns:android="http://schemas.android.com/apk/res/android" >
3.     <translate   android:duration="300"   android:fromXDelta="0.0"
4.                  android:toXDelta="100.0%p" />        </set>
```

步骤 2：该项目有两个界面，主界面 activity_main.xml 只有一个 Button，单击之后实现界面跳转，如源代码清单 4.13 所示。

源代码清单 4.13   \Ch04_Ex02_anim_translate\res\layout\activity_main.xml

```
1. <RelativeLayout xmlns:android="http://schemas.android.com/apk/res/android"
2.     xmlns:tools="http://schemas.android.com/tools"
3.     android:layout_width="match_parent"   android:layout_height="match_p
       arent" >
4.     <Button  android:id="@+id/button"
5.         android:layout_width="wrap_content"   android:layout_height="wrap
           _content"
6.         android:layout_centerHorizontal="true"  android:text="启动第二个活动界面"
7.         tools:context=".MainActivity" />     </RelativeLayout>
```

步骤 3：SecondActivity 界面包含一个 TextView 和一个 Button，项目的动画效果由 TextView 呈现，单击 Button 则返回上一界面，如源代码清单 4.14 所示。

源代码清单 4.14   \Ch04_Ex02_anim_translate\res\layout\activity_second.xml

```
1. <RelativeLayout xmlns:android="http://schemas.android.com/apk/res/android"
2.     xmlns:tools="http://schemas.android.com/tools"
3.     android:layout_width="match_parent"   android:layout_height="match_
       parent" >
4.     <TextView
5.         android:id="@+id/textView"
6.         android:layout_width="wrap_content"   android:layout_height="wrap
           _content"
7.         android:layout_alignParentTop="true"   android:layout_centerHori
           zontal="true"
8.         android:layout_centerVertical="true"   android:text="第二个活动界面" />
9.     <Button
10.        android:id="@id/button"
11.        android:layout_width="wrap_content"   android:layout_height="wrap_
           content"
12.        android:layout_below="@id/textView"   android:layout_centerHorizont
           al="true"
13.        android:layout_centerVertical="true"  android:padding="@dimen/paddi
           ng_medium"
14.        android:text="返回主活动界面" />        </RelativeLayout>
```

步骤 4：在 MainActivity 中为 Button 设置单击事件：单击 Button 时跳转到 SecondActivity，并调用 overridePendingTransition 方法，该方法包含两个参数，其中第一个参数是进入 SecondActivity 时的动画，第二个参数是退出 SecondActivity 时的动画，如源代码清单 4.15 所示。

源代码清单 4.15   \Ch04_Ex02_anim_translate\src\sziit\lihz\ch04_ex02_anim_translate\MainActivity.java

```
1. public class MainActivity extends Activity {      //从 Activity 基类派生子类
2.     public void onCreate(Bundle savedInstanceState) {//子类重写 onCreate 方法
3.         super.onCreate(savedInstanceState);       //基类调用 onCreate 方法
4.         setContentView(R.layout.activity_main);   //设置活动布局界面
```

```
5.        this.findViewById(R.id.button1).setOnClickListener(//设置按钮单击监听器
6.            new View.OnClickListener() {           //构建单击监听器接口
7.                public void onClick(View v) {      //实现单击接口方法
8.                    startActivity(new Intent(MainActivity.this,
9.                        SecondActivity.class));
                                          //启动意图，从主界面切换到第二个界面
10.                   overridePendingTransition(android.R.anim.slide_in_left,
11.                       android.R.anim.slide_out_right);//设置切换动画效果
12.       }});}}
```

步骤 5：在 SecondActivity 中，设置 Button 点击事件使之返回上一界面，如源代码清单 4.16 所示。

源代码清单 4.16　\Ch04_Ex02_anim_translate\src\sziit\lihz\ch04_ex02_anim_translate\SecondActivity.java

```
1. public class SecondActivity extends Activity { // 从 Activity 基类派生子类
2.    public void onCreate(Bundle savedInstanceState) {// 子类重写 onCreate 方法
3.        super.onCreate(savedInstanceState);// 基类调用 onCreate 方法
4.        setContentView(R.layout.activity_second);// 设置活动布局界面
5.        this.findViewById(R.id.button).setOnClickListener(//设置按钮单击监听器
6.            new View.OnClickListener() {           //构建单击监听器接口
7.                public void onClick(View v) {      //实现单击接口方法
8.                    onBackPressed();               //返回前一个活动界面
9.        }});}}
```

步骤 6：在 AndroidManifest.xml 中，注册 SecondActivity 活动，如源代码清单 4.17 所示。

源代码清单 4.17　\Ch04_Ex02_anim_translate\AndroidManifest.xml

```
1.  <?xml version="1.0" encoding="utf-8"?>
2.  <manifest xmlns:android="http://schemas.android.com/apk/res/android"
3.      package="sziit.lihz.ch04_ex02_anim_translate"
4.      android:versionCode="1"    android:versionName="1.0" >
5.      ...
6.      <application
7.          android:allowBackup="true"         android:icon="@drawable/ic_launcher"
8.          android:label="@string/app_name"   android:theme="@style/AppTheme" >
9.          <activity
10.             android:name="sziit.lihz.ch04_ex02_anim_translate.MainActivity"
11.             android:label="@string/MainActivity_Title" >
12.             <intent-filter>
13.                 <action android:name="android.intent.action.MAIN" />
14.                 <category android:name="android.intent.category.LAUNCHER" />
15.             </intent-filter>    </activity>
16.         <activity
17.             android:name="sziit.lihz.ch04_ex02_anim_translate.SecondActivity"
18.             android:label="@string/SecondActivity_Title" > </activity>
        </application> </manifest>
```

（3）运行结果

选择【Run As】→【Android Application】命令，运行结果如图 4.6 所示。

图 4.6　左进右出的程序过渡动画效果

4.2 利用 Drawable 资源文件美化程序 UI 界面

4.2.1 项目实训目标

- 掌握如何在程序中使用 Drawable 资源文件。
- 掌握如何编写及使用 StateListDrawable 资源文件。
- 掌握如何编写及使用 ShapeDrawable 资源文件。
- 掌握 Shared Preferences 数据存储方法。

4.2.2 知识准备

样式资源都是用于对 Android 应用程序进行"美化"的，只要充分利用 Android 应用的 Drawable 资源、StateListDrawable 资源、ShapeDrawable 资源和 Style 资源，开发者就可以开发出各种风格的应用。下面分别对上面提出的各种资源的知识点进行讲解。

1. Android Drawable 资源文件

Drawable 资源是 Android 应用中使用最广泛的资源，也是 Android 应用中最灵活的资源，它不仅可以直接使用*.png、*.jpg、*.gif 等图片作为资源，也可以使用多种 XML 文件作为资源。只要一份 XML 文件可以被系统编译成 Drawable 子类的对象，这份 XML 文件即可作为 Drawable 资源。

Drawable 资源通常保存在/res/drawable 目录，还可能保存在/res/drawable-ldpi、res/drawable-mdpi、res/drawable-hdpi 目录。

2. Android StateListDrawable 资源文件

StateListDrawable 用于组织多个 Drawable 对象。当使用 StateListDrawable 作为目标组件的背景图片时，StateListDrawable 对象所显示的 Drawable 对象会随目标组件形态的改变而自动切换。

源代码清单 4.18 是 Android 中的 StateListDrawable 文件。可以从此文件中了解到 StateListDrawable 文件的结构及各个组成部分。

源代码清单 4.18　StateListDrawable.xml 文件示意代码

```
1. <?xml version="1.0" encoding="utf-8"?>
2. <selector xmlns:android="http://schemas.android.com/apk/res/android">
3.    <item android:state_pressed="true" android:drawable="@drawable/shape_
      corner_round_login_pressed" />
4.    <item android:state_focused="true" android:drawable="@drawable/shape_
      corner_round_login_pressed" />
5.    <item android:state_focused="false" android:state_pressed="false" andr
      oid:drawable="@drawable/shape_corner_round_login_normal" />   </selector>
```

从代码中可以看到，StateListDrawable 对象的根元素为<selector/>，该元素可以包含多个<item/>元素，每个<item/>元素包含一种特定的状态，程序在运行过程中，当对应控件的状态发生改变时，例如触摸该控件导致 pressed 状态改变，就可以通过这种变化后的状态值去更新对应的 Drawable 对象。表 4.3 对 StateListDrawable 文件属性作了完整的描述。

表 4.3 StateListDrawable 文件属性

状态名称	含义描述
enable	是否处于可以单击状态，可以调用 setEnable()方法改变其状态
focused	是否处于聚焦（获得焦点）状态，一个窗口只能有一个视图拥有焦点，一般由用户交互导致，不需要应用程序直接改变
pressed	是否处于按下状态，一般由用户交互导致（通常是触摸 View 获取其状态值）
selected	是否处于被选择状态，程序可以调用 setSelected()方法改变其状态
window_focused	视图所在的窗口是否为当前交互窗口，即前台窗口（例如：Activity 处于 onRusume 状态，而不是 onPause/onStop 状态），该状态值由系统决定，应用程序不能改变其状态

需要说明的是，focused 不同于 selected，一个窗口只能有一个视图获得焦点（focus），但可以有多个视图处于 selected 状态。

3. Android ShapeDrawable 资源文件

ShapeDrawable 用于定义一个基本的几何图形(如矩形、圆形、线条等)，定义 ShapeDrawable 的 XML 文件的根元素是<shape/>元素，例如源代码清单 4.19。

源代码清单 4.19 ShapeDrawable.xml 文件示意代码

```
1. <?xml version="1.0" encoding="UTF-8"?>
2. <shape xmlns:android="http://schemas.android.com/apk/res/android">
3. <gradient
4.    android:startColor="#F67324" android:endColor="#F34324"   android:angle
      ="270.0"
5.    android:centerY="0.5"    android:centerColor="#F34324">  </gradient>
6. <corners android:radius="10px" />
7. <padding android:left="0dp" android:top="0dp" android:right="0dp"
   android:bottom="0dp" /></shape>
```

表 4.4 对<shape/>根元素下的子元素及其包含的属性进行了详细说明。

表 4.4 <shape/>根元素下的子元素及其包含的属性

子元素	含义	属性	作用
<corners/>	对该 shape 的角进行描述	radius	定义圆角半径
		topLeftRadius topRightRadius bottomLeftRadius bottomRightRadius	分别定义图形的四个角的半径
<gradient/>	定义这个 shape 的渐变颜色	startColor centerColor endColor	起始色、中间色、结束色
		useLevel	boolean 类型
		angle	渐变的角度，0 代表从 left 到 right，90 代表 bottom 到 top。必须是 45 的倍数，默认为 0
		type	渐变的模式，有三种：linear, 线性渐变；radial, 辐射渐变；sweep, 扫描线渐变
		centerX centerY	渐变中心的相对坐标，取值 0 到 1.0 之间
		gradientRadius	渐变的半径，只有在 android:type="radial"才使用

续表

子元素	含义	属性	作用
<padding/>	内容与视图边界的距离	left top right bottom	分别距左边、顶部、右边、底部的填充距离
<size/>	设置 shape 的大小	height	设置 shape 的高度
		width	设置 shape 的宽度
<solid/>	填充 shape 的纯色	color	颜色值，十六进制数或者一个 Color 资源
<stroke/>	shape 使用的笔画，当 android:shape="line" 时必须设置该元素	width	设置笔画的粗细
		color	设置笔画的颜色
		dashGap	每画一条线的间隔，只有当 android:dashWidth 也设置了才有效
		dashWidth	每画一条线的长度，只有当 android:dashGap 也设置了才有效

4.2.3 项目实践

（1）项目简介

如今每天都有成千上万的新 Android 应用程序出现在市场上，如何让你的应用程序脱颖而出？提供给用户一个友好的 UI 界面是一个非常重要的因素。仅仅依靠系统提供的基础控件显然达不到这一要求。下面通过一个用户登录界面的实例让读者对 UI 界面在程序中的重要性有一个基本的认识。

（2）实现步骤

实现要点：

- android:background="@drawable/login_bg"：设置布局背景颜色。
- android:src="@drawable/login_log"：设置 ImageView 显示图片来源。
- android:background="@drawable/selector_corner_round_top"：利用 selector 选择器美化 UI 界面，生成顶部带圆角、白色背景、灰色边框、无上边框的长方体。
- android:background="@drawable/selector_corner_round_bottom"：利用 selector 选择器美化 UI 界面，生成顶部带圆角、白色背景、灰色边框、无下边框的长方体。
- android:textColor="@color/gray"：设置 TextView 的字体颜色为自定义颜色。
- android:background="@drawable/metro_login_button_selector"：利用 selector 选择器美化按钮控件，可根据按钮的不同状态 android:state_pressed、android:state_focused 等设置不同的图片。
- void setOnCheckedChangeListener(OnCheckedChangeListener listener)：设置复选框监听器。
- interface OnCheckedChangeListener：多选框监听器接口，其包括一个接口方法 void onCheckedChanged(CompoundButton buttonView, boolean isChecked)。
- boolean isChecked()：检测复选框是否选中。
- SharedPreferences 偏好数据存储机制。

该实例中，运用以上所学知识点对"登录"按钮、用户名和密码输入框进行美化，使"登录"按钮在不同的状态（例如按下按钮）下显示不同的效果，并将按钮和输入框的四个角设置成圆角。

步骤1：创建 Ch04_Ex03_DrawableLogin 项目，新建 res/drawable 目录，用于存放该实例用到的一系列 Drawable 资源文件，下面是这些资源代码清单。

源代码清单4.20　\Ch04_Ex03_DrawableLogin\res\drawable\metro_login_button_selector.xml

```xml
1. <?xml version="1.0" encoding="utf-8"?>
2. <selector xmlns:android="http://schemas.android.com/apk/res/android">
3.    <item android:drawable="@drawable/shape_corner_round_login_pressed"
      android:state_pressed="true"/>
4.    <item android:drawable="@drawable/shape_corner_round_login_pressed"
      android:state_focused="true"/>
5.    <item android:drawable="@drawable/shape_corner_round_login_normal"
      android:state_focused="false" android:state_pressed="false"/>   </selector>
```

源代码清单4.21　\Ch04_Ex03_DrawableLogin\res\drawable\selector_corner_round_bottom_normal.xml

```xml
1. <?xml version="1.0" encoding="UTF-8"?>
2. <!-- 不带圆角 白色背景 灰色边框 无下边框 长方体 -->
3. <layer-list xmlns:android="http://schemas.android.com/apk/res/android" >
4.   <item android:top="-1px">
5.     <shape xmlns:android="http://schemas.android.com/apk/res/android" >
6.       <stroke  android:width="1px"  android:color="@color/corner_round_
          border"/>
7.       <solid  android:color="@color/corner_round_background_normal"/>
8.       <corners android:bottomLeftRadius="6dip" android:bottomRightRadius
          ="6dip"/>
9.     </shape>    </item>    </layer-list>
```

源代码清单4.22　\Ch04_Ex03_DrawableLogin \res\drawable\selector_corner_round_bottom_pressed.xml

```xml
1. <?xml version="1.0" encoding="UTF-8"?>
2. <!-- 不带圆角 白色背景 灰色边框 无下边框 长方体 -->
3. <layer-list xmlns:android="http://schemas.android.com/apk/res/android">
4.   <item android:top="-1px">
5.     <shape xmlns:android="http://schemas.android.com/apk/res/android">
6.       <stroke   android:width="1px"   android:color="@color/corner_round_
          border"/>
7.       <solid   android:color="@color/corner_round_background_pressed"/>
8.       <corners  android:bottomLeftRadius="6dip"  android:bottomRightRadius
          ="6dip"/>
9.     </shape>    </item>    </layer-list>
```

源代码清单4.23　\Ch04_Ex03_DrawableLogin\res\drawable\selector_corner_round_bottom.xml

```xml
1. <?xml version="1.0" encoding="UTF-8"?>
2. <!-- 顶部带圆角 白色背景 灰色边框 无下边框 长方体 -->
3. <selector xmlns:android="http://schemas.android.com/apk/res/android">
4.   <item android:drawable="@drawable/selector_corner_round_bottom_normal"
5.      android:state_pressed="false"/>  <!-- 默认背景 -->
6.   <item android:drawable="@drawable/selector_corner_round_bottom_pressed"
      android:state_pressed="true"/>  <!-- 单击时 -->   </selector>
```

源代码清单 4.24 \Ch04_Ex03_DrawableLogin\res\drawable\selector_corner_round_top.xml

```xml
1. <?xml version="1.0" encoding="UTF-8"?>
2. <!-- 顶部带圆角 白色背景 灰色边框 无下边框 长方体 -->
3. <selector xmlns:android="http://schemas.android.com/apk/res/android">
4.     <!-- 默认背景 -->
5.     <item android:state_pressed="false">
6.         <shape><stroke android:width="1px" android:color="@color/corner_round_border" />
7.         <solid android:color="@color/corner_round_background_normal" />
8.         <corners android:topLeftRadius="6dip" android:topRightRadius="6dip" />
9.         </shape></item>
10.     <!-- 单击时 -->
11.     <item android:state_pressed="true">
12.         <shape><stroke android:width="1px" android:color="@color/corner_round_border" />
13.         <solid android:color="@color/corner_round_background_pressed" />
14.         <corners android:topLeftRadius="6dip" android:topRightRadius="6dip" />
15.         </shape></item>    </selector>
```

源代码清单 4.25 \Ch04_Ex03_DrawableLogin\res\drawable\shape_corner_round_login_normal.xml

```xml
1. <?xml version="1.0" encoding="UTF-8"?>
2. <shape xmlns:android="http://schemas.android.com/apk/res/android" >
3.     <gradient android:angle="270.0"         android:centerColor="#F27324"  android:centerY="0.5"
4.     android:endColor="#E75B06" android:startColor="#FC8C40" > </gradient>
5.     <corners android:radius="10px"/>
6.     <padding android:bottom="0dp" android:left="0dp" android:right="0dp" android:top="0dp"/>
7. </shape>
```

源代码清单 4.26 \Ch04_Ex03_DrawableLogin \res\drawable\shape_corner_round_login_pressed.xml

```xml
1. <?xml version="1.0" encoding="UTF-8"?>
2. <shape xmlns:android="http://schemas.android.com/apk/res/android" >
3.     <gradient android:angle="270.0"         android:centerColor="#F34324" android:centerY="0.5"
4.     android:endColor="#F34324" android:startColor="#F67324" > </gradient>
5.     <corners android:radius="10px"/>
6.     <padding android:bottom="0dp" android:left="0dp" android:right="0dp" android:top="0dp"/>
7. </shape>
```

步骤 2：在程序界面中调用上面定义的各种动画资源文件。程序登录界面布局代码如源代码清单 4.27 所示。

源代码清单 4.27 \Ch04_Ex03_DrawableLogin\res\layout\activity_login.xml

```xml
1. <?xml version="1.0" encoding="utf-8"?>
2. <RelativeLayout xmlns:android="http://schemas.android.com/apk/res/android"
3.     android:layout_width="fill_parent"    android:layout_height="fill_parent"
```

```
4.    android:background="@drawable/login_bg" >
5.  <LinearLayout android:layout_width="wrap_content" android:layout_height
      ="wrap_content"
6.    android:layout_alignParentTop="true" android:layout_centerHorizontal=
      "true"
7.    android:layout_marginTop="50dp" android:background="@null" android:
      orientation="vertical">
8.    <LinearLayout android:layout_width="fill_parent" android:layout_
      height="fill_parent"
9.     android:orientation="vertical" android:paddingLeft="30dp" android:
      paddingRight="30dp">
10.    <ImageView android:id="@+id/imageView1" android:layout_width="252dp"
11.      android:layout_height="wrap_content" android:contentDescription=
      "@string/app_name"
12.      android:src="@drawable/login_log" />   </LinearLayout>
13.    <LinearLayout android:layout_width="fill_parent" android:layout_
      height="fill_parent"
14.     android:layout_marginTop="30dp"    android:orientation="vertical"
15.     android:paddingLeft="30dp"          android:paddingRight="30dp">
16.    <RelativeLayout android:id="@+id/rl_request_inteval" android:layout_
      width="match_parent"
17.      android:layout_height="wrap_content"
18.      android:background="@drawable/selector_corner_round_top"
19.      android:orientation="vertical"   android:paddingBottom="10dp"
20.      android:paddingLeft="25dp"    android:paddingTop="10dp">
21.     <EditText android:id="@+id/edt_userName" android:layout_width=
      "wrap_content"
22.       android:layout_height="wrap_content" android:layout_alignPaRent
      Right="true"
23.       android:layout_alignParentTop="true"
24.       android:layout_toRightOf="@+id/tv_request_inteval_lable"
25.       android:background="@null"      android:hint="@string/username"
26.       android:inputType="textPersonName" android:text="@string/ username"/>
27.     <TextView android:id="@+id/tv_request_inteval_lable"
28.       android:layout_width="wrap_content" android:layout_height=
      "wrap_content"
29.       android:layout_alignBaseline="@+id/edt_userName"
30.       android:layout_alignBottom="@+id/edt_userName"
31.       android:layout_alignParentLeft="true" android:text="@string/
      username_lable"
32.       android:textColor="@color/gray" />   </RelativeLayout>
33.     <RelativeLayout  android:id="@+id/rl_flashlight_set"
34.       android:layout_width="match_parent" android:layout_height=
      "wrap_content"
35.       android:background="@drawable/selector_corner_round_bottom"
36.       android:orientation="vertical"    android:paddingBottom="10dp"
37.       android:paddingLeft="25dp"       android:paddingTop="10dp" >
38.     <EditText android:id="@+id/edt_password" android:layout_width
          ="wrap_content"
```

```
39.            android:layout_height="wrap_content"    android:layout_align
               ParentRight="true"
40.            android:layout_alignParentTop="true"
41.            android:layout_toRightOf="@+id/tv_flashlight_set_lable"
42.            android:background="@null"      android:ems="10"
43.            android:hint="@string/password"   android:inputType="textPass
               word"
44.            android:text="@string/password"/>
45.        <TextView android:id="@+id/tv_flashlight_set_lable"
46.            android:layout_width="wrap_content"   android:layout_height=
               "wrap_content"
47.            android:layout_alignBaseline="@+id/edt_password"
48.            android:layout_alignBottom="@+id/edt_password"
49.            android:layout_alignParentLeft="true"   android:text="@string
               /password_lable"
50.            android:textColor="@color/gray" />   </RelativeLayout>
               </LinearLayout>
51.    <LinearLayout  android:layout_width="fill_parent"  android:layout_
       height="fill_parent"
52.        android:layout_marginTop="10dp" android:orientation="vertical"
           android:paddingLeft="30dp"
53.        android:paddingRight="30dp">
54.        <Button android:id="@+id/btn_login" android:layout_width="match_parent"
55.            android:layout_height="45dp"
56.            android:background="@drawable/metro_login_button_selector"
57.            android:text="@string/login"     android:textColor="#A42D03" />
               </LinearLayout>
58.        <RelativeLayout android:layout_width="fill_parent" android:layout_
           height="fill_parent"
59.        android:layout_marginTop="10dp"  android:orientation="horizontal"
60.        android:paddingLeft="30dp"      android:paddingRight="30dp">
61.        <CheckBox android:id="@+id/cb_mima"  android:layout_width="wrap_
           content"
62.            android:layout_height="wrap_content" android:checked="true"
63.            android:text="@string/save_password" android:textColor="#FFFFFF"
64.            android:textSize="13sp" />
65.        <CheckBox android:id="@+id/cb_auto"   android:layout_width="wrap_
           content"
66.            android:layout_height="wrap_content"  android:layout_alignParen
               tRight="true"
67.            android:layout_alignParentTop="true"   android:text="@string/
               auto_login"
68.            android:textColor="#FFFFFF"       android:textSize="13sp" />
69.        </RelativeLayout>    </LinearLayout>   </RelativeLayout>
```

步骤3：登录成功之后跳转的界面布局代码如源代码清单 4.28 所示。

源代码清单 4.28 \Ch04_Ex03_DrawableLogin\res\layout\activity_second.xml

```
1. <?xml version="1.0" encoding="utf-8"?>
2. <LinearLayout xmlns:android="http://schemas.android.com/apk/res/android"
3.     android:layout_width="match_parent"
```

```
4.         android:layout_height="match_parent"
5.         android:orientation="vertical" >
6.     <TextView
7.         android:id="@+id/textView1"
8.         android:layout_width="wrap_content"
9.         android:layout_height="wrap_content"
10.        android:text="登录成功,你已进入用户界面" />
11. </LinearLayout>
```

步骤 4：在该实例中，为了更接近实际工作需要，除对界面进行美化外，还增加了记住密码和自动登录功能，LoginActivity 代码如源代码清单 4.29 所示。

源代码清单 4.29　\Ch04_Ex03_DrawableLogin\src\sziit\lihz\ch04_ex03_drawablelogin\LoginActivity.java

```java
1. public class LoginActivity extends Activity {         //从 Activity 活动基类派生子类
2.     private EditText userName, passWord;              //声明 EditText 对象
3.     private CheckBox cb_pw, cb_auto_login;            //声明 CheckBox 对象
4.     private Button btn_login;                         //声明 Button 对象
5.     private String userNameValue, pwdValue;           //声明 String 对象
6.     private SharedPreferences sp;//声明 SharedPreferences 对象
7.     protected void onCreate(Bundle savedInstanceState) {//子类重写 onCreate 方法
8.         super.onCreate(savedInstanceState);           //基类调用 onCreate 方法
9.         this.requestWindowFeature(Window.FEATURE_NO_TITLE);//去除标题
10.        setContentView(R.layout.activity_login);//设置活动界面布局
11.        //获取实例对象
12.        sp = this.getSharedPreferences("userInfo", Context.MODE_WORLD_READABLE);
13.        userName = (EditText) findViewById(R.id.edt_userName);
14.        passWord = (EditText) findViewById(R.id.edt_password);
15.        btn_login = (Button) findViewById(R.id.btn_login);
16.        cb_pw = (CheckBox) findViewById(R.id.cb_mima);
17.        cb_auto_login = (CheckBox) findViewById(R.id.cb_auto);
18.        final Intent intent = new Intent(this, SecondActivity.class);
19.        if (sp.getBoolean("ISCHECK", true)) {      //判断记住密码多选框的状态
20.            userName.setText(sp.getString("USER_NAME", ""));
21.            passWord.setText(sp.getString("PASSWORD", ""));
22.            if (sp.getBoolean("AUTO_ISCHECK", false)) {//判断自动登录多选框的状态
23.                cb_auto_login.setChecked(true);       //设置默认是自动登录状态
24.                startActivity(intent);}}              //跳转界面
25.        sp.edit().putString("USER_NAME", "admin").commit();//设置默认用户名、密码
26.        sp.edit().putString("PASSWORD", "admin").commit();
27.        btn_login.setOnClickListener(new OnClickListener() {//登录按钮监听事件
28.          public void onClick(View v) { userNameValue = userName.getText().toString();
29.            pwdValue = passWord.getText().toString();
30.            if (userNameValue.equals("admin") & pwdValue.equals("admin")) {
31.                Toast.makeText(LoginActivity.this, "登录成功",Toast.LENGTH_SHORT).show();
32.                startActivity(intent);
33.                if (cb_pw.isChecked()) {//登录成功且保存密码复选框为选中状态时保存用户信息
34.                    Editor editor = sp.edit();        //记住用户名、密码
35.                    editor.putString("USER_NAME", userNameValue);
36.                    editor.putString("PASSWORD", pwdValue);editor.commit();  }
```

```
37.              } else { Toast.makeText(LoginActivity.this,"用户名或者密码错误,
38.                    请重新登录",Toast.LENGTH_SHORT).show();}}});
39.     //记住密码多选框监听事件
40.     cb_pw.setOnCheckedChangeListener(new OnCheckedChangeListener() {
41.         public void onCheckedChanged(CompoundButton buttonView, boolean
        isChecked) {
42.             if (cb_pw.isChecked()) {   System.out.println("记住密码已选中");
43.                 sp.edit().putBoolean("ISCHECK", true).commit();
44.             } else { System.out.println("记住密码没有选中");
45.                 sp.edit().putBoolean("ISCHECK", false).commit();}} });
46.     //自动登录多选框监听事件
47.     cb_auto_login.setOnCheckedChangeListener(new OnCheckedChangeListener() {
48.         public void onCheckedChanged(CompoundButton buttonView,
        boolean isChecked) {
49.             if (cb_auto_login.isChecked()) {   System.out.println("自动登
        录已选中");
50.                 sp.edit().putBoolean("AUTO_ISCHECK", true).commit();
51.             } else { System.out.println("自动登录没有选中");
52.                 sp.edit().putBoolean("", false).commit();   }}});}}
```

步骤5：SecondActivity界面代码如源代码清单4.30所示。

源代码清单4.30 \Ch04_Ex03_DrawableLogin\src\sziit\lihz\ch04_ex03_drawablelogin\SecondActivity.java

```
1. public class SecondActivity extends Activity {//从Activity活动基类派生子类
2.     protected void onCreate(Bundle savedInstanceState) {//子类重写onCreate方法
3.         super.onCreate(savedInstanceState);//基类调用onCreate方法
4.         setContentView(R.layout.activity_second);}}//设置活动界面布局
```

（3）运行结果

选择【Run As】→【Android Application】命令，运行程序，将会看到图4.7所示的用户登录界面。

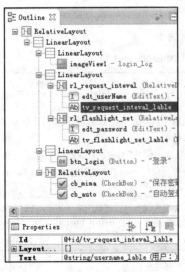
（a）设计轮廓

（b）运行界面

图4.7 用户登录界面

4.2.4 Android 数据存储 Shared Preferences

Shared Preferences 类似于常用的.ini 文件,用来保存应用程序中的一些属性设置,在 Android 平台下常用于存储简单的参数设置。就如上一节所讲到的,可以用来保存上一次用户所做的修改或者自定义参数设置,当再次启动程序时仍然保持原有的设置。下面通过一个实例来了解 SharedPreferences 的具体使用方法。这个实例中定义了两个 TextView 和两个 EditText,让用户输入用户名和密码。然后通过 Shared Preferences 的使用实现在下次启动程序时自动显示用户设置过的用户名和密码。实例初始运行状态如图 4.8(a)所示,在文本框中分别输入用户名和密码,按返回键,当再次进入程序时界面如图 4.8(b)所示。

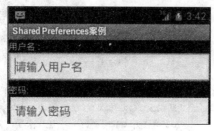 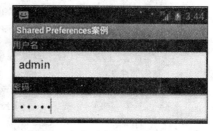

(a)程序初始运行截图　　　　　　　　　(b)再次启动程序截图

图 4.8　Shared Preferences 案例

使用 SharedPreferences 保存 key-value 对的步骤如下:

(1)使用 Activity 类的 getSharedPreferences()方法获得 SharedPreferences 对象,其中存储 key-value 的文件的名称由 getSharedPreferences()方法的第一个参数指定。

(2)使用 SharedPreferences 接口的 edit()方法获得 SharedPreferences.Editor 对象。

(3)通过 SharedPreferences.Editor 接口的 putXxx()方法保存 key-value 对。其中 Xxx 表示不同的数据类型。例如:字符串类型的 value 需要用 putString()方法。

(4)通过 SharedPreferences.Editor 接口的 commit()方法保存 key-value 对。Commit()方法相当于数据库事务中的提交操作。

表 4.5 中列出了两个获得 SharedPreferences 的方法。

表 4.5　获取 SharedPreferences 的方法

返回值	函数	备注
SharedPreferences	Context.getSharedPreferences (String name,int mode)	name 为本组件的配置文件名(如果想要与本应用程序的其他组件共享此配置文件,可以用这个名字来检索到这个配置文件)。 mode 为操作模式,默认的模式为 0 或 MODE_PRIVATE,还可以使用 MODE_WORLD_READABLE 和 MODE_WORLD_WRITEABLE
SharedPreferences	Activity.getPreferences(int mode)	配置文件仅可以被调用的 Activity 使用。mode 为操作模式,默认的模式为 0 或 MODE_PRIVATE,还可以使用 MODE_WORLD_READABLE 和 MODE_WORLD_WRITEABLE

除此之外还要介绍以下几个重要的方法:

● public abstract boolean contains(String key):检查是否已存在该文件,其中 key 是 XML 文件名。

- edit()：为 preferences 创建一个编辑器 Editor，通过创建的 Editor 可以修改 preferences 中的数据，但必须执行 commit() 方法。
- getAll()：返回 preferences 中的多余数据。
- getBoolean(String key, boolean defValue)：获取 Boolean 型数据。
- getFloat(String key, float defValue)：获取 Float 型数据。
- getInt(String key, int defValue)：获取 Int 型数据。
- getLong(String key, long defValue)：获取 Long 型数据。
- getString(String key, String defValue)：获取 String 型数据。
- registerOnSharedPreferenceChangeListener(SharedPreferences.OnSharedPreferenceChangeListener listener)：注册一个当 preference 发生改变时被调用的回调函数。
- unregisterOnSharedPreferenceChangeListener(SharedPreferences.OnSharedPreferenceChangeListener listener)：删除当前回调函数。

在理解了 SharedPreferences 的基本知识后来看一下程序的实现方法。创建 Ch04_Ex04_SharedPreferences 项目。布局文件 sp_layout.xml 如源代码清单 4.31 所示，程序中主程序源代码 SharedPreferencesActivity 如源代码清单 4.32 所示。

源代码清单 4.31 \Ch04_Ex04_SharedPreferences\res\layout\sp_layout.xml

```
1. ?xml version="1.0" encoding="utf-8"?>
2. <LinearLayout xmlns:android="http://schemas.android.com/apk/res/android"
3.   android:layout_width="fill_parent"  android:layout_height="fill_parent"
4.   android:orientation="vertical">
5.   <TextView   android:layout_width="fill_parent"
6.     android:layout_height="wrap_content"  android:text="用户名："/>
7.   <EditText  android:id="@+id/etUser"  android:layout_width="fill_parent"
8.     android:layout_height="wrap_content" android:hint="请输入用户名" />
9.   <TextView   android:layout_width="fill_parent"
10.    android:layout_height="wrap_content"  android:text="密码："/>
11.  <EditText   android:id="@+id/etPWD"   android:layout_width="fill_parent"
12.    android:layout_height="wrap_content"   android:hint="请输入密码"
13.    android:password="true" />      </LinearLayout>
```

源代码清单 4.32 \Ch04_Ex04_SharedPreferences\src\sziit\lihz\ch04_ex04_sharedproferences\SharedPreferencesActivity.java

```
1. public class SharedPreferencesActivity extends Activity {
                                           //从 Activity 基类派生子类
2.    public String userName, passWord;     //声明 String 对象
3.    public EditText etName, etPWD;        //声明 EditText 对象
4.    /** 当活动第一次创建时被调用. */
5.    public void onCreate(Bundle savedInstanceState) {//子类重写 onCreate 方法
6.       super.onCreate(savedInstanceState);    //设置活动界面布局
7.       setContentView(R.layout.sp_layout);    //基类调用 onCreate 方法
8.       etName = (EditText) this.findViewById(R.id.etUser);
                                           //查找布局文件中 EditText 对象
9.       etPWD = (EditText) this.findViewById(R.id.etPWD);
                                           //查找布局文件中 EditText 对象
```

```
10.     SharedPreferences sp = this.getPreferences(Activity.MODE_PRIVATE);
                                                                  //获得对象
11.     userName = sp.getString("username", "admin");     //获得用户名
12.     passWord = sp.getString("password", "admin");     //获得密码
13.     etName.setText(userName);                          //显示用户名
14.     etPWD.setText(passWord);}                          //显示密码
15. public boolean onKeyUp(int keyCode, KeyEvent event) {return false; }
16. public boolean onKeyDown(int keyCode, KeyEvent event) {//处理按键按下事件
17.     if (keyCode == KeyEvent.KEYCODE_BACK) {
18.       SharedPreferences sp = getPreferences(Activity.MODE_PRIVATE);
                                                                  //获得对象
19.       SharedPreferences.Editor editor = sp.edit();  //获得Editor对象
20.       editor.putString("username", etName.getText().toString());
                                                                  //保存用户名
21.       editor.putString("password", etPWD.getText().toString());//保存密码
22.       editor.commit();                               //提交数据
23.       this.finish();                                 //关闭活动
24.       return true;     }
25.     return super.onKeyDown(keyCode, event); }}
```

也许你想知道保存的数据到底存放到什么地方了，下面来看一下数据的保存。在 Eclipse 下切换到 DDMS 视图，选择 File Explorer 标签，找到 /data/data 目录中对应的项目文件夹下的 Shared_prefs 文件夹，我们所看到的 XML 文件就是保存数据的地方了，如图 4.9 所示。

图 4.9 Shared Preferences 数据存储目录

4.3 利用列表控件实现九宫格菜单

4.3.1 项目实训目标

掌握如何在程序中使用 GridView 组件。

4.3.2 知识准备

GridView 组件是 Android 应用程序 UI 界面的常用组件之一，是实现九宫格菜单的首选。下面将对 GridView 组件的相关知识进行介绍。

GridView 用于在界面上按行、列分布的方式来显示多个组件，需要通过 Adapter 来提供显示的数据。表 4.6 列出了 GridView 常用的 XML 属性。

表 4.6 GridView 常用的 XML 属性

XML 属 性	相 关 方 法	说　　明
android:conlumnWidth	setConlumnWidth(int)	设置列的宽度
android:gravity	setGravity(int)	设置对齐方式
android:horizontalSpacing	setHorizontalSpcing(int)	设置各元素之间的水平间距
android:numColumns	setNumColumns(int)	设置列数
android:stretchMode	setStretchMode(int)	设置拉伸模式
android:verticalSpacing	setVerticalSpacing(int)	设置各元素之间的垂直间距

需要注意的一点是，使用 GridView 时一般都应该指定 numColumns 大于 1，否则该属性的默认值为 1，则意味着该 GridView 只有一列，那么 GridView 就变成了 ListView。

表 4.6 所示的 android:stretchMode 属性支持如下几个属性值：

- NO_STRETCH：不拉伸。
- STRETCH_SPACING：仅拉伸元素之间的间距。
- STRETCH_SPACING_UNIFORM：表格元素本身、元素之间的间距一起拉伸。
- STRETCH_COLUMN_WIDTH：仅拉伸元素表格元素本身。

4.3.3　项目实践

1. 项目简介

九宫格菜单在程序 UI 界面非常实用，通过点击其中的一项，可以轻松实现界面跳转，接下来的实例就是利用 GridView 组件实现一个九宫格菜单界面。

2. 实现步骤

实现要点：

- OnItemClickListener：当 AdapterView 中某项被单击时的回调接口定义。
- void onItemClick(AdapterView<?> parent, View view, int position, long id)：当 AdapterView 中某项被单击时的接口方法。参数 parent 为被单击的 AdapterView 对象；view 为 AdapterView 内被单击的视图（由适配器提供）；position 为在适配器中视图位置；id 为被单击项的行 ID；调用 getItemAtPosition(position)可访问被选项的数据。
- SimpleAdapter(Context context, List<? extends Map<String, ?>> data, int resource, String[] from, int[] to)：简单适配器构造函数。
- void setAdapter(ListAdapter adapter)：设置适配器。
- void setTitle(CharSequence title)：设置活动标题。

该实例只有一个界面，当用户单击九宫格中的某一项时，将标题设置成该项的文本内容，具体步骤如下：

步骤 1：创建 Ch04_Ex05_GridView 项目，主界面包含一个 GridView，用于装载 item，如源代码清单 4.33 所示。

源代码清单 4.33　\Ch04_Ex05_GridView\res\layout\activity_main.xml

```
1. <?xml version="1.0" encoding="UTF-8"?>
2. <GridView xmlns:android="http://schemas.android.com/apk/res/android"
```

```xml
3.    xmlns:tools="http://schemas.android.com/tools"
4.    android:id="@+id/gridView"              android:layout_width="fill_parent"
5.    android:layout_height="fill_parent"     android:columnWidth="90dp"
6.    android:gravity="center"                android:horizontalSpacing="10dp"
7.    android:numColumns="auto_fit"           android:stretchMode="columnWidth"
8.    android:verticalSpacing="10dp" />
```

步骤2：新建item的布局文件cell.xml，如源代码清单4.34所示。

源代码清单4.34   \Ch04_Ex05_GridView\res\layout\cell.xml

```xml
1. <?xml version="1.0" encoding="utf-8"?>
2. <RelativeLayout xmlns:android="http://schemas.android.com/apk/res/android"
3.   android:layout_width="fill_parent"    android:layout_height="wrap_content">
4.   <ImageView android:id="@+id/itemImage" android:layout_width="wrap_content"
5.     android:layout_height="wrap_content" android:layout_centerHorizontal
       ="true"/>
6.   <TextView android:id="@+id/itemText" android:layout_width="wrap_content"
7.     android:layout_height="wrap_content"   android:layout_below="@+id/
       itemImage"
8.     android:layout_centerHorizontal="true" /> </RelativeLayout>
```

步骤3：实现主界面代码 GridViewActivity.java，并为 GridView 添加消息处理，如源代码清单4.35所示。

源代码清单4.35   \Ch04_Ex05_GridView\src\sziit\lihz\ch04_ex05_gridview\GridViewActivity.java

```java
1. public class GridViewActivity extends Activity {//从Activity活动基类派生子类
2.     GridView gridView;                          //声明GridView对象
3.     protected void onCreate(Bundle savedInstanceState) {//子类重写onCreate方法
4.         super.onCreate(savedInstanceState);     //基类调用onCreate方法
5.         setContentView(R.layout.activity_main); //设置活动界面布局
6.         //创建一个List对象,List对象的元素是Map
7.         List<Map<String, Object>> listItems = new ArrayList<Map<String, Object>>();
8.         for (int i = 0; i < 10; i++) {          //向listItems添加数据成员
9.             Map<String, Object> listItem = new HashMap<String, Object>();
                                                    //构建列表项
10.            listItem.put("itemImage", R.drawable.go);//添加图标
11.            listItem.put("itemText", "编号." + String.valueOf(i));//添加文本信息
12.            listItems.add(listItem);}           //将列表项添加到listItems中
13.        //创建一个SimpleAdapter
14.        SimpleAdapter adapter = new SimpleAdapter(this, listItems,
15.            R.layout.cell, new String[] { "itemImage", "itemText" },
16.            new int[] { R.id.itemImage, R.id.itemText });
17.        gridView = (GridView) findViewById(R.id.gridView);
                                                    //查找布局文件中的GridView对象
18.        gridView.setAdapter(adapter);           //为GridView设置Adapter
19.        gridView.setOnItemClickListener(new OnItemClickListener() {
                                                    //添加消息处理
20.            public void onItemClick(AdapterView<?> arg0, View arg1, int arg2,
21.                long arg3) {
22.                HashMap<String, Object> item = (HashMap<String, Object>) arg0
23.                    .getItemAtPosition(arg2);//访问被单击AdapterView对象中数据
24.                setTitle((String) item.get("itemText"));//设置活动标题
25.        }}); }}
```

3. 运行结果

选择【Run As】→【Android Application】命令，运行程序将会看到图 4.10 所示的用户登录界面。单击某项菜单后会将所选项设置为标题。

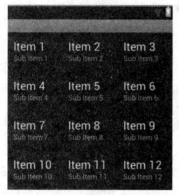 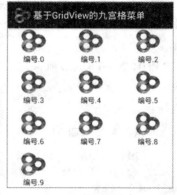

（a）GridView 布局示意图　　　　（b）程序运效果

图 4.10　九宫格菜单运行效果

4.4　利用列表控件实现 Metro UI 菜单

4.4.1　项目实训目标

- 掌握如何在程序中使用 GridView 组件。
- 掌握如何使用 StateListDrawable 资源文件。

4.4.2　知识准备

本项目主要目的在于巩固 GridView 组件和 StateListDrawable 资源文件这两个知识点。结合一个项目使用 StateListDrawable 资源文件来美化 GridView。

4.4.3　项目实践

1. 项目简介

Metro 是微软在 Windows Phone 7 中正式引入的一种界面设计语言，也是 Windows 8 的主要界面显示风格。接下来的程序中将使用 StateListDrawable 资源文件和 GridView 组件实现 Metro 风格菜单。

2. 实现步骤

实现要点：
- include：重用相同风格布局，此项目包含 activity_common_title_bar.xml 布局，重用标题栏。
- ImageView 的 android:src 属性：设置图像视图显示的图片来源。
- void startActivity(Intent intent)：启动意图所指定的活动。
- Intent(Context packageContext, Class<?> cls)：意图构造器。

- android:background="@drawable/main_default_bg"：设置布局背景图片。
- android:listSelector="@drawable/selector_list"：设置网格视图列表选择器属性，描述不同状态下列表项所显示图片。
- android:background="@drawable/title_back_selector"：设置文本视图在不同状态下的背景图片。

具体步骤如下：

步骤1：创建 Ch04_Ex06_Metro 项目，在项目的 res/drawable 目录下新建资源文件，如源代码清单 4.36 所示。通过选择器 selector 设置按钮在不同选择状态下所使用的图片资源，如源代码清单 4.37 所示。

源代码清单 4.36 \Ch04_Ex06_Metro\res\drawable\selector_list.xml

```
1.  <?xml version="1.0" encoding="utf-8"?>
2.  <selector xmlns:android="http://schemas.android.com/apk/res/android">
3.      <item android:drawable="@drawable/metro_home_blocks_hover"
        android:state_selected="true"/>
4.      <item android:drawable="@drawable/metro_home_blocks_hover"
        android:state_pressed="true"/>
5.      <item android:drawable="@drawable/metro_home_blocks_hover"
        android:state_focused="true"/>
6.      <item android:drawable="@drawable/metro_home_blocks"
        android:state_focused="false" android:state_pressed="false"/>
7.  </selector>
```

源代码清单 4.37 \Ch04_Ex06_Metro\res\drawable\title_back_selector.xml

```
1.  <?xml version="1.0" encoding="utf-8"?>
2.  <selector xmlns:android="http://schemas.android.com/apk/res/android">
3.      <item android:drawable="@drawable/back_btn_bg_hover"
        android:state_pressed="true"/>
4.      <item android:drawable="@drawable/back_btn_bg_hover"
        android:state_focused="true"/>
5.      <item android:drawable="@drawable/back_btn_bg" android:state_focused=
        "false" android:state_pressed="false"/>   </selector>
```

步骤2：实现该程序中的三个界面布局，如源代码清单 4.38、4.39、4.40 所示。

源代码清单 4.38 \Ch04_Ex06_Metro\res\layout\activity_main.xml（主活动界面）

```
1.  <?xml version="1.0" encoding="utf-8"?>
2.  <RelativeLayout xmlns:android="http://schemas.android.com/apk/res/android"
3.   android:layout_width="fill_parent"  android:layout_height="fill_parent"
4.   android:background="@drawable/main_default_bg" >
5.  <include //包含命令
6.      android:id="@id/bt_createtask_title_layout" layout="@layout/activity_
        common_title_bar"/>
7.  <RelativeLayout android:id="@+id/RelativeLayout1" android:layout_width
        ="match_parent"
8.
9.      android:layout_height="match_parent" android:layout_below="@id/bt_
        createtask_title_layout"
10.     android:layout_centerHorizontal="true" android:layout_marginLeft="15dp"
```

```
11.        android:layout_marginRight="15dp"    android:orientation="vertical"
12.        android:visibility="visible" >
13.    <LinearLayout android:id="@+id/LinearLayout1" android:layout_width=
       "match_parent"
14.        android:layout_height="match_parent"   android:layout_alignParentTop
           ="true"
15.        android:orientation="vertical" >
16.        <GridView android:id="@+id/gdv_main" android:layout_width="match_parent"
17.            android:layout_height="wrap_content" android:horizontalSpacing="10dp"
18.            android:listSelector="@drawable/selector_list" android:numColumns="2"
19.            android:verticalSpacing="10dp">    </GridView>
20.        </LinearLayout>    </RelativeLayout>    </RelativeLayout>
```

源代码清单 4.39　\Ch04_Ex06_Metro\res\layout\activity_common_title_bar.xml（标题栏）

```
1.  <?xml version="1.0" encoding="utf-8"?>
2.  <RelativeLayout xmlns:android="http://schemas.android.com/apk/res/android"
3.      android:id="@id/common_title_bar" android:layout_width="fill_parent"
4.      android:layout_height="54.0dip"  android:background="@android:color/
        transparent">
5.      <TextView android:id="@id/titlebar_left"   android:layout_width="48dp"
6.          android:layout_height="48dp"    android:layout_alignParentLeft="true"
7.          android:layout_centerVertical="true" android:layout_marginLeft="8.0dip"
8.          android:background="@drawable/title_back_selector"
9.          android:textSize="@dimen/bt_cm_textsize_17" />
10.     <TextView android:id="@id/titlebar_title" android:layout_width="wrap_
        content"
11.         android:layout_height="wrap_content"  android:layout_centerVertical
            ="true"
12.         android:layout_marginLeft="8.0dip"     android:layout_toRightOf="@id/
            titlebar_left"
13.         android:text="@string/back"        android:textColor="#ffffffff"
14.         android:textSize="@dimen/bt_cm_textsize_17" />
15.     <TextView android:id="@id/titlebar_right" android:layout_width="wrap_
        content"
16.         android:layout_height="wrap_content"   android:layout_alignParentRig
            ht="true"
17.         android:layout_centerVertical="true"    android:layout_marginRight=
            "8.0dip"
18.         android:background="@drawable/title_back_selector"    android:gravity
            ="center"
19.         android:visibility="gone" />   </RelativeLayout>
```

源代码清单 4.40　\Ch04_Ex06_Metro\res\layout\view_squared_item.xml（矩形视图布局）

```
1.  <?xml version="1.0" encoding="utf-8"?>
2.  <LinearLayout xmlns:android="http://schemas.android.com/apk/res/android"
3.      android:layout_width="fill_parent" android:layout_height="fill_parent"
4.      android:layout_weight="1.0"     android:background="@drawable/selector_list"
5.      android:gravity="center"      android:orientation="vertical">
6.      <ImageView android:id="@+id/itemImage" android:layout_width="wrap_content"
7.          android:layout_height="wrap_content" android:src="@drawable/metro_ho
            me_blacks_browser"/>
```

```
8.    <TextView android:id="@+id/itemText" android:layout_width="wrap_content"
9.        android:layout_height="wrap_content" android:clickable="false"
10.       android:paddingTop="10.0dip"      android:text="情景模式"
11.       android:textColor="#ffffffff"      android:textSize="14.0sp"/>
    </LinearLayout>
```

步骤3：实现程序的主界面，在该界面中为GridView添加事件处理，当用户单击其中的一项时将跳转至相应界面。如源代码清单4.41所示。

源代码清单4.41 \Ch04_Ex06_Metro\src\sziit\lihz\ch04_ex06_metro\MainActivity.java

```
1.  public class MainActivity extends Activity {       //从活动基类派生子类
2.      private TextView titlebar_left;               //声明TextView对象（左标题）
3.      private TextView titlebar_title;              //声明TextView对象（标题）
4.      protected void onCreate(Bundle savedInstanceState) {
                                                      //子类重写onCreate方法
5.         super.onCreate(savedInstanceState);        //基类调用onCreate方法
6.         setContentView(R.layout.activity_main);    //设置活动界面布局
7.         titlebar_left = (TextView) findViewById(R.id.titlebar_left);
                                                      //从布局查找TextView
8.         titlebar_left.setOnClickListener(new View.OnClickListener() {
                                                      //设置按钮单击监听器
9.            public void onClick(View v) {           //重写单击接口方法
10.              onBackPressed();}});                 //返回前页面
11.        titlebar_title = (TextView) findViewById(R.id.titlebar_title);
                                                      //从布局查找TextView
12.        titlebar_title.setText(R.string.app_name); //设置标题栏显示信息
13.        GridView gridView = (GridView) findViewById(R.id.gdv_main);
                                                      //从布局中查找GridView
14.        ArrayList<HashMap<String, Object>> lstImageItem = new ArrayList<HashMap<String, Object>>();                       //构造数组列表对象
15.        HashMap<String, Object> map = new HashMap<String, Object>();
                                                      //构造哈希映射对象
16.        map.put("ItemImage", R.drawable.icon_model);//向map中添加数据成员
17.        map.put("ItemText", "情景模式"); lstImageItem.add(map);
                                                      //向lstImageItem中添加数据
18.        map = new HashMap<String, Object>();map.put("ItemImage", R.drawable.icon_room);
19.        map.put("ItemText", "我的房间"); lstImageItem.add(map);
                                                      //向lstImageItem中添加数据
20.        map = new HashMap<String, Object>();map.put("ItemImage", R.drawable.icon_electric);
21.        map.put("ItemText", "家用电器"); lstImageItem.add(map);
                                                      //向lstImageItem中添加数据
22.        map = new HashMap<String, Object>();map.put("ItemImage", R.drawable.icon_camera);
23.        map.put("ItemText", "视频监控"); lstImageItem.add(map);
                                                      //向lstImageItem中添加数据
24.        map = new HashMap<String, Object>();map.put("ItemImage", R.drawable.icon_setting);
25.        map.put("ItemText", "定时任务"); lstImageItem.add(map);
                                                      //向lstImageItem中添加数据
```

```
26.     map = new HashMap<String, Object>();map.put("ItemImage", R.drawable.
        icon_setting);
27.     map.put("ItemText", "报警记录"); lstImageItem.add(map);
                                           //向lstImageItem中添加数据
28.     map = new HashMap<String, Object>();map.put("ItemImage", R.drawable.
        icon_setting);
29.     map.put("ItemText", "设置");      lstImageItem.add(map);
                                           //向lstImageItem中添加数据
30.     map = new HashMap<String, Object>();map.put("ItemImage", R.drawable.
        icon_add);
31.     map.put("ItemText", "预留");      lstImageItem.add(map);
                                           //向lstImageItem中添加数据
32.     SimpleAdapter saImageItems = new SimpleAdapter(this, lstImageItem,
33.         R.layout.view_squared_item, new String[] { "ItemImage","ItemText" },
34.         new int[] { R.id.itemImage, R.id.itemText });//构造简单适配器
35.     gridView.setAdapter(saImageItems);//设置网格视图适配器
36.     gridView.setOnItemClickListener(new OnItemClickListener() {
                                           //设置网格单击监听器
37.         public void onItemClick(AdapterView<?> arg0, View arg1, int arg2,
            long arg3) {
38.             switch (arg2) {
39.             case 0:                    //启动创建的匿名意图对象,实现页面切换
40.                 startActivity(new Intent(MainActivity.this,ModelListActivity.
                    class)); break;
41.             case 1:                    //启动创建的匿名意图对象,实现页面切换
42.                 startActivity(new Intent(MainActivity.this,RoomListActivity.
                    class)); break;
43.             case 2:                    //启动创建的匿名意图对象,实现页面切换
44.                 startActivity(new Intent(MainActivity.this,FunctionList
                    Activity.class));break;
45.             case 3:                    //启动创建的匿名意图对象,实现页面切换
46.                 startActivity(new Intent(MainActivity.this,VideoActivity.
                    class));        break;
47.             case 4:                    //启动创建的匿名意图对象,实现页面切换
48.                 startActivity(new Intent(MainActivity.this,TimingTask
                    Activity.class)); break;
49.             case 5:                    //启动创建的匿名意图对象,实现页面切换
50.                 startActivity(new Intent(MainActivity.this,AlarmRecord
                    Activity.class));break;
51.             case 6:                    //启动创建的匿名意图对象,实现页面切换
52.                 startActivity(new Intent(MainActivity.this,ConfigActivity.
                    class));    break;
53.         }} });
54.     gridView.setSelector(new ColorDrawable(Color.TRANSPARENT));}
55.     public boolean onKeyDown(int keyCode, KeyEvent event) {
56.         if (keyCode == KeyEvent.KEYCODE_BACK) { exitBy2Click();}
57.         return false; }
58.     private static Boolean isExit = false;
```

```
59.    private void exitBy2Click() {   Timer tExit = null;
60.       if (isExit == false) { isExit = true;
61.          Toast.makeText(this, "再按退出将程序", Toast.LENGTH_SHORT).show();
62.          tExit = new Timer();                    //启动定时器，延迟2秒后退出程序
63.          tExit.schedule(new TimerTask() {public void run() {isExit = false;
             }}, 2000);
64.       } else { finish();}}}                     //结束程序
```

步骤4：除主界面外，用户单击程序中每个item都将跳转至相应的Activity界面，这些界面不是该程序的重点，为了演示效果，其余Activity都只包含一个TextView，让它显示与主界面上的ItemText相对应的文字。例如当用户单击"情景模式"时将会跳转至ModelListActivity，ModelListActivity代码如源代码清单4.42所示，其余界面与之类似。注意在AndroidManifest.xml文件中注册所用到的所有活动。

源代码清单4.42   \Ch04_Ex06_Metro\src\sziit\lihz\ch04_ex06_metro\ModelListActivity.java

```
1. public class ModelListActivity extends Activity { //从活动基类派生子类
2.    protected void onCreate(Bundle savedInstanceState) {
                                                 //子类重写onCreate方法
3.       super.onCreate(savedInstanceState);     //基类调用onCreate方法
4.       TextView view=new TextView(this);       //构造文本视图对象
5.       view.setBackgroundColor(Color.WHITE);   //设置背景色
6.       view.setTextColor(Color.BLUE);          //设置字体颜色
7.       view.setTextSize(20);                   //设置字体大小
8.       view.setText("你已成功进入情景模式界面！");
9.       super.setContentView(view);}}           //设置活动界面
```

3. 运行结果

选择【Run As】→【Android Application】命令，运行程序，将会看到图4.11所示的用户登录界面。

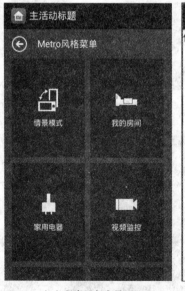

（a）程序运行主界面　　　　　（b）单击"情景模式"后的界面

图4.11　Metro风格菜单运行效果

4.5 获取 URL 地址上的资源

4.5.1 项目实训目标

- 了解 HTTP 通信编程相关知识。
- 掌握多线程编程的相关知识。

4.5.2 知识准备

1. HTTP 通信知识

HTTP（Hyper Text Transfer Protocol，超文本传输协议）用于传送 WWW 方式的数据。HTTP 采用请求/响应模型。客户端向服务器发送一个请求，请求头包含了请求的方法、URI、协议版本，以及包含请求修饰符、客户信息和内容的类似于 MIME 的消息结构。服务器以一个状态行作为响应，响应的内容包括消息协议的版本、成功或者错误编码，还包括服务器信息、实体元信息，以及可能的实体内容。

Google 以网络搜索引擎著称，自然而然也会使 Android SDK 拥有强大的 HTTP 访问能力。在 Android SDK 中，Google 集成了 Apache 的 HttpClient 模块。要注意的是，这里的 Apache HttpClient 模块是 HttpClient 4.0（org.apache.http.*），而不是 Jakarta Commons HttpClient 3.x（org.apache.commons.httpclient.*）。

在 HTTP 通信中，客户端发送的每次请求都需要服务器来响应，在请求结束后会主动释放连接。将从连接的建立到关闭的这个过程称为"一次连接"。要保持客户端在线的状态，就要不停地向服务器端发送连接请求。若服务器长时间无法收到客户端的请求，则认为客户端下线；若客户端长时间无法收到服务器的回复，则认为网络已经断开。在大多数情况下，需要服务器主动向客户端发送数据，保持客户端与服务器数据的同步，若通过 HTTP 建立连接，则服务器需要等到客户端发送一次请求后才能将数据回复给客户端，因此客户端定时向服务器发送连接请求不仅可以保持在线，同时也是在"询问"服务器是否有新的数据，如果有新数据就传送给客户端。但是，当需要多人同时在线联网时，HTTP 连接就已经不能很好地满足需要，所以就需要 Socket 通信。

2. Android 中的 HTTP 通信

（1）HttpGet 与 HttpPost

使用 Apache 提供的 HttpClient 接口同样可以进行 HTTP 操作。GET 和 POST 请求方法的操作有所不同。GET 方法的操作代码示例如下：

```
1. String httpUrl = "http://192.168.1.110:8080/httpget.jsp?par=HttpClient_
   android_Get";                                        //HTTP 地址
2. HttpGet httpRequest = new HttpGet(httpUrl);          //HttpGet 连接对象
3. HttpClient httpclient = new DefaultHttpClient();     //取得 HttpClient 对象
4. HttpResponse httpResponse = httpclient.execute(httpRequest);
                                   //请求 HttpClient，取得 HttpResponse
5. if (httpResponse.getStatusLine().getStatusCode() == HttpStatus.SC_OK){
                                                         //请求成功
6.    String strResult = EntityUtils.toString(httpResponse.getEntity());
```

```
                                                   //取得返回的字符串
7.    mTextView.setText(strResult); } else { mTextView.setText("请求错误!"); }
```

使用 POST 方法进行参数传递时，需要使用 NameValuePair 来保存要传递的参数，还需要设置所使用的字符集。代码示例如下：

```
1. String httpUrl = "http://192.168.1.110:8080/httpget.jsp";  //HTTP 地址
2. HttpPost httpRequest = new HttpPost(httpUrl);              //HttpPost 连接对象
3. //使用 NameValuePair 来保存要传递的 Post 参数
4. List<NameValuePair> params = new ArrayList<NameValuePair>();
5. params.add(new BasicNameValuePair("par", "HttpClient_android_Post"));
                                                   //添加要传递的参数
6. HttpEntity httpentity = new UrlEncodedFormEntity(params, "gb2312");
                                                   //设置字符集
7. httpRequest.setEntity(httpentity);              //请求 httpRequest
8. HttpClient httpclient = new DefaultHttpClient();//取得默认的 HttpClient
9. HttpResponse httpResponse = httpclient.execute(httpRequest);
                                                   //取得 HttpResponse
10. //HttpStatus.SC_OK 表示连接成功
11. if (httpResponse.getStatusLine().getStatusCode() == HttpStatus.SC_
    OK) {                                          //取得返回的字符串
12.     String strResult = EntityUtils.toString(httpResponse.getEntity());
        mTextView.setText(strResult);
13. } else { mTextView.setText("请求错误!"); }
```

HttpClient 实际上是对 Java 提供方法的一些封装，在 HttpURLConnection 的输入/输出流中操作，在这个接口中被统一封装成了 HttpPost(HttpGet)和 HttpResponse，这样就降低了操作的烦琐性。

需要注意的是，由于 Android 的很多操作都涉及权限问题，例如以前学习的打电话和发短信等，都需要权限。而 Android 尝试连续网络时，也需要权限。因此，在 AndroidManifest.xml 中加入网络连接权限：

```
<uses-permission android:name="android.permission.INTERNET" />
```

（2）HttpURLConnection

HttpURLConnection 是继承于 URLConnection 类，两者都是抽象类。其对象主要通过 URL 的 openConnection 方法获得。创建方法如下：

```
URL url = new URL("http://192.168.1.88:8888/abc/login.jsp");
HttpURLConnection conn = (HttpURLConnection) url.openConnection();
```

通过以下方法可以对请求的属性进行设置：

```
1. conn.setConnectTimeout(5*1000);conn.setRequestMethod("GET");
2. conn.setRequestProperty("Accept",
3.    "image/gif,image/jpeg,image/pjpeg,image/pjpeg,"
4.    +"application/x-shockwave-flash,application/xaml+xml,"
5.    +"application/vnd.ms-xpsdocument,application/x-ms-xbap,"
6.    +"application/x-ms-application,application/vnd.ms-excel,"
7.    +"application/vnd.ms-powerpoint,application/msword,*/*");
8. conn.setRequestProperty("Accept-Language","zh-CN");
9. conn.setRequestProperty("Charset", "UTF-8");
10.conn.setRequestProperty("Connection", "Keep-Alive");
```

HttpURLConnection 默认使用 Get 方式，例如下面这段代码：

```
1. //使用 HttpURLConnection 打开连接
2. HttpURLConnection urlConn = (HttpURLConnection) url.openConnection();
3. //得到读取的内容(流)
4. InputStreamReader in = new InputStreamReader(urlConn.getInputStream());
5. BufferedReader buffer = new BufferedReader(in);   //为输出创建 BufferedReader
6. String inputLine = null;
7. while (((inputLine = buffer.readLine()) != null)) {   //使用循环来读取获得的数据
8.    resultData += inputLine + "\n"; }           //在每一行后面加上一个"\n"来换行
9. in.close();                                    //关闭 InputStreamReader
10.urlConn.disconnect();                          //关闭 HTTP 连接
```

如果需要使用 POST 方式，则需要 setRequestMethod 设置。代码如下：

```
1. String httpUrl = "http://192.168.1.110:8080/httpget.jsp";
2. String resultData = "";           //获得的数据
3. URL url = null;
4. try {   url = new URL(httpUrl);}   //构造一个 URL 对象
5. catch (MalformedURLException e)  { Log.e(DEBUG_TAG, "MalformedURLExcepti
   on");  }
6. if (url != null)  {
7.  try {                           //使用 HttpURLConnection 打开连接
8.   HttpURLConnection urlConn = (HttpURLConnection) url.openConnection();
9.   urlConn.setDoOutput(true);     //因为这个是 post 请求, 设立需要设置为 true
10.  urlConn.setDoInput(true);
11.  urlConn.setRequestMethod("POST");//设置以 POST 方式
12.  urlConn.setUseCaches(false);    //Post 请求不能使用缓存
13.  urlConn.setInstanceFollowRedirects(true);
14.  //配置本次连接的 Content-type, 配置为 application/x-www-form-urlencoded 的连接
15.  urlConn.setRequestProperty("Content-Type","application/x-www-form-url
     encoded");
16.  //从 postUrl.openConnection()至此的配置必须要在连接之前完成
17.  //要注意的是 connection.getOutputStream 会隐式地进行连接
18.  urlConn.connect();
19.  DataOutputStream out = new DataOutputStream(urlConn.getOutputStream());
                                     //DataOutputStream 流
20.  String content = "par=" + URLEncoder.encode("ABCDEFG", "gb2312");
                                     //要上传的参数
21.  out.writeBytes(content);        //将要上传的内容写入流中
22.  out.flush();                    //刷新
23.  out.close();}}                  //关闭
```

（3）权限验证

首先需要明确的是，权限验证不同于参数传递。例如，在访问网页时，如果需要翻页，有的会把页数 id 这个参数传递过去，通过 id 指向对应的页面。而权限验证，比如是在路由器下，访问 192.168.0.1 时，会要求输入账号和密码，这就不能通过参数传递的方法。权限验证的方法如下：

```
DefaultHttpClient client = new DefaultHttpClient();
client.getCredentialsProvider().setCredentials(AuthScope.ANY,
  new UsernamePasswordCredentials("admin", "password"));
```

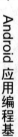

3. Android 中的多线程编程

在 Android 多线程编程中，经常要使用 Handler、Thread 和 Runnbale 这三个类，下面简要介绍它们之间的关系。

首先，Android 的 CPU 分配的最小单元是线程，Handler 一般是在某个线程里创建的，因而 Handler 和 Thread 就是相互绑定的，一一对应。而 Runnable 是一个接口，Thread 是 Runnbale 的子类。所以说，它们都算是一个线程。

HandlerThread 顾名思义就是可以处理消息循环的线程，它是一个拥有 Looper 的线程，可以处理消息循环。与其说 Handler 和一个线程绑定，不如说 Handler 是和 Looper 一一对应的。下面通过一段代码解释一下，以便于理解。在 UI 线程（主线程）中加入这段代码：

```
1. mHandler=new Handler();
2. mHandler.post(new Runnable(){ void run(){
3. //执行代码...
4. } });
```

这个线程其实是在 UI 线程之内运行的，并没有新建线程。常见的新建线程的方法是：

```
1. Thread thread = new Thread();   thread.start();
2. HandlerThread thread = new HandlerThread("string");  thread.start();
```

1）线程（Thread）

（1）Thread 概述

进程是计算机程序在特定数据集合上的一次执行过程，而线程是特定程序代码的一个执行路径。同一个进程的多个线程共享地址空间、全局变量、操作系统资源。线程是处理器调度的最小单位，每个线程有独立的调用栈及处理器资源。

Android 提供了多线程支持，在 Android 程序中，VM 采用抢占式调度模式对线程进行调度：基于线程优先级来决定 CPU 的使用权。下面从程序开发的角度介绍 Android Thread 的使用方法及其状态变化。

（2）如何创建/启动一个线程

Thread 类提供了 8 种构造函数来灵活创建 Thread 对象，但最常用的是以下 4 种方法：

方法 1：创建 Thread 实例。

```
1. Thread myThread = new Thread(){                    //构造 Thread 线程对象
2.   public void run() {                              //直接重写 run()方法
3.     Log.i("Thread", String.valueOf(getState()));   //日志信息
4.     //执行具体功能
5.   }}; myThread.start();                            //启动线程
```

此方法直接重写 Thread 的 run()函数来实现自己的线程，比较方便简单。

方法 2：创建 Thread 子类实例。

```
1. public class myThread extends Thread {             //从线程基类派生子线程类
2.   public void run() {                              //子线程类重写 run()方法
3.     Log.i("Thread", String.valueOf(getState()));   //日志信息
4.     //执行具体功能
5.   } } new myThread().start();                      //构造子线程对象，并启动该线程
```

此方法创建一个 Thread 子类来实现自己的线程，可以在子类中实现子线程特有的逻辑，便于多线程开发。

方法 3：使用 Runnable 实例化 Thread，多线程共享一个 Runnable。

```
1. Runnable  r = new Runnable(){        //构造 Runnable 对象
2.    private int count = 100;          //初始化计数器
3.    public void run() {               //重写 run()方法
4.       Log.i("Thread", String.valueOf(count));  //日志信息
5.       count--;                       //计数器减一
6.    //执行具体功能
7.    } };new Thread(r).start();        //以 Runnable 对象构造线程对象，并启动该线程
```

此方法使用同一个 Runnable 实例初始化线程，所有基于此 Runnable 的线程共享同一个 Runnable 实例，可实现多线程的资源共享，但没有 Thread 子类灵活，Runnable 实例只能有一个 run 接口。

方法 4：使用 Runnable 实例化 Thread，多线程独占 Runnable 实例。

```
1. Thread myThread = new Thread(new Runnable() {
                                       //以匿名 Runnable 对象作为参数构造线程
2.    public void run() {               //重写 run()方法
3.       Log.i("Thread", String.valueOf(getState()));
4.    //执行具体功能}
5.}); myThread.start();                 //启动线程
```

此方法使用 Runnable 实例初始化线程，但所有线程都有自己的 Runnable 实例。

如果以上几种方法不满足特定的应用场景，可选择 Thread.java 提供更多构造函数：

- Thread()，此函数不带参数。一般需要重载 run()函数来创建对象，否则该 Thread 是空线程，没有具体事情可做。默认新创建的 Thread 与创建者属于同一个 ThreadGroup，并自动分配 Thread 名字为 "Thread-" + id。
- Thread(Runnable runnable)，提供一个 Runnable 对象来创建 Thread 对象，于是 Thread 的执行体就是 Runnable 对象的 run()。默认新创建的 Thread 与创建者属于同一个 ThreadGroup，并自动分配 Thread 名字为 "Thread-" + id。
- Thread(Runnable runnable, String threadName)，提供一个 Runnable 对象来创建 Thread 对象，同时把新 Thread 命名为 threadName。Thread 的执行体就是 Runnable 对象的 run()。默认新创建的 Thread 与创建者属于同一个 ThreadGroup。
- Thread(String threadName)，创建名为 threadName 的 Thread 对象。一般需要重载 run()函数来创建对象，否则该 Thread 是空线程，没有具体事情可做。默认新创建的 Thread 与创建者属于同一个 ThreadGroup，并自动分配 Thread 名字为 "Thread-" + id。

2）句柄（Handler）

Android 的消息传递机制是另一种形式的"事件处理"，这种机制主要是为了解决 Android 应用的多线程问题——Android 平台不允许 Activity 新启动的线程访问该 Activity 里的界面组件，这样就会导致新启动的线程无法动态改变界面组件的属性值。但在实际 Android 应用开发中，尤其是涉及动画的游戏开发中，需要让新启动的线程周期性地改变界面组件的属性值，这就需要借助于 Handler 的消息传递机制来实现。

Handler 类的主要作用有两个：①在新启动的线程中发送消息；②在主线程中获取、处理消息。

上面的说法看上去很简单，似乎只要分成两步即可。但这个过程涉及一些问题：新启动的线程何时发送消息呢？主线程何时去获取并处理消息呢？这个时机显然不好控制。

为了让主线程能适时地处理新启动的线程所发送的消息，显然只能通过回调的方式来实现——开发者只要重写 Handler 类中处理消息的方法，当新启动的线程发送消息时，Handler 类中处理消息的方法被自动回调。Handler 类包含如下方法用于发送、处理消息。

- void handleMessage(Message msg)：处理消息的方法。该方法通常用于被重写。
- final boolean hasMessages(int what)：检查消息队列中是否包含 what 属性为指定值的消息。
- final boolean hasMessages(int what, Object object)：检查消息队列中是否包含 what 属性为指定值且 object 属性为指定对象的消息。
- 多个重载的 Message obtainMessage()：获取消息。
- sendEmptyMessage(int what)：发送空消息。
- final boolean sendEmptyMessageDelayed(int what, long delayMillis)：指定多少毫秒之后发送空消息。
- final boolean sendMessage(Message msg)：立即发送消息。

4.5.3 项目实践

实现要点：

- ImageView：图像视图控件，在 XML 布局文件中通过属性 android:src="@drawable/ic_launcher" 设置显示图片来源，在 Java 代码中通过 imageView.setImageBitmap(bitmap)设置显示图片。
- Thread：本项目通过直接构造线程对象，然后直接重写 run()实现进程执行体，然后调用 start()启动该线程。

该程序比较简单，访问一个 Web 网址获取图片并显示在界面上，步骤如下：

步骤 1：创建 Ch04_Ex07_URL 项目，程序界面只包含一个 ImageView 用于显示从 Web 站点上获取的图片。如源代码清单 4.43 所示。

源代码清单 4.43　\CH04\Ch04_Ex07_URL\res\layout\activity_main.xml

```
1.  <RelativeLayout xmlns:android="http://schemas.android.com/apk/res/android"
2.      xmlns:tools="http://schemas.android.com/tools"
3.      android:layout_width="match_parent" android:layout_height="match_parent"
4.      android:paddingBottom="@dimen/activity_vertical_margin"
5.      android:paddingLeft="@dimen/activity_horizontal_margin"
6.      android:paddingRight="@dimen/activity_horizontal_margin"
7.      android:paddingTop="@dimen/activity_vertical_margin"
8.      tools:context=".URLActivity">
9.      <ImageView  android:id="@+id/imageView"
10.         android:layout_width="wrap_content"
11.         android:layout_height="wrap_content"
12.         android:layout_centerHorizontal="true"
13.         android:contentDescription="@string/app_name"
14.         android:src="@drawable/ic_launcher"/>
15. </RelativeLayout>
```

步骤 2：界面部分，程序新开一个线程，在新开线程中访问 Web 网址获取图片并显示在界面中。如源代码清单 4.44 所示。

源代码清单 4.44　\CH04\Ch04_Ex07_URL\src\sziit\lihz\ch04_ex07_url\URLActivity.java

```
1. public class URLActivity extends Activity {        //从 Activity 活动基类派生子类
```

```
2.      private final int MESSAGE_SHOW = 0x123;     //声明自定义消息常量
3.      ImageView imageView;                         //声明 ImageView 对象
4.      Bitmap bitmap;                               //代表从网络下载得到的图片
5.      Handler handler = new Handler() {            //构造句柄对象
6.          public void handleMessage(Message msg) {//处理收到的消息
7.              if (msg.what == MESSAGE_SHOW) {      //接收到预定义消息
8.                  imageView.setImageBitmap(bitmap); //使用 ImageView 显示该图片
9.      }}};
10.     protected void onCreate(Bundle savedInstanceState) {
                                                     //子类重写 onCreate 方法
11.         super.onCreate(savedInstanceState);      //基类调用 onCreate 方法
12.         setContentView(R.layout.activity_main);  //设置活动界面布局
13.         imageView = (ImageView) findViewById(R.id.imageView);
                                                     //在布局中查找 ImageView 对象
14.         new Thread() {                           //构造进程
15.           public void run() {                    //实现进程执行体
16.             try {                                //捕获异常
17.             //定义一个 URL 对象
18.             URL url = new URL("http://www.sziit.com.cn/images2013/logo.jpg");
19.             InputStream is = url.openStream();//打开该 URL 对应的资源的输入流
20.             bitmap = BitmapFactory.decodeStream(is);
                                                     //从 InputStream 中解析出图片
21.             handler.sendEmptyMessage(MESSAGE_SHOW);
                                                     //发送消息、通知 UI 组件显示该图片
22.                 is.close();                      //关闭输入流
23.             } catch (Exception e) {              //处理异常
24.                 e.printStackTrace();}
25.         }.start();}}                             //启动进程
```

运行结果：

选择【Run As】→【Android Application】命令，运行程序，将会看到图 4.12 所示的用户登录界面。该程序调用的 URL 对象的 openStream()方法打开 URL 对应资源的输入流，并使用 BitmapFactory 的 decodeStream(InputStream in)方法来解析该输入的图片，使之在界面上进行显示。

（a）主界面布局轮廓

（b）程序运行效果

图 4.12 获取 URL 地址上的资源运行效果

4.6 利用天气预报接口编写天气查询软件

4.6.1 项目实训目标

掌握 JSON 数据的序列化和反序列化。

4.6.2 知识准备

1. 什么是 JSON?

JSON 的全称是 JavaScript Object Notation，即 JavaScript 对象符号，它是一种轻量级的数据交换格式。JSON 的数据格式既适合人来读/写，也适合计算机解析和生成。最初，JSON 是 JavaScript 语言的数据交换格式，后来慢慢发展成一种语言无关的数据交换格式，这一点类似于 XML。

相比于 XML 这种数据交换格式来说，解析 JSON 不需要像解析 XML 那样编写大段的代码，所以客户端和服务器的数据交换格式往往通过 JSON 来进行交换。尤其是对于 Web 开发来说，JSON 数据格式在客户端直接可以通过 JavaScript 来进行解析。

2. JSON 的数据类型

JSON 主要有如下两种数据结构。

（1）键值对（key-value）组成的数据结构。一个对象以"{"（左花括号）开始，以"}"（右花括号）结束。每个"名称"后跟一个":"（冒号）；'名称/值' 对"之间使用","（逗号）分隔，如图 4.13 所示。

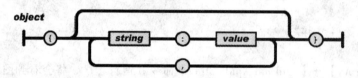

图 4.13　键值对

例如：{"name": "xiaoluo"}就是一个最简单的 JSON 对象，对于这种数据格式，key 值必须要是 string 类型，而对于 value，则可以是 string、number、object、array 等数据类型，如图 4.14 所示。

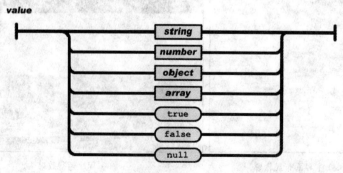

图 4.14　value 的类型

（2）有序集合。这种形式被称为是 jsonArray，数组是值（value）的有序集合。一个数组以"["（左方括号）开始，"]"（右方括号）结束。值之间使用","（逗号）分隔，如图 4.15 所示。

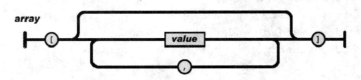

图 4.15　有序集合

更多的有关 JSON 数据格式可以参见 JSON 的官网 http://www.json.org。

3. JSON 数据的序列化和反序列化

这里将使用两种 JSON 封装库来对 JSON 数据格式进行解析，以及生成 JSON 数据格式。

1）Jackson

首先，从 http://jackson.codehaus.org/1.7.6/jackson-all-1.7.6.jar 下载 Jackson 的 jar 包。

把 jackson-all-1.7.6.jar 复制到 libs（在项目根目录新建一个 libs 文件夹）中。Jackson 框架提供以下几种方式处理 JSON：

（1）流模型，具有最好的性能（最低的开销，最快的读写速度；其他两种方式都是基于它的）。

（2）数据绑定，通常最方便。

（3）树模型，最灵活。

下面分别讲解几个实例来学习如何在项目中使用 Jackson 框架。

（1）完全数据绑定（POJO）。

org.codehaus.jackson.map.ObjectMapper 用于将 JSON 数据映射成普通的 Java 对象（plain old Java objects，POJOs）。因此，"完全数据绑定"就是将 JSON 内容完整地映射成 Java 对象，它们一一对应。这是最简单的情况。例如，对于给定的 JSON 数据：

```
1. { "name": {  "first": "Joe", "last": "Sixpack" }, "gender": "MALE",
2.      "verified": false,"userImage": "Rm9vYmFyIQ==" }
```

用两行 Java 代码就可以把它转换成一个 User 实例：

```
1. ObjectMapper mapper = new ObjectMapper();   //可重用，全局共享
2. User user = mapper.readValue(new File("user.json"), User.class);
```

User 类的定义如下所示。

```
1. public class User {
2.   public enum Gender { MALE, FEMALE };
3.   public static class Name { private String _first, _last;
4.      public String getFirst() { return _first; }
5.      public String getLast() { return _last; }
6.      public void setFirst(String s) { _first = s; }
7.      public void setLast(String s) { _last = s; } }
8.   private Gender _gender;         private Name _name;
9.   private boolean _isVerified;    private byte[] _userImage;
10.  public Name getName() { return _name; }
```

```
11.    public boolean isVerified() { return _isVerified; }
12.    public Gender getGender() { return _gender; }
13.    public byte[] getUserImage() { return _userImage; }
14.    public void setName(Name n){ _name = n; }
15.    public void setVerified(boolean b) { _isVerified = b; }
16.    public void setGender(Gender g) { _gender = g; }
17.    public void setUserImage(byte[] b) { _userImage = b; }
18.    public String toString() { return "name=" + this._name._last + " " +
19.            this._name._first + ",gender=" + this._gender;  }}
```

将 User 对象转换成 JSON，并保存成名为 user-modified.json 的文件，如下所示。

```
mapper.writeValue(new File("user-modified.json"), user);
```

对于某些数据绑定（例如，把格式化的日期编排成 java.util.Date），Jackson 提供注解自定义重排的处理。

总结：完全数据绑定是最简单的情况。我们很容易想到，一个是 Java 系统，另一个是.NET 系统，显然它们的类结构会不同（除非当初设计好了），即便是同一个框架下的不同子系统，要让它们通信，一个系统序列化成 JSON，另一个系统反序列化 JSON 时完全不知道是什么。最好的办法是，直接序列化成另一个系统能识别的 JSON。千万不能将序列化/反序列化理解得太狭隘，不是仅调用序列化/反序列化这两个函数。

（2）Raw 数据绑定示例。

Raw 数据绑定也称为"非类型"，有时称为简单数据绑定。在一些情况下，没有明确的 Java 类（也不想这么做）去绑定 JSON，那么非类型的数据绑定是最好的方法。它的使用与完全数据绑定一样，只是简单地规定把 Object.class（或是 Map.class、List.class、String[].class 等）作为绑定类型。User 的 JSON 绑定如下所示。

```
Map<String, Object> userData = mapper.readValue(new File("user.json"),Map.class);
```

userData 可以如下列代码显示构造：

```
1. Map<String, Object> userData = new HashMap<String, Object>();
2. Map<String, String> nameStruct = new HashMap<String, String>();
3. nameStruct.put("first","Joe");           nameStruct.put("last","Sixpack");
4. userData.put("name", nameStruct);         userData.put("gender", "MALE");
5. userData.put("verified", Boolean.FALSE); userData.put("userImage", "Rm9v
   YmFyIQ==");
```

如果构造一个 Map，或从 JSON 构造，并可能进行修改，那么可以跟之前一样写成 JSON 文件：

```
mapper.writeValue(new File("user-modified-map.json"), userData);
```

Jackson 用于简单数据绑定的具体 Java 类型如表 4.7 所示。

表 4.7　Jackson 用于简单数据绑定的 Java 类型

JSON 类型	Java 类型
object	LinkedHashMap<String,Object>
array	ArrayList

续表

JSON 类型	Java 类型
string	String
number（非小数）	Integer、Long 或 BigInteger (smallest applicable)
number（小数）	Double (configurable to use BigDecimal)
true|false	Boolean
null	Null

（3）用泛型数据绑定。

除了绑定 POJO 和简单类型外，还可以绑定泛型。这种情况需要特殊处理，这是由于所谓的类型擦除（Type Erasure）（Java 以向后兼容的方式实现泛型），这会阻止使用类似 Collection<String>.class（它不会被编译）。因此，如果想绑定数据到 Map<String,User>，需要使用：

```
Map<String, User> userData = mapper.readValue(new File("user-modified-generic.json"),new TypeReference<Map<String, User>>() {});
```

其中，只能通过 TypeReference 传递泛型定义（在这种情况下，是通过 anynomous 内部类），最重要的部分是<Map<String,User>>，它定义绑定到的类型。如果不这样做，而只是使用 Map.class，那么调用等价于绑定到 Map<?,?>（例如，"untyped" Map），正如前面说明的。

2）Gson

首先，从 code.google.com/p/google-gson/downloads/list 下载 GsonAPI：

google-gson-1.7.1-release.zip

把 gson-1.7.jar 复制到 libs（在项目根目录新建一个 libs 文件夹）中。

可以使用以下两种方式来解析 JSON 数据。

（1）通过流模式解析 JSON 数据。

```
1.  String jsonData = "[{\"username\":\"arthinking\",\"userId\":001},
    {\"username\":\"Jason\",\"userId\":002}]";
2.  try{ JsonReader reader = new JsonReader(new StringReader(jsonData));
3.      reader.beginArray();
4.      while(reader.hasNext()){ reader.beginObject();
5.          while(reader.hasNext()){ String tagName = reader.nextName();
6.              if(tagName.equals("username")) {
7.                  System.out.println(reader.nextString());}
8.              else if(tagName.equals("userId")) {
9.                  System.out.println(reader.nextString());} }
10.         reader.endObject(); } reader.endArray();}
11. catch(Exception e){ e.printStackTrace();}
```

（2）通过绑定模式解析 JSON 数据。

第一步，创建 JSON 数据对应的一个 POJO 对象 User.java：

```
1.  public class User {
2.      private String username ;    private int userId ;
3.      public String getUsername() {   return username;}
4.      public void setUsername(String username) {this.username = username; }
```

```
5.    public int getUserId() {  return userId;}
6.    public void setUserId(int userId) { this.userId = userId;}}
```

第二步，使用 Gson 对象获取 User 对象数据进行相应的操作。

```
1. Type listType = new TypeToken<LinkedList<User>>(){}.getType();
2. Gson gson = new Gson();
3. LinkedList<User> users = gson.fromJson(jsonData, listType);
4. for (Iterator iterator = users.iterator(); iterator.hasNext();) {
5.    User user = (User) iterator.next();
6.    System.out.println(user.getUsername());
7. System.out.println(user.getUserId());}
```

如果要处理的 JSON 字符串只包含一个 JSON 对象，则可以直接使用 fromJson 获取一个 User 对象。

```
1. String jsonData = "{\"username\":\"arthinking\",\"userId\":001}";
2. Gson gson = new Gson();
3. User user = gson.fromJson(jsonData, User.class);
4. System.out.println(user.getUsername());
5. System.out.println(user.getUserId());
```

4.6.3 项目实践

（1）利用天气预报接口编写天气查询软件项目简介

在一些基于 Web 系统的 Android 客户端程序开发过程中，通常从 Web 服务接口获取的数据是以 JSON 这种数据格式传输的，需要对 JSON 格式的数据进行解析（反序列化）才能在 Android 程序中使用，下面的实例是从天气预报的 Web 服务接口获取各个地方的天气预报信息（JSON 格式），使用 Jackson 解析 JSON 数据，编写一个天气查询的小软件。

（2）利用天气预报接口编写天气查询软件实现步骤

实现要点：

- android:ems="12"：设置 TextView 或者 EditText 的宽度为 12 个字符。
- 如何对 JSON 格式的数据进行解析。

步骤 1：创建 Ch04_Ex08_Weather 项目，程序界面包含 4 个 EditText 用于显示天气相关信息，如源代码清单 4.45 所示。

源代码清单 4.45　\CH04\Ch04_Ex08_Wheather\res\layout\wheather_layout.xml

```
1. <LinearLayout xmlns:android="http://schemas.android.com/apk/res/android"
2.    xmlns:tools="http://schemas.android.com/tools"
3.    android:layout_width="fill_parent"  android:layout_height="fill_parent"
4.    android:orientation="vertical" >
5.    <TextView   android:id="@+id/textView1" android:layout_width="wrap_content"
6.       android:layout_height="wrap_content"   android:text="城 市： " />
7.    <EditText   android:id="@+id/city"  android:layout_width="match_parent"
8.       android:layout_height="wrap_content"   android:ems="12">
         <requestFocus/>  </EditText>
9.    <TextView   android:id="@+id/textView2" android:layout_width="wrap_content"
```

```
10.        android:layout_height="wrap_content"  android:text="天 气: "/>
11.    <EditText  android:id="@+id/weather"  android:layout_width="match_parent"
12.        android:layout_height="wrap_content"  android:ems="12"/>
13.    <TextView android:id="@+id/textView3" android:layout_width="wrap_content"
14.        android:layout_height="wrap_content"  android:text="最高温度: "/>
15.    <EditText  android:id="@+id/highTemp"  android:layout_width="match_parent"
16.        android:layout_height="wrap_content"  android:ems="12"/>
17.    <TextView android:id="@+id/textView4" android:layout_width="wrap_content"
18.        android:layout_height="wrap_content"  android:text="最低温度: " />
19.     <EditText  android:id="@+id/lowTemp"  android:layout_width="match_parent"
20.        android:layout_height="wrap_content"  android:ems="12" /> </Linear
       Layout>
```

步骤 2：主界面部分，所用知识点有线程、网络通信、JSON 数据解析，重点是 JSON 数据解析，如源代码清单 4.46 所示。

源代码清单 4.46　\Ch04_Ex08_Weather\src\sziit\lihz\ch04_ex08_weather\WeatherActivity.java

```
1. public class WeatherActivity extends Activity {//从Activity活动基类派生子类
2.    private EditText city, weather, highTemp, lowTemp;//声明EditText控件
3.    protected void onCreate(Bundle savedInstanceState) {
                                                    //子类重写onCreate方法
4.      super.onCreate(savedInstanceState);        //基类调用onCreate方法
5.      setContentView(R.layout.weather_layout);//设置活动界面布局
6.      city = (EditText) findViewById(R.id.city);//从布局文件查找EditText对象
7.      weather = (EditText) findViewById(R.id.weather);
                                              //从布局文件查找EditText对象
8.      highTemp = (EditText) findViewById(R.id.highTemp);
                                              //从布局文件查找EditText对象
9.      lowTemp = (EditText) findViewById(R.id.lowTemp);
                                              //从布局文件查找EditText对象
10.     //新开线程，读取网络数据
11.     new Thread() {                              //构造匿名线程
12.       public void run() {                       //重写run()方法
13.         List<Map<String, String>> list = getWeatherInfo();//查询天气信息
14.         Map<String, String> map = new HashMap<String, String>();
                                              //构造哈希映射对象
15.         map = list.get(0);//从天气信息列表中取得数据
16.         city.setText(map.get("city"));          //获得城市信息并显示
17.         weather.setText(map.get("weather"));    //获得天气信息并显示
18.         highTemp.setText(map.get("highTemp"));  //获得最高温度并显示
19.         lowTemp.setText(map.get("lowTemp"));    //获得最大温度并显示
20.     }}.start();                                 //启动线程 }
21.   /***连接网络* @throws MalformedURLException*/
22.   private static String getConnection(String path)
23.       throws MalformedURLException {            //获得网络连接
24.     URL url = new URL(path);                    //构造URL对象
```

```
25.    try {
26.        HttpURLConnection conn = (HttpURLConnection) url.openConnection();
                                                                    //打开连接
27.        conn.setConnectTimeout(5 * 1000);           //设置连接超时
28.        conn.setRequestMethod("GET");               //设置请求方法
29.        InputStreamReader in =new InputStreamReader(conn.getInputStream());
                                                                    //构造输入流
30.        BufferedReader br = new BufferedReader(in);//构造带传冲区的读对象
31.        String line = br.readLine().toString();    //读一行数据
32.        br.close();                                 //关闭读写器
33.        in.close();                                 //关闭输入流
34.        return line;                                //返回连接
35.    } catch (IOException e) {e.printStackTrace(); } return null;}
36.    /*** JSON 获取数据
37.     * 北京: http://www.weather.com.cn/data/cityinfo/101010100.html
38.     * 上海: http://www.weather.com.cn/data/cityinfo/101020100.html
39.     * 深圳: http://www.weather.com.cn/data/cityinfo/101280601.html */
40.    public static List<Map<String, String>> getWeatherInfo() {//查询天气信息
41.      List<Map<String, String>> list = new ArrayList<Map<String, String>>();
                                                                    //构造列表对象
42.      String json = null; Map<String, String> map;  //声明映射对象
43.      try { String line =getConnection("http://www.weather.com.cn/data/
44.    cityinfo/101280601.html");
45.        if (line != null) { json = new String(line);
46.        JSONObject item1 = new JSONObject(json);    //JSON 数据解析
47.        JSONObject item = item1.getJSONObject("weatherinfo");
48.        String city = item.getString("city");
49.        String highTemp = item.getString("temp1");
50.        String lowTemp = item.getString("temp2");
51.        String weather = item.getString("weather");
52.        map = new HashMap<String, String>();
53.        map.put("city", city);
54.        map.put("highTemp", highTemp);
55.        map.put("lowTemp", lowTemp);
56.        map.put("weather", weather);
57.        list.add(map);
58.           return list;   }
59.    } catch (MalformedURLException e) {e.printStackTrace();
60.    } catch (JSONException e) { e.printStackTrace();}
61.    return null;   }}
```

(3) 运行结果

选择【Run As】→【Android Application】命令,运行程序,将会看到图 4.16 所示的用户登录界面。

（a）天气预报界面布局轮廓　　　　（b）程序运行效果

图 4.16　利用天气预报接口编写天气查询软件运行效果

本章小结

本章主要讲解了如何编写 Android 动画资源文件、Shared Preferences 数据存储、HTTP 通信、JSON、Drawable 资源文件和多线程编程等基本知识，通过几个基本实训项目的设计与实现，让学生掌握利用动画资源文件实现程序过渡动画、利用 Drawable 资源文件美化程序 UI 界面、利用列表控件实现九宫格菜单、获取 URL 地址上的资源和利用天气预报接口编写天气查询软件等应用编程技巧。

强化练习

1. 填空题

（1）Android 的动画由四种类型组成，其中，alpha 负责（　　　　）的改变，scale 负责对图片进行（　　　　）改变，translate 负责对图片进行（　　　　）变换，rotate 负责对图片进行（　　　　）。

（2）程序中使用动画资源文件有两种方式，分别是在（　　　　）代码中引用和在（　　　　）文件中引用。

（3）class AnimationUtils 是实现（　　　　）功能的通用工具类。

（4）Animation loadAnimation(Context context, int id)用于从资源文件（　　　　）动画。

（5）void setAnimationListener(AnimationListener listener)用于设置动画（　　　　）。

（6）动画监听器接口 interface AnimationListener 包含（　　　　）个接口方法，void onAnimationStart(Animation animation)方法在动画（　　　　）时调用，void onAnimationEnd(Animation animation)方法在动画（　　　　）时调用。

（7）GridView 控件的 android:columnWidth 属性用于设置（　　　　）的宽度。

（8）GridView 控件的 android: numColumns 属性用于设置（　　　　）数。

（9）ImageView 的（　　　　）属性用来设置图像视图显示的图片来源。

（10）android:background="@drawable/main_default_bg"用来设置布局（　　　　）。

（11）超文本传输协议的简称为（　　　　）。

（12） sendEmptyMessage(int what)用于发送（　　　　）。
（13） JavaScript 对象符号（JavaScript Object Notation）简称为（　　　　）。
（14） android:ems="14"属性设置 TextView 或者 Edittext 的宽度为（　　　　）个字符。
（15） 给列表视图（ListView）设置适配器的方法是（　　　　）。
（16） 在 Android 应用程序中可通过（　　　　）方法来设置布局的背景颜色。
（17） 对按钮控件单击事件进行监听的方法是（　　　　）。
（18） Android 系统为用户提供了事件的处理机制。在处理单击事件时可以调用（　　　　）方法，其回调方法为（　　　　）。
（19） 在 Android 中使用 SharedPreferences 数据存储机制，采用（　　　　）方式来存储数据。
（20） 线性布局、相对布局、帧布局和绝对布局是直接继承自（　　　　）类。

2. 单选题

（1） Android 布局方式不包括下面的（　　）。
　　A. 线性布局　　　　　　　　B. 相对布局
　　C. 多维布局　　　　　　　　D. 单帧布局

（2） Android 线性布局分为两种方式：纵向和横向，设置该方式的属性为（　　）
　　A. android: orientation　　　　B. android:layout_gravity
　　C. android:layout_width　　　D. android:layout_height

（3） 在 Android 应用程序开发中用户要用到一些开发工具，但其中不包括（　　）。
　　A. Eclipse　　　　　　　　　B. VC6.0
　　C. ADT　　　　　　　　　　D. JDK

（4） Android 开发中用户经常用到打印日志的方式来进行调试，其中日志类型不包括下面的（　　）
　　A. Log.v　　　　　　　　　B. Log.c
　　C. Log.e　　　　　　　　　D. Log.d

（5） 在 Android 程序开发中，用户经常用到文本框控件，该控件在布局文件中的标签为（　　）。
　　A. EditText　　　　　　　　B. Text
　　C. TextView　　　　　　　　D. Label

（6） 在调用对话框时，需要最后调用（　　）方法来显示对话框。
　　A. onLongClick()　　　　　　B. onClick()
　　C. onTouch()　　　　　　　　D. show()

（7） 列表控件在布局文件中的标签为（　　）。
　　A. ListArray　　　　　　　　B. List
　　C. ListAdapter　　　　　　　D. ListView

（8） 在 Android 中，进度条对话框（ProgressDialog）必须要在后台程序运行完毕前，以（　　）方法来关闭所取得的焦点。
　　A. finish()　　　　　　　　　B. close()

  C．dimiss()   D．以上都不是

（9）在使用系统自带的 TabHost 时需要注意：TabHost 的 ID 必须为（ ）。

  A．@android:id/tabhost   B．@android:id/tabs

  C．@android:id/tabcontent   D．@android:id/tabframework

（10）为了在 Android 程序中能够自适应手机屏幕的大小，所选择的布局方式一般是（ ）。

  A．线性布局   B．相对布局

  C．单帧布局   D．绝对布局

3．多选题

（1）活动的状态包括（ ）。

  A．睡眠状态   B．暂停状态

  C．停止状态   D．运行状态

（2）下面关于句柄（Handler）的说话中正确的是（ ）。

  A．它是采用栈的方式来组织任务的

  B．它可以属于一个新的线程

  C．它是不同线程间通信的一种机制

  D．它避免了新线程操作用户界面（UI）组件

（3）下面属于视图（View）的子类组件包是（ ）。

  A．TextView   B．Service

  C．ViewGroup   D．Activity

（4）在布局文件中，定义一个视图组件时，哪两个属性必须写（ ）。

  A．android:id="@+id/button"   B．android:text

  C．android:layout_width   D．android:layout_height

（5）在列表视图或 Spinner 控件中，经常使用的适配器类包括（ ）。

  A．SimpleAdapter   B．ArrayDdapter

  C．SimpleCursorAdapter   D．SimpleCursorsAdapter

4．问答题

（1）活动的生命周期事件回调函数包括哪些？

（2）Android 软件框架结构自上而下可分为哪些层？

（3）单帧布局的特点是什么？

（4）Android 应用程序的四大组件是什么？

（5）Android 的动画由哪几种类型组成？

（6）简要描述 Android Drawable 资源文件。

（7）Android Service 与 Activity 的区别是什么？

（8）简述使用 SharedPreferences 存储数据的基本步骤。

（9）Android SQLite 数据库的操作方法有哪些？

5．编程题

（1）编写一个图 4.17 所示的应用程序，演示主要 UI 组件的基本用法。

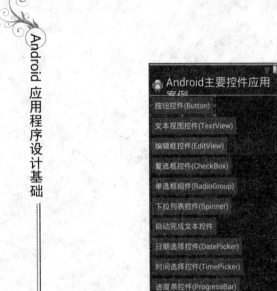

(a)界面布局　　　　　　　　　　(b)界面轮廓

图 4.17　常用 UI 组件功能演示界面布局

（2）参考 QQ 和微信 UI 界面，动手编写一款社交软件，要求操作界面简单、流畅，且能提供聊天、交友、娱乐、购物等主流功能。

第 5 章 基于 Android 未来超市系统项目实训

【知识目标】
- 掌握界面编程与视图组件知识；
- 掌握布局管理器知识；
- 掌握 TextView 及其子类知识；
- 掌握 ImageView 及其子类知识；
- 掌握 AdapterView 及其子类知识；
- 掌握 ViewAnimator 及其子类知识；
- 掌握对话框和菜单知识；
- 掌握基于监听的事件处理知识；
- 掌握基于回调的事件处理知识；
- 掌握响应的系统设置的事件。

【能力目标】
- 掌握界面的设计和实现方法；
- 掌握页面与页面之间数据的交互技术；
- 掌握数据库的设计和实现方法；
- 掌握 Handle 消息传递机制；
- 掌握基于 TCP 的网络通信；
- 掌握使用 URL 访问网络资源。

【重点、难点】
- 使用 WebServer 进行网络编程；
- 数据存储与访问；
- 嵌入式关系型 SQLite 数据库存储数据；
- 视频采集。

5.1 项目概述

物联网未来超市系统是基于 RFID、Zigbee、WIFI 等技术的可用于真实项目的系统。整个系统具有五大子系统：环境监控系统、安防报警系统、客户流量系统、货架系统、查询系统。

无论何时何地，都可以通过手持设备或者计算机监控超市环境状况、客户流量状况、货架状况等功能，同时还支持火灾报警等功能，集成了目前智能仓储中众多声光电因素，能够很好地体现现代超市管理的安全化和智能化。

未来超市实训台高度结合了物联网工程技术与行业体系架构，还原了行业真实环境，将感知层、网络层和应用层三个区域分开、区域清晰；预留扩展接口，方便用户二次开发调试；烧写接口集中管控，方便烧写；集成远程教学功能；智能断电系统，过电流可自动切断电源。

5.2 项目设计

5.2.1 项目总体功能需求

1. 环境监控系统

在手持设备或平板电脑上能查到温度、湿度、光照度、空气质量等传感器的状况。

2. 安防报警系统

（1）防盗报警：系统检测到有人入侵时，启动报报警器，同时，手持设备或平板电脑能显示报警信息。

（2）防灾报警：有烟雾异常时，启动报报警器，打开通风设备，同时，手持设备或平板电脑能显示报警信息。

3. 客户流量系统

（1）通过远红外探测器自动检测出入超市客户流量信息，客户流量信息经智能网关定时上传到系统服务器中。

（2）在客户流量系统通过选择不同的时间段，可以直观查看每天、每周、每月货物入库信息的曲线图。

4. 货架系统

（1）在货架系统中通过选择具体的货架，可以查看该货架上的详情：货物名称、货物价格、货物数量、货物产地。

（2）在货架系统中通过选择具体的货架上的货物，可以查看该货物换货时间频率的曲线图。

5. 查询系统

在查询系统中通过输入具体的货物，可以查看该货物的总库存数据、入库总数量、入库价格。

5.2.2 项目总体设计

现在全球在进行信息化建设，而信息化建设最终能否落地，移动终端上的应用开发将起到决定性的作用，Android 是目前市场占有率最高的移动设备操作系统，同时它是开源的，这样有利于其上应用程序的开发。因此，本系统的上位机应用程序选择在 Android 操作系统上进行开发。本项目总体设计如图 5.1 所示。

从安全性和便利性方面考虑，本系统将采用云与端的方式进行开发。

本系统采用常规的用户与密码的方式登录系统，系统主页面中有六个采用扁平化设计的功能选项：环境监控、安防报警、客户流量、货架、查询和更多。

在"环境监控"中包括：温湿度、光照度和空气质量；用户还可以根据超市的情况添加新的探测器，同时也可以对探测器进行删除操作。在界面中查看到简单的探测器状况信息。

在"安防报警"中包括：烟雾探测器、报警器和通风扇；用户还可以根据实际情况添加新的设备,同时也可以对设备进行删除操作。在界面中查看到简单的设备状况信息。对于可控制的设备,单击具体的设备后,可以查看设备具体信息,并可以对设备进行控制。

在"客户流量"中包括:日记录图、周记录图和月记录图;用户还可根据需要对功能进行扩展。

在"货架"中包括:货架现况、货物数量和换货时间频率;用户还可根据需要对功能进行扩展。"

"查询"主要用于货物的查询。"更多"选项用于后期功能的扩展。

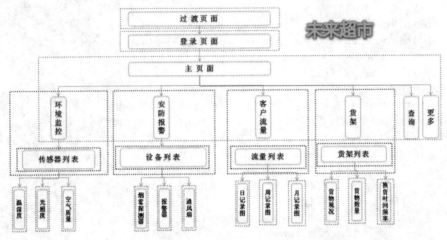

图 5.1 项目整体设计框架图

5.3 项目实施

5.3.1 登录页面

实现账号和密码输入文本框、实现登录按键功能、实现账号和密码记录功能。页面布局如图 5.2 所示。

当用户输入账号和密码,单击登录按键时,系统读取账号和密码文本框信息,并把账号和密码信息经网络传输给智能网关,与智能网关中的数据库中的账号和密码数据信息进行比对,如果账号和密码同时存在于智能网关中的数据库中,则允许用户登录系统。如果账号或密码不正确,则根据返回的标记码的不同,提示"账号不存在"或"密码不正确"。

当用户选中"自动登录"和"保存密码"时,系统会把账号和密码保存在系统相关的配置文件中,当下次登录系统时,则不需要输入账号和密码。如果用户只选中"自动登录"或"保存密码",则系统只保存账号或密码。程序的界面布局代码如源代码清单 5.1 所示。

图 5.2 登录界面

源代码清单 5.1　\CH05_Supermarket\src\com\zigcloud\supermarket\activity\LoginActivity.java

```
1. package com.zigcloud.supermarket.activity;
2. public class LoginActivity extends BaseActivity{  /*** 用户登录**/
```

```
3.    public void onCreate(Bundle savedInstanceState) {
4.        super.onCreate(savedInstanceState);
5.        setContentView(R.layout.activity_login);
6.        MainApplication.getInstance().addActivity(this);
7.        findViewById(R.id.btn_login).setOnClickListener(new View.OnClickL
istener() {
8.            public void onClick(View v) {
9.                startActivity(new Intent(LoginActivity.this,MainActivity.class));
10.           }
11.       }); }
12.   public void onBackPressed() {   super.onBackPressed();}
13.   public boolean onKeyDown(int keyCode, KeyEvent event) {
14.       if(keyCode == KeyEvent.KEYCODE_BACK)
15.         { exitBy2Click(); }
16.       return false; }
17.   private static Boolean isExit = false;
18.   private void exitBy2Click() {/*** 双击后退按钮退出**/
19.       Timer tExit = null;
20.       if (isExit == false) { isExit = true;
21.         Toast.makeText(this, getResources().getString(R.string.exit_dia
22.         log_message),Toast.LENGTH_SHORT).show();
23.         tExit = new Timer();
24.         tExit.schedule(new TimerTask() {
25.           public void run() {isExit = false; }
26.       }, 2000); } else { MainApplication.getInstance().exit();}}}
```

5.3.2 主页面

实现"返回"、"报警信息"和六项图示功能，页面布局如图 5.3 所示。

图 5.3 主页面

"返回"功能的作用是返回上一级页面；"环境监控"功能的作用是查看当前仓库中的传感器设备信息。其他功能类似。程序的界面布局代码如源代码清单 5.2、5.3 所示。

源代码清单 5.2 \CH05_Supermarket\res\layout\activity_main.xml

```
1. <?xml version="1.0" encoding="utf-8"?>
```

```
2. <RelativeLayout    xmlns:android="http://schemas.android.com/apk/res/android"
3.     android:layout_width="fill_parent"  android:layout_height="fill_parent"
4.     android:background="@drawable/main_default_bg" >
5.     <include
6.         android:id="@id/bt_createtask_title_layout" layout="@layout/activity_common_title_bar"/>
7.     <RelativeLayout
8.         android:id="@+id/RelativeLayout1"
9.         android:layout_width="match_parent" android:layout_height="match_parent"
10.        android:layout_below="@id/bt_createtask_title_layout"
11.        android:layout_centerHorizontal="true"  android:layout_marginLeft="15dp"
12.        android:layout_marginRight="15dp"    android:orientation="vertical"
13.        android:visibility="visible" >
14.        <LinearLayout
15.            android:id="@+id/LinearLayout1"
16.            android:layout_width="match_parent" android:layout_height="match_parent"
17.            android:layout_alignParentTop="true" android:orientation="vertical" >
18.            <GridView
19.                android:id="@+id/gdv_main"
20.                android:layout_width="match_parent" android:layout_height="wrap_content"
21.                android:listSelector="@drawable/selector_list" android:horizontalSpacing="10dp"
22.                android:numColumns="2"   android:verticalSpacing="10dp" >
23.            </GridView>
24.            <include
25.                android:id="@+id/view_loading_error" android:layout_width="match_parent"
26.                android:layout_height="match_parent" layout="@layout/view_loading_error"
27.                android:visibility="gone" />
28.            <include
29.                android:id="@+id/view_loading"    android:layout_width="match_parent"
30.                android:layout_height="match_parent" layout="@layout/view_loading"
31.                android:visibility="gone" />
32.    </LinearLayout>   </RelativeLayout> </RelativeLayout>
```

源代码清单5.3 \CH05_Supermarket\src\com\zigcloud\supermarket \activity\MainActivity.java

```
1. package com.zigcloud.supermarket.activity;
2. public class MainActivity extends BaseActivity{ /*** 主界面**/
3.     private TextView titlebar_left;
4.     private TextView titlebar_title;
5.     public void onCreate(Bundle savedInstanceState) {
6.         super.onCreate(savedInstanceState);
7.         setContentView(R.layout.activity_main);
```

```
8.        MainApplication.getInstance().addActivity(this);
9.        titlebar_left = (TextView) findViewById(R.id.titlebar_left);
10.       titlebar_left.setOnClickListener(new View.OnClickListener() {
11.          public void onClick(View v) { onBackPressed();    }});
12.       titlebar_title = (TextView) findViewById(R.id.titlebar_title);
13.       titlebar_title.setText(R.string.app_name);
14.       GridView gridView = (GridView) findViewById(R.id.gdv_main);
15.       ArrayList<HashMap<String, Object>> lstImageItem=new ArrayList<HashMap<String, Object>>();
16.       HashMap<String, Object> map = new HashMap<String, Object>();
17.       map.put("ItemImage", R.drawable.icon_environment);
18.       map.put("ItemText", "环境监控");
19.       lstImageItem.add(map);
20.       map = new HashMap<String, Object>();
21.       map.put("ItemImage", R.drawable.icon_alarm);
22.       map.put("ItemText", "安防报警");
23.       lstImageItem.add(map);
24.       map = new HashMap<String, Object>();
25.       map.put("ItemImage", R.drawable.icon_import);
26.       map.put("ItemText", "物品入库");
27.       lstImageItem.add(map);
28.       map = new HashMap<String, Object>();
29.       map.put("ItemImage", R.drawable.icon_outport);
30.       map.put("ItemText", "物品出库");
31.       lstImageItem.add(map);
32.       map = new HashMap<String, Object>();
33.       map.put("ItemImage", R.drawable.icon_search);
34.       map.put("ItemText", "物品查询");
35.       lstImageItem.add(map);
36.       map = new HashMap<String, Object>();
37.       map.put("ItemImage", R.drawable.icon_setting);
38.       map.put("ItemText", "设置");
39.       lstImageItem.add(map);
40.       map = new HashMap<String, Object>();
41.       map.put("ItemImage", R.drawable.icon_add);
42.       map.put("ItemText", "预留");
43.       lstImageItem.add(map);
44.       map = new HashMap<String, Object>();
45.       map.put("ItemImage", R.drawable.icon_add);
46.       map.put("ItemText", "预留");
47.       lstImageItem.add(map);
48.       SimpleAdapter saImageItems = new SimpleAdapter(this, lstImageItem,R.layout.view_squared_item, new String[] {"ItemImage",
          "ItemText"}, new int[] {R.id.itemImage,R.id.itemText});
49.       gridView.setAdapter(saImageItems);
50.       gridView.setOnItemClickListener(new OnItemClickListener() {
51.         public void onItemClick(AdapterView<?> arg0, View arg1, int arg2, long arg3) {
52.           switch(arg2){
53.           case 0: startActivity(new Intent(MainActivity.this,EnvironmentMonitorActivity.class));break;
```

```
54.         case 1: break;
55.         case 2: startActivity(new Intent(MainActivity.this,GoodsImport
                 Activity.class));break;
56.         case 3: startActivity(new Intent(MainActivity.this,
                 GoodsOutportActivity.class));break;
57.         case 4:startActivity(new Intent(MainActivity.this,
                 GoodsSearchActivity.class));break;
58.     }}});
59.     gridView.setSelector(new ColorDrawable(Color.TRANSPARENT)); }
60.     public boolean onKeyDown(int keyCode, KeyEvent event) {
61.     if(keyCode == KeyEvent.KEYCODE_BACK) { exitBy2Click();}return false;}
62.     private static Boolean isExit = false;
63.     private void exitBy2Click() {Timer tExit = null;
64.         if (isExit == false) {
65.         isExit = true;
66.         Toast.makeText(this,getResources().getString(R.string.exit_dial
                og_message), Toast.LENGTH_SHORT).show();
67.         tExit = new Timer();
68.         tExit.schedule(new TimerTask() {public void run() {isExit = fals
                e; }}, 2000);
69.         } else { MainApplication.getInstance().exit();}}}
```

5.3.3 环境监控页面

实现查看仓库中各个区域的传感器信息,如火焰传感器、温度传感器、湿度传感器等环境信息。环境监控页面图如图 5.4 所示。

图 5.4 环境监测页

源代码清单 5.4 \CH05_Supermarket\res\layout\ activity_environmentmonitor.xml

```
1.  <?xml version="1.0" encoding="utf-8"?>
2.  <RelativeLayout xmlns:android="http://schemas.android.com/apk/res/android"
3.      android:layout_width="fill_parent" android:layout_height="fill_parent"
4.      android:background="@drawable/main_default_bg" >
5.      <include
6.          android:id="@id/bt_createtask_title_layout" layout="@layout/activity
            _common_title_bar"/>
```

```
7.      <RelativeLayout
8.          android:id="@+id/RelativeLayout1"
9.          android:layout_width="match_parent"    android:layout_height="match_
            parent"
10.         android:layout_below="@id/bt_createtask_title_layout"
11.         android:layout_centerHorizontal="true"
12.         android:layout_marginLeft="15dp"    android:layout_marginRight="15dp"
13.         android:background="@drawable/content_default_bg"
14.         android:orientation="vertical"    android:visibility="visible" >
15.         <com.zigcloud.warehousing.widget.Gallery3D
16.             android:id="@+id/gal_rooms"    android:layout_width="fill_parent"
17.             android:layout_height="wrap_content"    android:layout_align
                ParentBottom="true"
18.             android:layout_alignParentLeft="true"    android:spacing="30dp"
19.             android:unselectedAlpha="128" />
20.         <LinearLayout
21.             android:id="@+id/LinearLayout1"    android:layout_width=
                "match_parent"
22.             android:layout_height="match_parent"    android:layout_above=
                "@+id/gal_rooms"
23.             android:layout_alignParentTop="true"    android:orientation=
                "vertical" >
24.             <include
25.                 android:id="@+id/view_loading_error"    android:layout_width
                    ="match_parent"
26.                 android:layout_height="match_parent"    layout="@layout/view_
                    loading_error"
27.                 android:visibility="gone" />
28.             <include
29.                 android:id="@+id/view_loading"    android:layout_width=
                    "match_parent"
30.                 android:layout_height="match_parent"    layout="@layout/view
                    _loading"
31.                 android:visibility="gone" />
32.             <GridView
33.                 android:id="@+id/gdv_equipments"    android:layout_width=
                    "match_parent"
34.                 android:layout_height="wrap_content"    android:numColumns="1"
35.                 android:padding="10dp"    android:verticalSpacing="10dp"
36.                 android:visibility="visible" >
37.         </GridView></LinearLayout>    </RelativeLayout></RelativeLayout>
```

源代码清单 5.5 \CH05_Supermarket\src\com\zigcloud\supermarket\activity\
EnvironmentMonitorActivity.java

```
1. package com.zigcloud.supermarket.activity;
2. @SuppressLint("HandlerLeak")
3. public class EnvironmentMonitorActivity extends BaseActivity{ /*** 环境
   监控* */
4.     private TextView titlebar_left;
```

```
5.    private TextView titlebar_title;
6.    private GridView equipmentsGridView;
7.    private ArrayList<HashMap<String, Object>> equipmentsArrayList = new
      ArrayList<HashMap<String, Object>>();
8.    private HashMap<String, Object> equipmentsHashMap = new HashMap<String,
      Object>();
9.    private SimpleAdapter equipmentsAdapter;
10.   /*** 获取设备列表**/
11.   private EquipmentListHttpRequestTask equipmentListHttpRequestTask=new
      EquipmentListHttpRequestTask();
12.   public void onCreate(Bundle savedInstanceState) {super.onCreate(savedIn
      stanceState);
13.      setContentView(R.layout.activity_environmentmonitor);
14.      MainApplication.getInstance().addActivity(this);
15.      titlebar_left = (TextView) findViewById(R.id.titlebar_left);
16.      titlebar_left.setOnClickListener(new View.OnClickListener() {
17.        public void onClick(View v) {   onBackPressed(); }});
18.      titlebar_title = (TextView) findViewById(R.id.titlebar_title);
19.      titlebar_title.setText(R.string.title_environment_monitor);
20.      initialRoomsGallery();   initialEquipmentsGridView(); }
21.   private void initialEquipmentsGridView(){    /*** 初始化设备信息列表**/
22.      equipmentsGridView=(GridView) findViewById(R.id.gdv_equipments);
23.      equipmentsAdapter = new SimpleAdapter(getApplicationContext(),
         equipmentsArrayList, R.layout.activity_equipment_list_item, new
         String[] {"ItemImage","ItemNodeName","ItemDataValue","ItemArea"}, new
         int[] {R.id.img_image,R.id.tv_name,R.id.tv_datavalue,R.id.tv_area});
24.      equipmentsGridView.setAdapter(equipmentsAdapter);
25.      equipmentsGridView.setOnItemClickListener(new OnItemClickListener() {
26.        public void onItemClick(AdapterView<?> arg0, View arg1, int arg2,
           long arg3) {
27.        HashMap<String, Object> item=( HashMap<String, Object>) arg0.getItem
           AtPosition(arg2);
28.        String nodeId=item.get("ItemNodeId")==null ?null:item.get("ItemNode
           Id").toString();
29.        String nodeName=item.get("ItemNodeName")==null ?null:item.get
           ("ItemNodeName").toString();
30.        String nodeTypeId=item.get("ItemNodeTypeId")==null ?null:item
           .get("ItemNodeTypeId").toString();
31.        String nodeTypeName=item.get("ItemNodeTypeName")==null ?null:
           item.get("ItemNodeTypeName").toString();
32.        String dataValueString=item.get("ItemDataValueString")==null ?
           null:item.get("ItemDataValueString").toString();
33.        String updateTimeString=item.get("ItemUpdateTimeString")==null
           ?null:item.get("ItemUpdateTimeString").toString();
34.        Intent intent =new Intent(getApplicationContext(),EquipmentActivity
           .class);
35.        intent.putExtra("nodeId", nodeId);      intent.putExtra("nodeName",
           nodeName);
36.        intent.putExtra("nodeTypeId", nodeTypeId);
```

```
37.            intent.putExtra("nodeTypeName", nodeTypeName);
38.            intent.putExtra("dataValueString", dataValueString);
39.            intent.putExtra("updateTimeString",updateTimeString);
40.            startActivity(intent);} });
41.       equipmentListHttpRequestTask.execute(); }
42.     private EquipmentDAO mEquipmentDAO=new EquipmentDAO(); /*** 设备业务控制类**/
43.     private Handler mHandler=new Handler(){
44.        public void handleMessage(Message msg) {
45.            super.handleMessage(msg);
46.            switch(msg.what){
47.               case 10001: //正在加载
48.                   findViewById(R.id.gdv_equipments).setVisibility(8);
49.                   findViewById(R.id.view_loading).setVisibility(0);
50.                   findViewById(R.id.view_loading_error).setVisibility(8);
51.                   break;
52.               case 10002: //加载成功
53.                   findViewById(R.id.gdv_equipments).setVisibility(0);
54.                   findViewById(R.id.view_loading).setVisibility(8);
55.                   findViewById(R.id.view_loading_error).setVisibility(8);
56.                   break;
57.               case 10003: //加载失败
58.                   findViewById(R.id.gdv_equipments).setVisibility(8);
59.                   findViewById(R.id.view_loading).setVisibility(8);
60.                   findViewById(R.id.view_loading_error).setVisibility(0);
61.                   break;   }} };
62.     private class EquipmentListHttpRequestTask extends AsyncTask<String, Integer, List<EquipmentJson>>{/*** 获取 equipment 列表**/
63.        protected void onPreExecute() {super.onPreExecute();
64.           if(!(equipmentsAdapter!=null&&equipmentsAdapter.getCount()>0)){
65.              mHandler.sendEmptyMessage(10001);}}
66.        protected List<EquipmentJson> doInBackground(String... params) {
67.           return mEquipmentDAO.getAll(); }
68.        protected void onProgressUpdate(Integer... values) {
69.           super.onProgressUpdate(values); }
70.        protected void onPostExecute(List<EquipmentJson> result) {
71.           if(result!=null){
72.              EquipmentJson equipmentEntity=null;
73.              BaseEquipmentEntity equipment=null;
74.              for(int i=0;i<result.size();i++){
75.                 equipmentEntity= result.get(i);
76.                 equipment= BaseEquipmentEntity.parse(equipmentEntity);
77.                 if(equipment!=null){
78.                    equipmentsHashMap=new HashMap<String, Object>();
79.                    equipmentsHashMap.put("ItemImage", equipment.getIconRes());
80.                    equipmentsHashMap.put("ItemNodeName", equipment.getName());
81.                    equipmentsHashMap.put("ItemDataValue", equipment.getDataValueString());
82.                    equipmentsHashMap.put("ItemArea", "未知区域");
```

```
83.                equipmentsHashMap.put("ItemNodeId",equipment.getNodeId());
84.                equipmentsHashMap.put("ItemNodeTypeId",equipment.get
                   NodeTypeId());
85.                equipmentsHashMap.put("ItemNodeTypeName",equipment.get
                   TypeName());
86.                equipmentsHashMap.put("ItemDataValueString",equipment.
                   getDataValueString());
87.                equipmentsHashMap.put("ItemUpdateTimeString",equipment.
                   getUpdateTimeString());
88.                equipmentsArrayList.add(equipmentsHashMap);       }}
89.          equipmentsAdapter.notifyDataSetChanged(); mHandler.sendEmpty
             Message(10002);
90.       }else{ mHandler.sendEmptyMessage(10003);}} };
91.    private void initialRoomsGallery(){
92.      Gallery3D  gallery = (Gallery3D) findViewById(R.id.gal_rooms);
93.      ArrayList<HashMap<String, Object>> lstImageItem = new ArrayList<Hash
         Map<String, Object>>();
94.      HashMap<String, Object> map = new HashMap<String, Object>();
95.      map.put("ItemImage", R.drawable.scene_gallery_0);
96.      map.put("ItemText", "仓库1");    map.put("ItemContent", "仓库1");
97.      lstImageItem.add(map);           map = new HashMap<String, Object>();
98.      map.put("ItemImage", R.drawable.scene_gallery_1);
99.      map.put("ItemText", "仓库2");    map.put("ItemContent", "仓库2");
100.     lstImageItem.add(map);           map = new HashMap<String, Object>();
101.     map.put("ItemImage", R.drawable.scene_gallery_2);
102.     map.put("ItemText", "仓库3");    map.put("ItemContent", "仓库3");
103.     lstImageItem.add(map);           map = new HashMap<String, Object>();
104.     map.put("ItemImage", R.drawable.scene_gallery_3);
105.     map.put("ItemText", "仓库4");    map.put("ItemContent", "仓库4");
106.     lstImageItem.add(map);
107.     SimpleAdapter saImageItems = new SimpleAdapter(this, lstImageItem,
108.        R.layout.view_grallery3d_item, new String[] {"ItemImage","Item
            Text","ItemContent"},
109.        new int[] {R.id.itemImage,R.id.itemText,R.id.itemContent});
110.     gallery.setFadingEdgeLength(0);   gallery.setAdapter(saImageItems);
111.     gallery.setOnItemSelectedListener(new OnItemSelectedListener() {
112.     public void onItemSelected(AdapterView<?> parent, View view, int position,
         long id) {
113.       Toast.makeText(RoomListActivity.this, "img " + (position+1) +
           " selected", Toast.LENGTH_SHORT).show();}
114.     public void onNothingSelected(AdapterView<?> parent) {}});
115.     gallery.setOnItemClickListener(new OnItemClickListener() {
116.       public void onItemClick(AdapterView<?> parent, View view, int position,
           long id) {
117.         Toast.makeText(RoomListActivity.this, "img " + (position+1) +
            " selected", Toast.LENGTH_SHORT).show();}});}}
```

5.3.4 物品入库页面

实现货物入库的功能。页面布局如图5.5所示。

图 5.5 物品入库页面

源代码清单 5.6 \CH05_Supermarket\res\layout\ activity_goods_import.xml

```xml
1.  <?xml version="1.0" encoding="utf-8"?>
2.  <RelativeLayout xmlns:android="http://schemas.android.com/apk/res/android"
3.    android:layout_width="fill_parent"   android:layout_height="fill_parent"
4.    android:background="@drawable/main_default_bg" >
5.    <include
6.      android:id="@+id/bt_createtask_title_layout" layout="@layout/activity_common_title_bar"/>
7.    <RelativeLayout
8.      android:id="@+id/RelativeLayout1"   android:layout_width="match_parent"
9.      android:layout_height="match_parent"  android:layout_below="@id/bt_createtask_title_layout"
10.     android:layout_centerHorizontal="true"  android:layout_marginLeft="15dp"
11.     android:layout_marginRight="15dp"   android:background="@drawable/content_default_bg"
12.     android:orientation="vertical"      android:visibility="visible" >
13.     <TableLayout
14.       android:layout_width="wrap_content"    android:layout_height="wrap_content"
15.       android:layout_alignParentBottom="true"  android:layout_alignParentLeft="true"
16.       android:layout_alignParentRight="true"   android:layout_alignParentTop="true" >
17.       <TableRow
18.         android:id="@+id/tableRow1"    android:layout_width="wrap_content"
19.         android:layout_height="wrap_content" >
20.         <TextView
21.           android:id="@+id/textView1"    android:layout_width="wrap_content"
```

```
22.            android:layout_height="wrap_content" android:text="@string/
               goods_cardid" />
23.        <EditText
24.            android:id="@+id/edt_goods_cardId" android:layout_width="wrap_
               content"
25.            android:layout_height="wrap_content" android:ems="10" >
26.            <requestFocus />        </EditText>
27.        </TableRow>
28.      <TableRow
29.            android:id="@+id/tableRow2" android:layout_width="wrap_content"
30.            android:layout_height="wrap_content" >
31.        <TextView
32.            android:id="@+id/textView2" android:layout_width="wrap_
               content"
33.            android:layout_height="wrap_content" android:text="@string
               /goods_name" />
34.        <EditText
35.            android:id="@+id/edt_goods_name" android:layout_width=
               "wrap_content"
36.            android:layout_height="wrap_content" android:ems="10" />
               </TableRow>
37.      <TableRow
38.            android:id="@+id/tableRow3" android:layout_width="wrap_content"
39.            android:layout_height="wrap_content" >
40.        <TextView
41.            android:id="@+id/textView3"      android:layout_width="wrap_
               content"
42.            android:layout_height="wrap_content" android:text="@string
               /goods_address"/>
43.        <EditText
44.            android:id="@+id/edt_goods_address" android:layout_width=
               "wrap_content"
45.            android:layout_height="wrap_content" android:ems="10" />
               </TableRow>
46.      <TableRow
47.            android:id="@+id/tableRow4"    android:layout_width="wrap_content"
48.            android:layout_height="wrap_content" >
49.        <TextView
50.            android:id="@+id/textView4"         android:layout_width="wrap_
               content"
51.            android:layout_height="wrap_content" android:text="@string
               /operate_option"/>
52.        <Button
53.            android:id="@+id/btn_ok" android:layout_width="wrap_content"
54.            android:layout_height="wrap_content"  android:text="@string
               /ok"/></TableRow>
55.     </TableLayout>     </RelativeLayout>   </RelativeLayout>
```

源代码清单 5.7　\CH05_Supermarket\src\com\zigcloud\supermarket\activity\GoodsImportActivity.java

```
1. package com.zigcloud.supermarket.activity;
```

```java
2.  public class GoodsImportActivity extends BaseActivity{/*** 物品入库**/
3.      private TextView titlebar_left; private TextView titlebar_title;
4.      private GoodsDAO mGoodsDAO=new GoodsDAO(); /*** 物品操作业务类**/
5.      /*** 物品入库任务类**/
6.      private GoodsImportTask mGoodsImportTask=new GoodsImportTask();
7.      protected void onCreate(Bundle savedInstanceState) {
8.          super.onCreate(savedInstanceState);
9.          setContentView(R.layout.activity_goods_import);
10.         MainApplication.getInstance().addActivity(this);
11.         titlebar_left = (TextView) findViewById(R.id.titlebar_left);
12.         titlebar_left.setOnClickListener(new View.OnClickListener(){
13.             public void onClick(View v) {onBackPressed(); } });
14.         titlebar_title = (TextView) findViewById(R.id.titlebar_title);
15.         titlebar_title.setText(R.string.title_goods_import);
16.         /*** 商品入库按钮事件**/
17.         findViewById(R.id.btn_ok).setOnClickListener(new OnClickListener() {
18.             public void onClick(View v) {
19.                 EditText edt_goods_cardId=(EditText)findViewById(R.id.edt_goods_cardId);
20.                 EditText edt_goods_name=(EditText)findViewById(R.id.edt_goods_name);
21.                 EditText edt_goods_address=(EditText)findViewById(R.id.edt_goods_address);
22.                 mGoodsImportTask=new GoodsImportTask();
23.                 mGoodsImportTask.execute(edt_goods_cardId.getText().toString(),
                        edt_goods_name.getText().toString(), edt_goods_address.getText().toString(),"admin");}});}
24.     /*** 物品入库任务类**/
25.     private class GoodsImportTask extends AsyncTask<String, Integer, ControlResultJson>{
26.         protected ControlResultJson doInBackground(String... params) {
27.             if(params!=null&&params.length>3){
28.                 String cardNum=params[0];        String dataName=params[1];
29.                 String originPlace=params[2];    String staffName=params[3];
30.                 if(cardNum!=null&&dataName!=null&&originPlace!=null&&staffName!=null)
31.                     return mGoodsDAO.goodsImport(cardNum, dataName, originPlace, staffName); }
32.             return null;    }
33.         protected void onPostExecute(ControlResultJson result) {
34.             super.onPostExecute(result);
35.             String resStr=null;
36.             if(result!=null){resStr=result.flag==0?"添加成功! ":"添加失败";}
37.             else{resStr="添加失败"; }
38.             Toast.makeText(GoodsImportActivity.this, resStr, Toast.LENGTH_SHORT).show();}
39.         protected void onPreExecute() { super.onPreExecute();}
40.         protected void onProgressUpdate(Integer... values) {super.onProgressUpdate(values);} }
```

5.3.5 具体设备页面

实现"返回"、"报警信息"、设备名称显示、设备图片呈现、"开启"、"关闭"功能。页面布局如图 5.6 所示。

"返回"功能的作用是返回上一级页面;"报警信息"功能的作用是报警信息的提示。在"返回"和"报警信息"之间显示当前设备的名称。设备图片采用设备实物图;"开启"按键和"关闭"按键分别实现对设备的开关控制。

图 5.6 设备页面控制图

源代码清单 5.8 \CH05_Supermarket\res\layout\activity_equipment.xml

```
1.  <?xml version="1.0" encoding="utf-8"?>
2.  <RelativeLayout xmlns:android="http://schemas.android.com/apk/res/android"
3.      android:layout_width="fill_parent"   android:layout_height="fill_parent"
4.      android:background="@drawable/main_default_bg" >
5.      <include
6.         android:id="@id/bt_createtask_title_layout" layout="@layout/activity
           _common_title_bar"/>
7.      <RelativeLayout
8.         android:id="@+id/RelativeLayout1"   android:layout_width="match_parent"
9.         android:layout_height="match_parent"   android:layout_below="@id/bt_
           createtask_title_layout"
10.        android:layout_centerHorizontal="true"   android:layout_marginLeft="15dp"
11.        android:layout_marginRight="15dp"   android:background="@drawable/
           content_default_bg"
12.        android:orientation="vertical"   android:visibility="visible">
13.        <com.zigcloud.warehousing.widget.Gallery3D
14.            android:id="@+id/gal_rooms"   android:layout_width="fill_parent"
15.            android:layout_height="wrap_content"   android:layout_alignParent
               Bottom="true"
16.            android:layout_alignParentLeft="true"   android:spacing="30dp"
17.            android:unselectedAlpha="128" />
18.        <LinearLayout
```

```xml
19.        android:id="@+id/LinearLayout1"        android:layout_width="match
           _parent"
20.        android:layout_height="match_parent"   android:layout_above="@+
           id/gal_rooms"
21.        android:layout_alignParentTop="true"   android:orientation=
           "vertical">
22.        <include
23.          android:id="@+id/view_loading_error" android:layout_width=
             "match_parent"
24.          android:layout_height="match_parent" layout="@layout/view_
             loading_error"
25.          android:visibility="gone" />
26.        <include
27.          android:id="@+id/view_loading"       android:layout_width="match_
             parent"
28.          android:layout_height="match_parent" layout="@layout/view_
             loading"
29.          android:visibility="gone" />
30.        <LinearLayout
31.          android:id="@+id/ll_equipment"       android:layout_width=
             "match_parent"
32.          android:layout_height="match_parent" android:orientation=
             "vertical">
33.        </LinearLayout>
34.      </LinearLayout>    </RelativeLayout>    </RelativeLayout>
```

源代码清单 5.9 \CH05_Supermarket\src\com\zigcloud\supermarket\activity\EquipmentActivity.java

```java
1.  package com.zigcloud.supermarket.activity;
2.  public class EquipmentActivity extends BaseActivity{  /*** 设备* */
3.    private TextView titlebar_left; private TextView titlebar_title;
      private String nodeId;         private String nodeName;  private String
      nodeTypeId;    private String nodeTypeName;
4.    private String dataValueString;    private String updateTimeString;
5.    protected void onCreate(Bundle savedInstanceState) {super.onCreate(save
      dInstanceState);
6.    setContentView(R.layout.activity_equipment);
7.    MainApplication.getInstance().addActivity(this);
8.    nodeId=getIntent().getStringExtra("nodeId");
9.    nodeName=getIntent().getStringExtra("nodeName");
10.   nodeTypeId=getIntent().getStringExtra("nodeTypeId");
11.   nodeTypeName=getIntent().getStringExtra("nodeTypeName");
12.   dataValueString=getIntent().getStringExtra("dataValueString");
13.   updateTimeString=getIntent().getStringExtra("updateTimeString");
14.     titlebar_left = (TextView) findViewById(R.id.titlebar_left);
15.     titlebar_left.setOnClickListener(new View.OnClickListener() {
16.       public void onClick(View v) { onBackPressed(); }});
17.     titlebar_title = (TextView) findViewById(R.id.titlebar_title);
18.     titlebar_title.setText(nodeName); initalEquipment();}
19.   private void initalEquipment(){ /*** 初始化设备信息**/
```

```
20.        LinearLayout ll_equipment=(LinearLayout) findViewById(R.id.ll_equip
           ment);
21.        if(ll_equipment!=null){
22.           if(nodeTypeId!=null){BaseEquipmentWidget equipmentWidget=null;
23.             if(nodeTypeId.equals("20")){
24.                equipmentWidget=new CurtainWidget(EquipmentActivity.this);}
25.             else if(nodeTypeId.equals("24")){
26.                equipmentWidget=new LampWidget(EquipmentActivity.this);}
27.             else if(nodeTypeId.equals("25")){
28.                equipmentWidget=new BedLampWidget(EquipmentActivity.this);}
29.             else if(nodeTypeId.equals("26")){
30.                equipmentWidget=new WallLampWidget(EquipmentActivity.this);}
31.             else{equipmentWidget=new BaseEquipmentWidget(EquipmentActivi
           ty.this);}
32.             equipmentWidget.setNodeId(nodeId);
33.             equipmentWidget.setNodeTypeName(nodeTypeName);
34.             equipmentWidget.setNodeDataString(dataValueString);
35.             equipmentWidget.setUpdateTimeString(updateTimeString);
36.             equipmentWidget.addSendCmdListener(new BaseEquipmentWidget.
           Listener(){
37.                public void sendCmd(String nodeId, String stateFlag, String
           stateValue) {
38.                  if(mSendCmdHttpRequestTask.getStatus()!=Status.RUNNING){
39.                    mSendCmdHttpRequestTask=new SendCmdHttpRequestTask();
40.                    mSendCmdHttpRequestTask.execute(nodeId, stateFlag,
           stateValue) ;}}});
41.             ll_equipment.addView(equipmentWidget,new LayoutParams(Layout
           Params.MATCH_PARENT,LayoutParams.MATCH_PARENT)); }}
42.  protected EquipmentDAO mEquipmentDAO=new EquipmentDAO();/*** 设备控制业
     务类**/
43.  /*** 异步发送命令任务类**/
44.  protected SendCmdHttpRequestTask mSendCmdHttpRequestTask=new SendCmd
     HttpRequestTask();
45.  /*** 发送控制指令**/
46.  public class SendCmdHttpRequestTask extends AsyncTask<String, Integer,
     ControlResultJson>{
47.      protected ControlResultJson doInBackground(String... params) {
48.        if(params!=null&&params.length>2){
49.          mEquipmentDAO.sendCmd(params[0], params[1], params[2]);}
             return null; }}
50.  protected EquipmentHttpRequestTask mEquipmentHttpRequestTask
51.                       =new EquipmentHttpRequestTask();
52.  /*** 获取equipment**/
53.  protected class EquipmentHttpRequestTask extends AsyncTask<String,
     Integer, EquipmentJson>{
54.      protected void onPreExecute() { super.onPreExecute();}
55.      protected void onProgressUpdate(Integer... values) {super.onProgress
     Update(values);}
56.      protected EquipmentJson doInBackground(String... arg0) {
57.        String nodeId=arg0!=null&&arg0.length>0?arg0[0]:null;
```

```
58.         return mEquipmentDAO.getByNodeId(nodeId);}
59.     protected void onPostExecute(EquipmentJson result) {
60.         super.onPostExecute(result);
61.         BaseEquipmentEntity baseEquipment=BaseEquipmentEntity.parse(result);
62.         if(baseEquipment!=null){ }}}}
```

5.3.6 物品出库页面

实现物品出库功能。页面布局如图 5.7 所示。

图 5.7 物品出库页面列表图

源代码清单 5.10 \CH05_Supermarket\res\layout\activity_goods_outport.xml

```
1. <?xml version="1.0" encoding="utf-8"?>
2. <RelativeLayout xmlns:android="http://schemas.android.com/apk/res/android"
3.  android:layout_width="fill_parent"  android:layout_height="fill_parent"
4.  android:background="@drawable/main_default_bg" >
5.  <include
6.   android:id="@+id/bt_createtask_title_layout" layout="@layout/activity
    _common_title_bar"/>
7.  <RelativeLayout
8.    android:id="@+id/RelativeLayout1"   android:layout_width="match_parent"
9.    android:layout_height="match_parent"  android:layout_below="@id/bt_
      createtask_title_layout"
10.   android:layout_centerHorizontal="true"  android:layout_marginLeft="15dp"
11.   android:layout_marginRight="15dp"   android:background="@drawable/
      content_default_bg"
12.   android:orientation="vertical"       android:visibility="visible">
13.   <GridView
14.     android:id="@+id/gridView1"         android:layout_width="match_parent"
15.     android:layout_height="wrap_content"  android:layout_alignParent
        Left="true"
16.     android:layout_alignParentTop="true"    android:horizontalSpacing
        ="10dp"
17.     android:numColumns="1"         android:verticalSpacing="10dp"
18.     android:visibility="visible" >
```

```
19.      </GridView>
20.      <include
21.         android:id="@+id/view_loading_error"     android:layout_width="match_
            parent"
22.         android:layout_height="match_parent"     android:layout_alignParent
            Left="true"
23.         android:layout_below="@+id/gdv_models"   layout="@layout/view_
            loading_error"
24.         android:visibility="gone" />
25.   </RelativeLayout>    </RelativeLayout>
```

源代码清单 5.11 \CH05_Supermarket\src\com\zigcloud\supermarket
\activity\GoodsOutportActivity.java

```
1. package com.zigcloud.supermarket.activity;
2. public class GoodsOutportActivity extends BaseActivity{/*** 物品出库* */
3.    private TextView titlebar_left; private TextView titlebar_title; privat
      e GridView goodsGridView;
4.    private ArrayList<HashMap<String, Object>> goodsArrayList = new ArrayL
      ist<HashMap<String, Object>>();
5.    private HashMap<String, Object> goodsHashMap = new HashMap<String,
      Object>();
6.    private SimpleAdapter goodsAdapter;
7.    private GoodsDAO mGoodsDAO=new GoodsDAO(); /*** 物品业务处理类**/
8.    /*** 获取物品信息任务类**/
9.    private GoodsRequestTask mGoodsRequestTask=new GoodsRequestTask();
10.   /*** 获取物品出库**/
11.   private GoodsOutportTask mGoodsOutportTask=new GoodsOutportTask();
12.   protected void onCreate(Bundle savedInstanceState) {super.onCreate(save
      dInstanceState);
13.    setContentView(R.layout.activity_goods_outport);
14.    MainApplication.getInstance().addActivity(this);
15.      titlebar_left = (TextView) findViewById(R.id.titlebar_left);
16.      titlebar_left.setOnClickListener(new View.OnClickListener() {
17.         public void onClick(View v) { onBackPressed(); }});
18.      titlebar_title = (TextView) findViewById(R.id.titlebar_title);
19.      titlebar_title.setText(R.string.title_goods_outport);
20.      initialGoodsGridView();    }
21.   private void initialGoodsGridView(){  /*** 初始化情景模式列表**/
22.      goodsGridView=(GridView) findViewById(R.id.gridView1);
23.      goodsAdapter = new SimpleAdapter(getApplicationContext(), goods
24.          ArrayList,R.layout.activity_goods_list_item, new String[]
              {"Image","goodsCardId","goodsName"},
25.          new int[] {R.id.itemImage,R.id.tv_goods_cardid,R.id.
              tv_goods_name});
26.     goodsGridView.setAdapter(goodsAdapter);
27.     goodsGridView.setOnItemClickListener(new OnItemClickListener() {
28.        @SuppressWarnings({ "unchecked" })
29.        public void onItemClick(AdapterView<?> arg0, View arg1, int arg2,
           long arg3) {
```

```
30.            HashMap<String, Object> item=( HashMap<String, Object>) arg0.
               getItemAtPosition(arg2);
31.            final String goodsCardId= item.get("goodsCardId").toString();
32.            AlertDialog.Builder builder=new AlertDialog.Builder(GoodsOut
               portActivity.this)
33.                .setIcon(R.drawable.ic_launcher)
34.                .setItems(new String[]{"出库"},new DialogInterface.OnClick
                    Listener() {
35.                    public void onClick(DialogInterface dialog, int which) {
36.                        switch(which){
37.                        case 0:
38.                            mGoodsOutportTask=new GoodsOutportTask();
39.                            mGoodsOutportTask.execute(goodsCardId);
40.                            break; } }});
41.            builder.create().show();} });
42.        mGoodsRequestTask.execute();}
43.    /*** 获取物品信息任务类* */
44.    private class GoodsRequestTask extends AsyncTask<String, Integer, List
        <GoodsJson>>{
45.        protected List<GoodsJson> doInBackground(String... params) {
46.            return mGoodsDAO.getAll();}
47.        protected void onPreExecute() { super.onPreExecute();}
48.        protected void onProgressUpdate(Integer... values) {    super.onPro
            gressUpdate(values);}
49.        protected void onPostExecute(List<GoodsJson> result) { super.onPost
            Execute(result);
50.            if(result!=null){
51.                goodsArrayList.clear();
52.                GoodsJson goodsEntity=null;
53.                for(int i=0;i<result.size();i++){
54.                    goodsEntity= result.get(i);
55.                    if(goodsEntity!=null){
56.                        goodsHashMap=new HashMap<String, Object>();
57.                        goodsHashMap.put("Image", R.drawable.metro_home_blacks
                            _scan_code);
58.                        goodsHashMap.put("goodsCardId", goodsEntity.cardNum);
59.                        goodsHashMap.put("goodsName", goodsEntity.dataName);
60.                        goodsHashMap.put("goodsAddress",goodsEntity.originPlace);
61.                        goodsHashMap.put("goodsUser", goodsEntity.staffName);
62.                        goodsArrayList.add(goodsHashMap);}}
63.                goodsAdapter.notifyDataSetChanged();
64.            }else{   findViewById(R.id.view_loading_error).setVisibility(0);}} }
65.    /*** 物品出库* */
66.    private class GoodsOutportTask extends AsyncTask<String, Integer, Contro
        lResultJson>{
67.        protected ControlResultJson doInBackground(String... params) {
68.            if(params!=null&&params.length>0) return mGoodsDAO.goodsOutpo
                rt(params[0]);
69.            return null;   }
70.        protected void onPreExecute() { super.onPreExecute();}
```

```
71.     protected void onProgressUpdate(Integer... values) {  super.onPro
        gressUpdate(values);    }
72.     protected void onPostExecute(ControlResultJson result) { super.on
        PostExecute(result);
73.         String resStr=null;
74.         if(result!=null){resStr=result.flag==0?"出库成功!":"出库失败";}
75.         else{resStr="出库失败";  }
76.         mGoodsRequestTask=new GoodsRequestTask();
77.         mGoodsRequestTask.execute();
78.         Toast.makeText(GoodsOutportActivity.this, resStr, Toast.LENGTH_
        SHORT).show();}}}
```

具体代码实现请见本教材所附系统源码。

本章小结

本章通过未来超市项目，让学生了解企业项目的整个开发流程：项目立项、需求分析、总体设计、详细设计、编码实现、系统调试。同时通过这个项目理解以前章节中学习到的知识并灵活应用到要实现的项目。

强化练习

1. 填空题

（1）Timer 类是用来管理（　　　　）的，其会按照设置好的时间间隔，定时触发任务。

（2）TimerTask 类描述定时需要执行的任务，它是一个抽象类，需要实现其中的（　　　　）方法，添加定时任务所需要执行的功能代码。

（3）Timer 启动后会周期性地触发 TimerTask 类的 run()方法，在 run()方法中会向 UI 线程的句柄（　　　　）对象发送消息（　　　　），从而触发 Handle 对象的（　　　　）方法，完成 UI 线程中控件的更新操作。

（4）定时器（　　　　）提供了开启定时器、取消定时器等方法，其中 void schedule(TimerTask task, long delay, long period)方法的功能为（　　　　）；void cancel()方法的功能为（　　　　）和所有的定时任务。

（5）一个定时器（　　　　）一旦调用了 Cancel()方法，它就不能再执行（　　　　）方法，否则会抛出异常。

（6）定义线性布局（　　　　）时至少设置的三个属性为（　　　　）、（　　　　）和（　　　　）。

（7）布局文件 XML 的命名中不能出现（　　　　）字母。

（8）设置文本视图（　　　　）控件字体的属性是（　　　　）。

（9）Android 的常用四大组件是（　　　　）、（　　　　）、（　　　　）和（　　　　）。

（10）Android 的所有基本 UI 都是由（　　　　）类或者其子类实现的。

2. 单选题

（1）Android 的基本 UI 控件和高级 UI 控件均继承自（　　　　）类。

A. TextView　　　　　　　　　　B. Button
C. GridView　　　　　　　　　　D. View

（2）Android UI 包括几类布局管理器：线性布局、相对布局、帧布局、表格布局和绝对布局等，它们均继承自（　　）类。

A. Activity　　　　　　　　　　B. Service
C. ViewGroup　　　　　　　　　D. SurfaceView

（3）ListView 是 Android 中最常用的控件之一，ListView 中一个重要的概念就是（　　），它是控件与数据源之间的桥梁。

A. 适配器　　　　　　　　　　B. 视图
C. 数据源　　　　　　　　　　D. 事件监听器

（4）在启动 Activity 的方法中，不包括以下（　　）方法。

A. startActivity　　　　　　　　B. startActivityFromChild
C. startActivityForResult　　　　D. startActivityForFragment

（5）Activity 的生命周期方法不包括下列（　　）方法。

A. onPause　　　　　　　　　　B. onCreate
C. onNewIntent　　　　　　　　D. onRestart

（6）设置编辑框（EditText）提示信息的属性为（　　）。

A. android:inputType　　　　　　B. android:hint
C. android:digits　　　　　　　　D. android:text

（7）Android 意图的作用在于（　　）。

A. 是连接四大组件的纽带，可实现界面间的切换，可包含动作和动作数据
B. 实现应用程序间的数据共享
C. 可保持应用在后台运行，而不会因为切换页面而消失
D. 处理一个应用程序的后台工作

（8）句柄是线程与活动通信的桥梁，若线程处理不当，会导致机器变得越来越慢，可调用（　　）方法销毁该线程。

A. onStop()　　　　　　　　　　B. onFinish()
C. onDestroy()　　　　　　　　　D. onClear()

（9）下面属于 View 子类的是（　　）。

A. Activity　　　　　　　　　　B. Service
C. ViewGroup　　　　　　　　　D. Content Provider

（10）当活动被销毁时，应该实现它在什么方法来保存其原来的状态（　　）。

A. onResume()　　　　　　　　　B. onSaveInstance()
C. onInstanceState()　　　　　　　D. onSaveInstanceState()

3. 问答题

（1）Android 软件框架结构自上而下可分为哪些层？
（2）Android 应用程序的 4 大组件是什么？
（3）Android 应用工程文件结构有哪些？

（4）Android 开发应用程序最有可能使用到的应用框架部分是哪些？

4. 编程题

（1）编写一个简单 Android 应用程序，演示常用日志包 android.util.Log 中包含的日志方法 Log.d()、Log.e()、Log.i()和 Log.w()的使用方法，并用 DDMS 观察记录活动生命周期日志信息（参见 Ch05_Coding01_Log.rar）。

（2）编写图 5.8 所示的程序，按下"普通""小型""切换""华盖"和四个位图按钮后，分别会在屏幕的不同位置用 Toast 短时间或长时间提示信息，并相应设置活动标题信息。分别按下/释放上面四个按钮时，会更新位图背景位图、用 Toast 向用户给出提示信息并设置活动标题。分别按下上面四个按钮并移动时，也会更新位图背景位图、用 Toast 向用户给出提示信息并设置活动。注意：按"普通"按钮时弹出自定义风格的提示信息（参见 Ch05_Coding02_Toast.rar）。

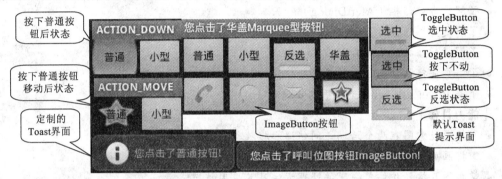

图 5.8 Toast 显示提示信息范例练习

（3）本程序运行结果如图 5.9 所示。单击"发送通知"按钮后，出现带图标通知信息（参见 Ch05_Coding03_Notification.rar）。

图 5.9 利用 Notification 实现状态栏提示功能练习

（4）编程练习图 5.10 所示的自定义 Toast 提示信息框（参见 Ch05_Coding04_CustomedToast.rar）。

图 5.10 自定义 Toast 练习

博物馆智能导览系统项目综合实训

【知识目标】
- 了解利用 JNI（Java Native Interface）接口实现串口通信的方法；
- 了解警告对话框（AlertDialog）的基本知识；
- 了解定时器任务（TimerTask）的基本知识；
- 了解视频播放的基本知识；
- 了解广播接收动态注册的基本知识；
- 了解 BroadcastReceiver 的基本知识。

【能力目标】
- 掌握通过意图在两个活动之间传递和接收数据的方法；
- 掌握画廊控件的使用方法及技巧；
- 掌握图片按钮控件单击事件处理方法及其应用技巧；
- 掌握网格视图控件的使用方法及技巧；
- 掌握媒体播放器（MediaPlayer）的使用方法及技巧。

【重点、难点】
- SQLiteDatabase 数据库的应用编程方法；
- SurfaceView 的使用方法及其应用技巧。

6.1 项目概述

6.1.1 项目简要介绍

博物馆物联网智能导览系统是基于 Zigbee 定位技术的可用于真实项目的非精确定位系统。系统包含一套 Cortex A9 智能终端、智能插座系统，智能终端和插座系统的连接通过上位机基于 Android 的智能导览应用软件实现。本系统可用于实际的项目。用户可以在 Cortex A9 智能终端上体验真实的应用场景，例如博物馆展品展示、校园实验室展板介绍、商场智能导购等。

6.1.2 项目背景

（1）实用性：近些年来，随着社会科技、文化和经济的发展，特别是物联网技术和人工智能的大发展，人类社会从工业社会向信息社会过渡的趋势越来越明显，人们对信息化生活的意识和追求愿望越来越强。通过智能终端应用软件，用户可以很容易地查询和浏览信息资源。然而在平时的生活中，我们不管是在博物馆浏览展品还是在大商场购物，对于展品和产品的信息缺乏自主或直观的了解。大多还是服务人员或者静态标牌指示，这些还是满足不了用户多方面的便捷阅览需求，大大影响了浏览效率。试想未来的商场只要给消费者配发一张存储卡，消费者安装在智能移动平台上通过智能导览的应用软件就能自主地浏览、选择适合自己的产品，那是件多么惬意的事。所以"智能导览"应用是在人们日益丰富的物质文化追求下满足人们自主便捷的阅览需求而定制的，通过移动智能终端让用户直观地、感性地知道自己所处阅览位置以及即时了解展品或产品信息。该智能终端的这款应用将是未来信息化生活的必备元素。

（2）产品化："智能导览"的一大核心是基于 Android 系统的智能终端平台的一款应用软件，在如今各类智能平台发展普遍化和大众化大前提下，将可执行程序安装在智能终端平台上，便可在相应场合轻松实现产品化。

基于物联网发展和人工智能发展的理念之上，平台上的 Android 应用软件是结合硬件系统的一款实用性软件，该软件接收底层硬件（自由组网的 ID 结点）上传数据进行解码处理，自动分析 ID 结点位置的关系及远近信息。然后根据解码后的数据显示最近 ID 结点信息（信息动态更新），然后可以启动对应媒体阅览功能，做到"你的地盘我做主""信谁我动"的自由阅览。

6.2 目标分析与运行环境

6.2.1 目标分析

（1）ID 结点的位置关系主要是远近关系

Zigbee 网络是自组网，网络中数据传输的路径并不是预先设定的，在传输数据前对网络当时可利用的所有路径进行搜索，通过信号强度进行算法过滤取得最近的结点信息并在智能终端平台上显示。

（2）实时更新 ID 信息，可通过信号强度判断目标距离

导览系统不需要精确的经纬度定位，判断出当前所在区域，再根据界面显示的信息值进行信息及媒体资源浏览，而根据信号强度又可以初步判断 ID 结点的远近。

（3）呈现给用户简洁美观的访问浏览界面

智能导览应用软件的开发环境是 Eclipse，应用在带 Android 系统的智能终端平台上。浏览界面包括动态文字、图片自主浏览以及对应视频观赏。

6.2.2 运行环境

智能终端平台。

（1）系统要求：Android4.2.2 版本。

（2）硬件配置：处理器采用 ARM Cortex A9 系列（Amlogic Cortex-A9 AML8726-M3 处理器）及以上；主频不低于 1 GHz；内存推荐使用 1 GB 以上；存储器提供 8 GB 的内置 Flash 和一个 MiroSD 卡插槽；Zigbee 板载芯片型号为主流 TI 的 CC2530 芯片或以上。

6.3 需求分析

6.3.1 功能需求

本项目主要功能需求，分成数据处理（见表 6.1）、界面显示（见表 6.2）和插座灯控制（见表 6.3）等三个部分。

表 6.1 数据处理部分功能需求

需 求 编 号	001
名称	ID 结点数据接收
描述	上层需要接收底层结点上传的数据帧并解码
输入	不断变化的十六进制数据帧
处理	读取数据帧并解码过滤
输出	结点对应的相关信息（结点号、状态、RSSI 值等）
需 求 编 号	002
名称	结点对应图标信息动态显示
描述	解码过滤后的结点信息实时送显
输入	过滤后的个结点信息
处理	选择图标或文字形象的显示结点信息
输出	结点对应图标和文字信息

表 6.2 界面显示部分功能需求

需 求 编 号	003
名称	媒体信息浏览
描述	对应结点的文字图片浏览及视频欣赏
输入	单击最近结点对应的图标
处理	根据此图标对应结点的信息对应媒体浏览界面
输出	文字图片浏览界面及视频浏览界面
需 求 编 号	004
名称	临近结点位置更新后智能提醒
描述	用户正在浏览媒体信息时可能进入到另一个结点附近
输入	更新后的结点信息（主要是结点号和信息强度）
处理	最新结点信息与原信息进行比较判断
输出	信息提示框

表 6.3 插座灯控制部分功能需求

需 求 编 号	005
名称	小夜灯开关控制
描述	点亮临近区域的小夜灯，关闭较远的

续表

需求编号	005
输入	开关命令语句
处理	根据定义的控制命令协议向对应结点发送开关命令数据帧
输出	小夜灯的点亮和熄灭

6.3.2 性能需求

1. 位置精确度

在实际应用中需要准确判断出结点位置的远近关系（相对于智能终端平台），准确显示出最近区域的结点信息，同时准确控制智能插座对应小夜灯的点亮与熄灭。

2. 时间特性要求

小夜灯的点亮和界面的信息显示时间基本保持一致；时差 1 s 左右，关闭较远结点的小夜灯延迟 3 s 左右；结点信息（主要是信号强度）实时更新。

6.4 总体设计

6.4.1 总体结构

本项目总体结构按照功能归类，其主要系统元素划分如表 6.4 所示。本项目以上模块之间的结构关系如图 6.1 所示。

表 6.4 主要系统元素划分

模块名称	标识符	模块描述
数据帧读取模块	SerialPort	读取串口数据帧
数据解码模块	ZigbeePackage	分解出数据帧包含的各类结点信息
数据过滤模块	Rssi_max	过滤出最近结点数据
数据转换模块	ConvertHelper	提供各类数据类型的相互转换
滚动字幕模块	MarqueeText	实现 Allshow 界面的滚动字幕
图片加载模块	ImageAdapter	从 SDcard 加载图片文件
图片处理模块	GalleryFlow	实现图片拨动浏览以及美化显示效果
视频处理模块	Medisplay	从 SDcard 加载视频文件并做控制处理
反向控制模块	CommandBuilder	实现智能插座的开关控制
退出应用模块	onKeyDown	处于主界面时单击返回或 Home 键时提示退出主程序
列表建立模块	BuildList	建立信息列表，为信息显示做准备
主界面显示模块	IntelligentBroswing	动态刷新并显示出结点图标
图文显示模块	Allshow	实现单击主界面图标后出现文字和图片浏览界面，不同的结点对应不同的浏览信息
视频显示模块	Allmovie	实现单击 Allshow 界面的纪录片图标后呈现视频欣赏界面，不同结点对应不同视频信息

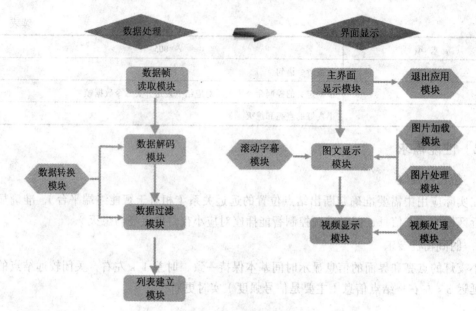

图 6.1 模块之间结构关系

6.4.2 处理流程设计

1. 基本设计概念

四个 ID 结点按实际场合需要分布在四个不同的位置,移动智能终端上安装有智能导览的应用软件。用户手持移动平台,移动终端不断地向外广播信号数据包,所有结点接受信号包过滤处理后在向移动平台反馈相应信号包。平板接收到信号包后再进行解码分析,过滤分析以及数据转换,然后在应用软件的主界面上显示出结点信息。最后达到的效果是走到最近 ID 结点处点亮小夜灯的同时在屏幕界面上显示出对应的信息,对主界面的信息图标执行单击事件从而开始对应媒体的浏览,如图 6.2 所示。

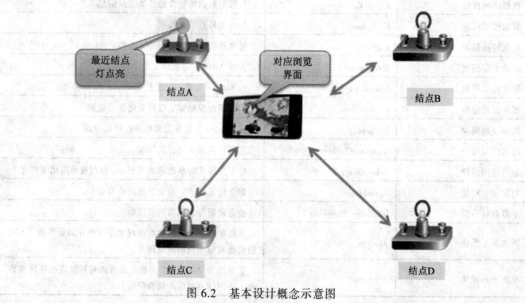

图 6.2 基本设计概念示意图

2. 主要处理流程

（1）Activity 跳转（主线程）处理：主界面（IntelligentBrowsing）通过单击结点图标跳转到对应图文浏览界面（Allshow），再通过单击事件跳转到对应视频信息的浏览界面（Allmovie），如图 6.3 所示。

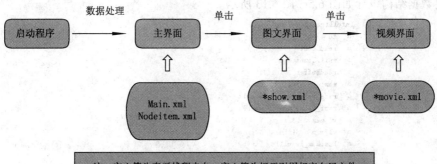

图 6.3 Activity 跳转（主线程）处理

（2）数据处理线程：这个线程设置在 IntelligentBrowsing 模块中，进行实时的串口数据读取，然后实现解码和过滤操作。把处理的数据结果建立列表（BuildList），即为主界面的结点信息显示做准备。

（3）设备休眠处理线程：用户结束浏览离开展区并超过一定的时间，通过设置一定时间达到设备休眠的效果。

（4）信息数据广播线程：处理后的结点信息数据也是跟随读取的数据不断变化的，由于用户的位置是动态的，可能在浏览某结点对应媒体文件时已经临近另一个结点，这时候需要接收广播来的动态数据，根据前后结点信息的比较判断是否在界面上显示提示框，用户根据提示进行页面的切换或者继续浏览，如图 6.4 所示。

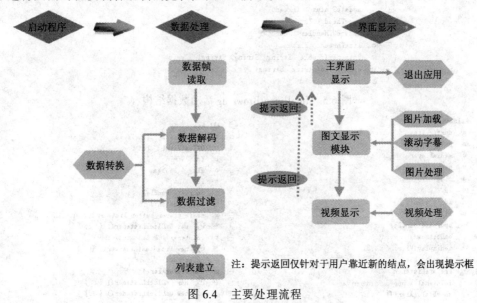

图 6.4 主要处理流程

6.5 详细设计

6.5.1 主要数据结构设计

主要数据结构设计如图 6.5～图 6.13 所示。

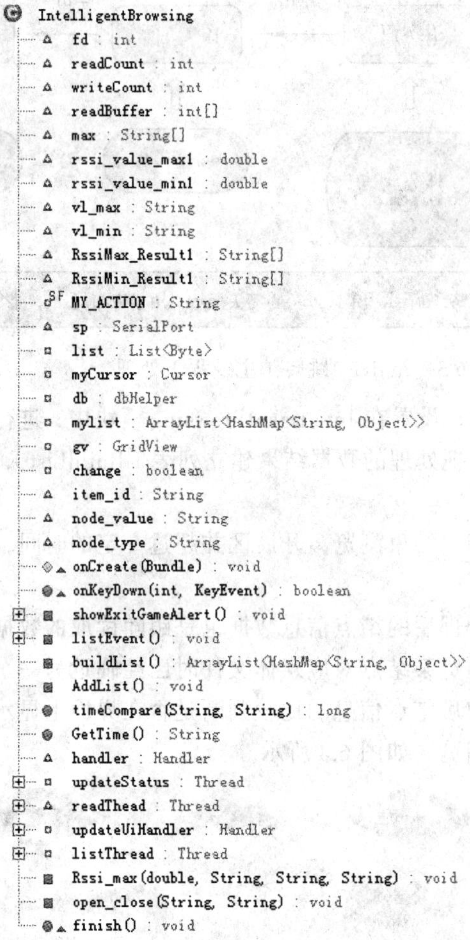

```
IntelligentBrowsing
 △  fd : int
 △  readCount : int
 △  writeCount : int
 △  readBuffer : int[]
 △  max : String[]
 △  rssi_value_max1 : double
 △  rssi_value_min1 : double
 △  vl_max : String
 △  vl_min : String
 △  RssiMax_Result1 : String[]
 △  RssiMin_Result1 : String[]
 ᵝᶠ MY_ACTION : String
 △  sp : SerialPort
 □  list : List<Byte>
 □  myCursor : Cursor
 □  db : dbHelper
 □  mylist : ArrayList<HashMap<String, Object>>
 □  gv : GridView
 □  change : boolean
 △  item_id : String
 △  node_value : String
 △  node_type : String
 ◇△ onCreate(Bundle) : void
 ●△ onKeyDown(int, KeyEvent) : boolean
 ⊞■ showExitGameAlert() : void
 ⊞■ listEvent() : void
    ■ buildList() : ArrayList<HashMap<String, Object>>
    ■ AddList() : void
    ● timeCompare(String, String) : long
    ● GetTime() : String
 △  handler : Handler
 ⊞□ updateStatus : Thread
 ⊞△ readThead : Thread
 ⊞□ updateUiHandler : Handler
 ⊞■ listThread : Thread
    ■ Rssi_max(double, String, String, String) : void
    ■ open_close(String, String) : void
 ●△ finish() : void
```

图 6.5 IntelligentBrowsing 活动数据结构

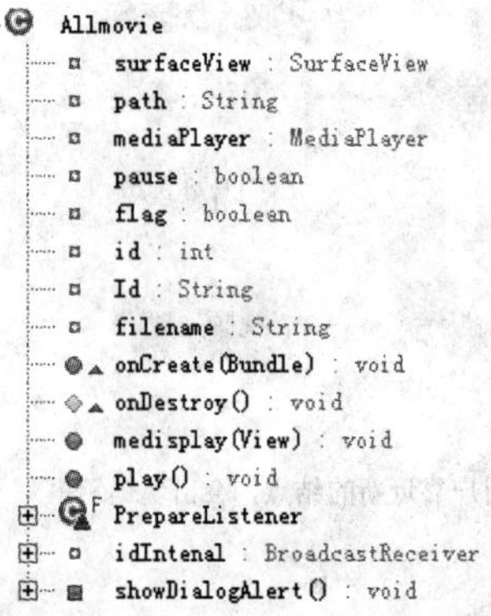

```
Allmovie
 □ surfaceView : SurfaceView
 □ path : String
 □ mediaPlayer : MediaPlayer
 □ pause : boolean
 □ flag : boolean
 □ id : int
 □ Id : String
 □ filename : String
 ●△ onCreate(Bundle) : void
 ◇△ onDestroy() : void
 ● medisplay(View) : void
 ● play() : void
 ᶠ PrepareListener
 ⊞■ idIntenal : BroadcastReceiver
 ⊞■ showDialogAlert() : void
```

图 6.6 Allmovie 活动数据结构

```
Allshow
 □ TCTextView1 : TextView
 □ QTTextView1 : TextView
 □ Id : String
 △ flag : boolean
 □ id : int
 ◇△ onCreate(Bundle) : void
 ■ image_button() : void
 ⊞■ media_player_button_listener : OnClickListener
   ⊞Ⓖⁿⁿⁿⁿⁿⁿⁿⁿⁿⁿⁿ new OnClickListener() {...}
 ⊞■ idIntenal : BroadcastReceiver
   ⊞Ⓖ new BroadcastReceiver() {...}
 ◇△ onDestroy() : void
 ⊞■ showDialogAlert() : void
   ⊞Ⓖ new OnClickListener() {...}
   ⊞Ⓖ new OnClickListener() {...}
```

图 6.7 Allshow 活动数据结构

```
ConvertHelper
  byte2HexString(byte[]) : String
  hexString2ByteArray(String) : byte[]
  byteArrayJoin(byte[], byte[]) : byte[]
  byteArray2Integer(byte[], boolean) : int
  byteArray2Long(byte[], boolean) : long
  short2ByteArray(short, boolean) : byte[]
  int2ByteArray(int, boolean) : byte[]
  long2ByteArray(long, boolean) : byte[]
  byteArray2Short(byte[], boolean) : short
  getDoubleValue(String) : double
  getIntValue(String) : int
```

图 6.8 ConvertHelper 数据结构

```
GalleryFlow
  mCamera : Camera
  mMaxRotationAngle : int
  mMaxZoom : int
  mCoveflowCenter : int
  GalleryFlow(Context)
  GalleryFlow(Context, AttributeSet)
  GalleryFlow(Context, AttributeSet, int)
  getMaxRotationAngle() : int
  setMaxRotationAngle(int) : void
  getMaxZoom() : int
  setMaxZoom(int) : void
  getCenterOfCoverflow() : int
  getCenterOfView() : int
  getChildStaticTransformation(View, Transformation) : boolean
  onSizeChanged(int, int, int, int) : void
  transformImageBitmap(ImageView, Transformation, int) : void
```

图 6.9 GalleryFlow 数据结构

```
ImageAdapter
  mGalleryItemBackground : int
  mContext : Context
  mImages : ImageView[]
  lis : List<String>
  ImageAdapter(Context, List<String>)
  createReflectedImages() : boolean
  getCount() : int
  getItem(int) : Object
  getItemId(int) : long
  getView(int, View, ViewGroup) : View
  getScale(boolean, int) : float
```

图 6.10 ImageAdapter 数据结构

```
MarqueeText
  MarqueeText(Context)
  MarqueeText(Context, AttributeSet)
  MarqueeText(Context, AttributeSet, int)
  isFocused() : boolean
  onFocusChanged(boolean, int, Rect) : void
```

图 6.11 MarqueeText 数据结构

```
SerialPort
  S {...}
  EM7910_Open() : int
  EM7910_Close(int) : int
  EM7910_read(int, int, int[]) : int
  EM7910_write(int, byte[], int) : int
```

图 6.12 SerialPort 数据结构

```
ZigbemPackage
  byteArray : int[]
  byteArrayCount : int
  data : double
  get_data() : double
  get_status() : int
  get_id() : int
  get_type() : String
  hex : String
  getHex() : String
  ZigbemPackage(int[], int)
```

图 6.13 ZigbeePackage 数据结构

6.5.2 关键或难点技术的实现

1）结点上传数据的过滤

这里的过滤主要是对信号强度而言的，结点反馈的数据帧包含信号强度值。

难点及实现：由于硬件或者其他外界干扰结点反馈的信号强度不稳定，例如偶尔远的结点反馈的信号反而比较近结点的信号要强，所以不能用绝对过滤的方法（即设定一个中间强度值，反馈的信号强度与此值进行比较）。经过测试，较远结点的信号强度在宏观时间段来讲大体小于较远结点，仅仅偶尔出现异常，所以这里利用相对比较过滤法，即不断地将所有结点的 RSSI 值进行比较，保留最大值。与此同时保留绝对过滤法，由于数据动态变化，两种过滤法结合后基本上就剔除了异常。

2）数据解码以及再合成

上传的数据是按照特定协议组成的数据帧，数据帧包含了所有结点的相关信息（结点类型，结点状态，结点 ID 号以及 RSSI 值等）。程序需要对这些数据进行分解处理。分解处理结束还需要再次合成可供显示的列表信息。

难点及实现：数据的分解首先不能有一位出错，而且解码的同时需要数据类型的转换，解码后进行 RSSI 过滤，过滤后的再次合成所以信息必须不失同步性（结点之间的信息不能交叉）。这里建立一个数据库类，数据读取进来解码后把每一个结点的所有信息整体存入数据库（数据库的每一列均代表某一结点的所有信息），过滤时动态更新数据库。

3）临近新结点时提示返回

移动终端平台的位置是不断变化的，当用户靠近新的结点时正在浏览原结点的媒体信息，需要给予信息提示。

难点及实现：考虑到程序的稳定性和内存暂用，不可能在每一个 Activity 中都新建一个串口数据读取的线程。这里利用 Android 的广播（Broadcast）与接收功能，在主界面串口数据读取处理的同时发送广播信息（这里只需要结点 ID 号），在图文和视频界面实现程序中仅需要接收这个广播信息和原结点 ID 号比较，然后根据判断结果提示返回。

6.5.3 软件主要功能的使用说明

本项目主要功能的使用步骤如下：

步骤 1：对底层上传的结点信息进行数据处理，获取结点号以及结点的其他信息，对应到上位机的图标并显示。只有最近的结点才能在主界面上显示出来。

步骤 2：图 6.14 中显示出最近的结点 4 的信息，此处假设 ID 号为 4 的结点对应的是"玉器展区"。单击图标进入玉器展示界面，如图 6.15 所示，其中的展示界面分为标题、图片、字幕介绍和视频按钮四个部分。其中标题为结点对应展区的信息，图片和字幕是玉器区的相关介绍，单击视频按钮进入玉器视频浏览界面。

图 6.14　最近的结点 4 的信息

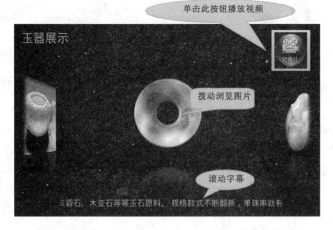

图 6.15　玉器展示界面

步骤 3：单击图 6.15 中的视频播放按钮进入视频播放界面，如图 6.16 所示。

黑色区域为视频播放区，左边的四个按钮用来控制视频播放，最上面的按钮为开始播放按钮，单击后开始出现视频播放页面；紧接着为暂停按钮，暂停后再次单击暂停按钮可以继续播放；再接着为重播按钮，单击后视频从头开始播放；最下面的是停止按钮，单击后视频停止。

步骤 4：在浏览玉器展区时靠近了另一个展区，此时显示图 6.17 所示的提示框。

图 6.16　视频播放界面

图 6.17　新入展区提示信息对话框

6.6 项目实施

6.6.1 IntelligentBrowsing 活动

IntelligentBrowsing 活动的界面布局包括：一个图像视图控件、一个网格视图控件和两个文本视图控件，其 XML 布局源代码清单如 6.1 所示。Java 源代码清单如 6.2 所示。

源代码清单 6.1　\ch06_com.emdoor.smartBrowsing\res\layout\main.xml

```xml
1. <?xml version="1.0" encoding="utf-8"?>
2. <LinearLayout xmlns:android="http://schemas.android.com/apk/res/android"
3.   android:layout_width="fill_parent"  android:layout_height="fill_parent"
4.   android:background="#000000"    android:orientation="vertical" >
5.   <RelativeLayout android:layout_width="match_parent" android:layout_
     height="wrap_content">
6.     <ImageView  android:id="@+id/imageView1" android:layout_width="wrap
       _content"
7.       android:layout_height="wrap_content"     android:layout_alignParent
         Left="true"
8.       android:layout_marginLeft="5dp"  android:src="@drawable/emdoor"/>
9.     <TextView   android:id="@+id/textView1"   android:textColor="#3366FF"
10.      android:textSize="12pt"          android:layout_width="wrap_content"
11.      android:layout_height="wrap_content"    android:layout_alignParent
         Right="true"
12.      android:layout_alignParentTop="true"    android:layout_marginRight
         ="32dp"
13.      android:text="@string/welcome" />   </RelativeLayout>
14.    <GridView  android:id="@+id/gridView"  android:layout_width="match_
       parent"
15.      android:layout_height="wrap_content" android:numColumns="4">
       </GridView>
16.    <TextView  android:id="@+id/touchScreen"   android:textSize="15pt"
17.      android:gravity="center_horizontal"  android:layout_width="fill_
         parent"
18.      android:layout_height="500dp" android:shadowColor="#666666"
19.      android:shadowRadius="1.0"    android:shadowDx="1.0"
20.      android:shadowDy="1.0" >    </TextView>   </LinearLayout>
```

源代码清单 6.2　\ch06_com.emdoor.smartBrowsing\src\com\emdoor\smartBrowsing

```java
1. public class IntelligentBrowsing extends Activity{
2.    int fd = -1;   /*文件句柄*/
3.    int readCount=0;
4.    int[] readBuffer=new int[128]; /*读取数据缓冲区*/
5.    boolean updateUiThreadFlag=false; /*数据采集线程是否开启*/
6.    boolean fdOpenFlag=false; /*设备是否打开*/
7.    boolean fdReadFlag=false; /*是否读取设备数据*/
8.    boolean ListTreadflag = true; /*执行命令的标识*/
9.    boolean doCmdThreadFlag=false;
10.   double rssi_value_max1=0.0; /*RSSI 过滤相关定义*/
11.   double rssi_value_min1=0.0;
```

```java
12.     String vl_max = new String(Double.toString(rssi_value_max1));
13.     String vl_min = new String(Double.toString(rssi_value_min1));
14.     String[] RssiMax_Result1 =new String[]{vl_max,null,null,null};
15.     String[] RssiMin_Result1 =new String[]{"0.0",null,null,null};
16.     SerialPort sp=new SerialPort();
17.     private Cursor myCursor;
18.     private dbHelper db;
19.     private ArrayList<HashMap<String, Object>> mylist;
20.     private GridView gv = null;
21.     private boolean change =false;//为闪烁文字设置循环条件值
22.     String item_id = "id = null;";
23.     String node_value = "value = 0;";
24.     String node_type = "unknown";
25.     private static final String MY_ACTION = "MY_ACTION";//定义一个Action常量
26.     protected void onCreate(Bundle savedInstanceState) {
27.         super.onCreate(savedInstanceState);
28.         setContentView(R.layout.main);
29.         gv = (GridView) findViewById(R.id.gridView);
30.         gv.setVisibility(-1);
31.         db=new dbHelper(IntelligentBrowsing.this);
32.     updateUiThreadFlag=true;         fd= sp.EM7910_Open();
33.     if(fd!=-1)fdOpenFlag=true;       fdReadFlag=true;
34.     db.deleteAll("Node");            readThead.start();
35.     listThread.start(); }
36.     /*以下这段代码为点击pad的返回键，确定是否退出该应用程序*/
37.     public boolean onKeyDown(int keyCode, KeyEvent event) {
                                                        //判断是否是返回键
38.         if((keyCode == KeyEvent.KEYCODE_BACK )|| (keyCode == KeyEvent.KEYCODE_HOME)){
39.             showExitGameAlert();}
40.         return super.onKeyDown(keyCode, event);}
41.     private void showExitGameAlert() {
42.     AlertDialog alertDialog = new AlertDialog.Builder(IntelligentBrowsing.this)
43.         .setTitle("退出").setMessage("确定要退出吗? ").setIcon(R.drawable.emdoor)
44.         .setPositiveButton("确定", new DialogInterface.OnClickListener() {
45.             public void onClick(DialogInterface dialog, int id) {
                                                //跳转到桌面，退出该系统
46.                 Intent startMain = new Intent(Intent.ACTION_MAIN);
47.                 startMain.addCategory(Intent.CATEGORY_HOME);
48.                 startMain.setFlags(Intent.FLAG_ACTIVITY_NEW_TASK);
49.                 startActivity(startMain);   finish();}})
50.         .setNegativeButton("取消", new DialogInterface.OnClickListener() {
51.             public void onClick(DialogInterface dialog, int id) {
52.                 dialog.cancel(); }}).setCancelable(true).show();}
53.     private void listEvent(){/* ListEvent部分(视图显示)*/
54.         gv.setOnItemClickListener(new OnItemClickListener(){
55.         public void onItemClick(AdapterView<?> arg0, View arg1, final int arg2, long arg3){
```

```
56.          HashMap<String, Object> _hm=mylist.get(arg2); item_id=_hm.get
             ("NODE_ID").toString();
57.          Intent intent = new Intent();
58.             switch(arg2){
59.                case  0: intent.putExtra("Id",item_id);
60.                   intent.setClass(IntelligentBrowsing.this, Allshow.class);
61.                   IntelligentBrowsing.this.startActivity(intent);break;
62.                default: break; }}});
63.     mylist = buildList();
64.     /*一些关于文字闪烁的定义*/
65.     final TextView touchScreen =(TextView)findViewById(R.id.touchScreen);
66.     Timer SStimer = new Timer();
67.     if(mylist != null){ touchScreen.setText("欢迎参观此展区！");
68.     MySimpleAdapter mSchedule = new MySimpleAdapter(this,
69.         mylist,//数据来源
70.         R.layout.nodeitem,//ListItem 的 XML 实现动态数组与 ListItem 对应的子项
71.         new String[] {"NODE_ID","status","logo","detail"},
72.           //ListItem 的 XML 文件里面的两个 TextView ID
73.           new int[] {R.id.listview_item_tv,R.id.listview_item_status,
              R.id.listview_item_logo,R.id.lisview_item_detail});
74.     gv.setAdapter(mSchedule); gv.setVisibility(1);}
75.     else{touchScreen.setText("附近没有展区，感谢您的光临！");gv.setVisibili
        ty(-1);}
76.     /*此处添加一段代码，为主界面的 TextView 实现文字闪烁*/
77.     TimerTask task =new TimerTask(){
78.        public void run(){ runOnUiThread(new Runnable(){
79.        public void run(){
80.            if(change){ change=false;
81.                touchScreen.setTextColor(Color.TRANSPARENT);//透明，看不到文字
82.            }else{change=true;touchScreen.setTextColor(Color.parseColor
               ("#3366FF"));}}});} };
83.        SStimer.schedule(task,1000);}//一秒闪一次
84.  private ArrayList<HashMap<String, Object>> buildList() {  /*buildList,
     添加 list 内容*/
85.     ArrayList<HashMap<String, Object>> mylist = new ArrayList<HashMap
        <String, Object>>();
86.     myCursor=db.selectNode("Node","");
87.     if(myCursor.getCount() > 0){
88.     myCursor.moveToLast();
89.     do{ HashMap<String, Object> map = new HashMap<String, Object>();
90.         map.put("NODE_ID",myCursor.getString(1));
91.         map.put("status", R.drawable.using);
92.         map.put("logo",R.drawable.main_logo);
93.         map.put("detail", "RSSI 值:"+myCursor.getString(3).substring
            (4));   mylist.add(map);
94.        }while(myCursor.moveToPrevious()); }
95.     else{mylist = null; }
96.     return mylist;}
97.  /*读取数据分解过滤并存入数据库的进程*/
98.  Thread readThead=new Thread(){
```

```
99.  public void run(){
100.    try{  while(updateUiThreadFlag){
101.       if(fd !=-1 && fdReadFlag ) {
102.          readCount = sp.EM7910_read(fd, 128, readBuffer);
103.          if(readCount > 0) { Message msg = new Message();
104.             msg.what = 100; updateUiHandler.sendMessage(msg); }}
105.          else  {Thread.sleep(100); }}}
106.       /*线程被阻塞时(Thread.sleep(),Thread.wait()方法会造成阻塞),出现中断异常*/
107.       catch(InterruptedException e)
108.          {Thread.currentThread().interrupt(); }  } };
109.  private Handler updateUiHandler = new Handler() {
110.     public void handleMessage(Message msg) {
111.     switch (msg.what) {
112.        case 100:ZigbeePackage zigbemPackage=new ZigbeePackage(readBuffer,
          readCount);
113.        double value = zigbemPackage.get_data();
114.        String nodeID = "结点 ID: " + zigbemPackage.get_id();
115.        String nodeType = zigbemPackage.get_type();
116.        String nodeStatus = "";
117.        if(zigbemPackage.get_status()>0){nodeStatus = "是"; }
118.        else{nodeStatus = "否";}
119.        Rssi_max(value,nodeID,nodeType,nodeStatus);
120.        Intent intent = new Intent();  intent.setAction(MY_ACTION);
121.        intent.putExtra("msg", RssiMax_Result1[1].trim());
122.        sendBroadcast(intent); /*广播动态结点信息*/
123.        double valuemax_convert=ConvertHelper.getDoubleValue(RssiMax_
          Result1[0]);
124.        myCursor=db.select("Node", "Node_id='"+RssiMax_Result1[1].trim()
          +"'");
125.        if((myCursor.getCount()==0)&&(valuemax_convert>25.0)){db.delete
          All("Node");
126.          db.insertNode("Node", RssiMax_Result1[1].trim(),RssiMax_Resul
          t1[3].trim(),RssiMax_Result1[2].trim()+valuemax_convert); }
127.        else{ db.is_used_update("Node",RssiMax_Result1[3],
128.          "node_id='"+RssiMax_Result1[1].trim()+"'");
129.          db.value_update("Node", RssiMax_Result1[2].trim()+valuemax_
          convert,"node_id='"+RssiMax_Result1[1].trim()+"'");} break;
130.        default: break;}}};
131.  Handler handler=new Handler();     //定义了一个 handler 对象
132.  private Thread listThread = new Thread() {
133.    public void run() { listEvent();   //do something
134.        handler.postDelayed(this, 1000*1);}};
135./*建立一个 rssi 值处理方法,过滤出最大 RSSI 值对应的那个结点*/
136.private void Rssi_max( double value, String node_id, String node_type,
  String node_status){
137.  String vl = new String(Double.toString(value));
148.  String[] temp =new String[]{vl,node_id,node_type,node_status};
139.  if(value>=rssi_value_max1){RssiMax_Result1=temp;}
140.  else {RssiMin_Result1=temp;}}
```

```
141.   public void finish() { super.finish(); updateUiThreadFlag=false;
142.     try {readThead.join();                //等待线程完成
143.     } catch (InterruptedException e) {throw new RuntimeException(e);}
144.     listThread.interrupt();
145.     try { listThread.join();               //等待线程完成
146.     } catch (InterruptedException e) {
147.     throw new RuntimeException(e);}
148.     android.os.Process.killProcess(android.os.Process.myPid());
149.     System.exit(0);} }
```

6.6.2 数据帧读取模块 SerialPort

数据帧读取模块 SerialPort 负责读取串口数据帧，如源代码清单 6.3 所示。

源代码清单 6.3　\ch06_com.emdoor.smartBrowsing\src\com\JNIEmdoor\SerialPort.java

```
1. public class SerialPort {
2.   static {System.loadLibrary("JNIEmdoor_SerialPort");}
3.   public native int EM7910_Open();public native int EM7910_Close(int fd);
4.   public native int EM7910_read(int fd, int num, int[] buff);
5.   public native int EM7910_write(int fd, byte[] src, int sum);}
```

6.6.3 数据解码模块 ZigbeePackage

数据解码模块 ZigbeePackage 负责分解出数据帧包含的各类结点信息，如源代码清单 6.4 所示。

源代码清单 6.4　\ch06_com.emdoor.smartBrowsing\src\com\emdoor\smartBrowsing\port\ZigbeePackage.java

```
1.  package com.emdoor.smartBrowsing.port;
2.  /*** Zigbee 数据包**/
3.  public class ZigbeePackage {
4.    private int[] byteArray; /*当前字节数组*/
5.    private int byteArrayCount; /*当前字节数组长度*/
6.    private double data;
7.    /*** 取智能插座数据**/
8.    public double get_data() {
9.      if (byteArray != null) {
10.       byte[] dataArray = new byte[] {getByte(byteArray[byteArrayCount
11.        - 4]),   getByte(byteArray[byteA rrayCount - 3]) };
12.       data = ConvertHelper.byteArray2Integer(dataArray, false);}
13.     return this.data;}
14.   public int get_status() {int status;
15.     if (byteArray != null) {
16.       byte[] dataArray = new byte[] { getByte(byteArray[23]),getByte
            (byteArray[24]) };
17.       status = ConvertHelper.byteArray2Integer(dataArray, false);
18.     } else { status = -1;}
19.     return status;}
20.   public int get_id() {int id = 0;
21.     if (byteArray != null) {
22.       byte[] dataArray = new byte[] { getByte(byteArray[17]), getByte
            (byteArray[18]) };
23.       id = ConvertHelper.byteArray2Integer(dataArray, false);   }
```

```
24.     return id;}
25. public String get_type() {String node_type = "未知类型";
26.     int int_type;
27.     if (byteArray != null) {
28.         byte[] type = new byte[] {getByte(byteArray[13]), getByte(byteArray[14])};
29.         int_type = ConvertHelper.byteArray2Integer(type, false);
30.     } else { int_type = 0;}
31.     switch (int_type) {
32.     case 131:node_type = "智能插座";break;
33.     case 133:node_type = "智能雨棚";break;
34.     case 134:node_type = "智能灯光";break;
35.     case 136:node_type = "红外控制";break; }
36.     return node_type;}
37. private String hex;
38. public String getHex() {//取hex字符串
39.     hex = "";
40.     for (int i = 0; i < byteArrayCount; i++) {
41.         hex += new String(String.format("%02x", byteArray[i]));}
42.     return this.hex;}
43. /** * @param b     :内容字节数组 * @param c     :内容字节数组的数量 * */
44. public ZigbeePackage(int[] b, int c) {this.byteArray = b;this.byteArrayCount = c; }
45. private byte getByte(int i) {Integer in = new Integer(i);return in.byteValue();}}
```

6.6.4 数据转换模块 ConvertHelper

数据转换模块 ConvertHelper 负责提供各类数据类型的相互转换,如源代码清单 6.5 所示。

源代码清单 6.5 Ch06_com.emdoor.smartBrowsing\src\com\emdoor\smartBrowsing\port\ConvertHelper.java

```
1. package com.emdoor.smartBrowsing.port;
2. /** 数据类型转换辅助类* */
3. public class ConvertHelper {
4.     /** * byte 数组转 String * */
5.     public final static String byte2HexString(byte[] b) {
6.         char[] hex = {'0', '1', '2', '3', '4', '5', '6', '7', '8', '9', 'A', 'B',
                       'C', 'D', 'E', 'F'};
7.         char[] newChar = new char[b.length * 2];
8.         for(int i = 0; i < b.length; i++) {
9.             newChar[2 * i] = hex[(b[i] & 0xf0) >> 4];
10.            newChar[2 * i + 1] = hex[b[i] & 0xf]; }
11.        return new String(newChar); }
12. /*** String 转 byte 数组* */
13. public final static byte[] hexString2ByteArray(String hexString) {
14.     if(hexString.length() % 2 != 0) {
15.         //throw new IllegalArgumentException("error");
16.         return null;}
17.     char[] chars = hexString.toCharArray(); byte[] b = new byte[chars.length / 2];
18.     for(int i = 0; i < b.length; i++) {
19.         int high = Character.digit(chars[2 * i], 16) << 4;
20.         int low = Character.digit(chars[2 * i + 1], 16); b[i] = (byte)(high |
```

```
          low); }
21.    return b; }
22. /***连接byte 数组**/
23. public final static byte[] byteArrayJoin(byte[] b1,byte[] b2){
24.   byte[] bRes=new byte[b1.length  +  b2.length];
25.   System.arraycopy(b1, 0, bRes, 0, b1.length);
26.   System.arraycopy(b2, 0, bRes, b1.length,b2.length);
27.   return bRes; }
28. /*字节数组转int asc（与desc对应）：如果值是false则为小端模式，反之为大端模式*/
29. public final static int byteArray2Integer(byte[] buf, boolean asc) {
30.   if (buf == null) { throw new IllegalArgumentException("byte array is null!"); }
31.   if (buf.length > 4) { throw new IllegalArgumentException("byte array size > 4 !"); }
32.   int r = 0;
33.   if (asc)  for (int i = buf.length - 1; i >= 0; i--) { r <<= 8; r |= (buf[i] & 0x000000ff);}
34.   else   for (int i = 0; i < buf.length; i++) { r <<= 8; r |= (buf[i] & 0x000000ff); }
35.   return r; }
36. /*** byteArray转long**/
37. public final static long byteArray2Long(byte[] buf, boolean asc) {
38.   if (buf == null) { throw new IllegalArgumentException("byte array is null!"); }
39.   if (buf.length > 8) {throw new IllegalArgumentException("byte array size > 8 !"); }
40.   long r = 0;
41.   if (asc)  for (int i = buf.length - 1; i >= 0; i--) { r <<= 8; r |= (buf[i] & 0x00000000000000ff); }
42.   else   for (int i = 0; i < buf.length; i++) { r <<= 8; r |= (buf[i] & 0x00000000000000ff); }
43.   return r; }
44. /*** short 转 byteArray**/
45. public final static byte[] short2ByteArray(short s, boolean asc) {  byte[] buf = new byte[2];
46.   if (asc)  for (int i = buf.length - 1; i >= 0; i--) { buf[i] = (byte) (s & 0x00ff); s >>= 8; }
47.   else   for (int i = 0; i < buf.length; i++) { buf[i] = (byte) (s & 0x00ff); s >>= 8; }
48.   return buf; }
49. /***int 转 byteArray**/
50. public final static byte[] int2ByteArray(int s, boolean asc) { byte[] buf = new byte[4];
51.   if (asc) for (int i = buf.length - 1; i >= 0; i--) {buf[i] = (byte) (s & 0x000000ff);s >>= 8;}
52.   else   for (int i = 0; i < buf.length; i++) {buf[i] = (byte) (s & 0x000000ff);s >>= 8;}
53.   return buf; }
54. /*** long 转 byteArray**/
55. public final static byte[] long2ByteArray(long s, boolean asc) { byte[] buf = new byte[8];
```

```
56.    if (asc) for (int i = buf.length - 1; i >= 0; i--) {buf[i] = (byte) (s
       & 0x00000000000000ff); s >>= 8;}
57.    else    for (int i = 0; i < buf.length; i++) {buf[i] = (byte) (s & 0x00
       000000000000ff);s >>= 8;}
58.    return buf;}
59. /*** byteArray 转 short**/
60. public final static short byteArray2Short(byte[] buf, boolean asc) {
61.    if (buf == null) {throw new IllegalArgumentException("byte array is n
       ull!");}
62.    if (buf.length > 2) { throw new IllegalArgumentException("byte array size
       > 2 !");}
63.    short r = 0;
64.    if (asc) for (int i = buf.length - 1; i >= 0; i--) {r <<= 8;r |= (buf[i]
       & 0x00ff);}
65.    else    for (int i = 0; i < buf.length; i++) {r <<= 8; r |= (buf[i] & 0x00ff);}
66.    return r; }
67. /***解析字符串获得双精度型数值**/
68. public  static double getDoubleValue(String str)  { double d = 0;
69.    if(str!=null && str.length()!=0) {StringBuffer bf = new StringBuffer();
70.       char[] chars = str.toCharArray();
71.       for(int i=0;i<chars.length;i++){char c = chars[i];
72.          if(c>='0' && c<='9') { bf.append(c); }
73.          else if(c=='.'){if(bf.length()==0) { continue; }
74.          else if(bf.indexOf(".")!=-1)  { break; }
75.          else { bf.append(c); }}
76.          else  { if(bf.length()!=0) { break; } }}
77.       try { d = Double.parseDouble(bf.toString());}
78.       catch(Exception e) {}}
79.       return d; }
80. /*** 解析 str,获得其中的整数 @param str   @return**/
81. public static int getIntValue(String str)  {int r = 0;
82.    if(str!=null && str.length()!=0) {StringBuffer bf = new StringBuffer();
83.       char[] chars = str.toCharArray();
84.       for(int i=0;i<chars.length;i++) {char c = chars[i];
85.          if(c>='0' && c<='9') { bf.append(c); }
86.          else if(c==',') {continue;}
87.          else{if(bf.length()!=0) { break; } } }
88.       try { r = Integer.parseInt(bf.toString()); }
89.       catch(Exception e) {} }
90.       return r; }}
```

6.6.5 滚动字幕模块 MarqueeText

滚动字幕模块 MarqueeText 负责实现 Allshow 界面的滚动字幕,如源代码清单 6.6 所示。

源代码清单 6.6 ch06_com.emdoor.smartBrowsing\src\com\emdoor\smartBrowsing\text\MarqueeText.java

```
1. public class MarqueeText extends TextView {
2.    public MarqueeText(Context con) {super(con);}
3.    public MarqueeText(Context context, AttributeSet attrs) {super(context,
      attrs);}
4.    public MarqueeText(Context context, AttributeSet attrs, int defStyle)
```

```
        {super(context, attrs, defStyle);}
5.   @Override
6.   public boolean isFocused() {return true;}
7.   @Override
8.   protected void onFocusChanged(boolean focused, int direction,Rect pre
     viouslyFocusedRect) {}
```

6.6.6 图片处理模块 GalleryFlow

图片处理模块 GalleryFlow 负责实现图片拨动浏览以及美化显示效果，如源代码清单 6.7 所示。

源代码清单 6.7　\ch06_com.emdoor.smartBrowsing\src\com\emdoor\smartBrowsing\show\GalleryFlow.java

```
1. public class GalleryFlow extends Gallery {
2.     private Camera mCamera = new  Camera(); private int mMaxRotationAngle = 60;
3.     private int mMaxZoom = -120;    private int mCoveflowCenter;
4.     public GalleryFlow(Context context) {super(context);this.setStaticTra
       nsformationsEnabled(true);}
5.     public GalleryFlow(Context context, AttributeSet attrs) {super(context,
       attrs);
6.        this.setStaticTransformationsEnabled(true);}
7.     public GalleryFlow(Context context, AttributeSet attrs, int defStyle)
       { super(context, attrs, defStyle);
8.        this.setStaticTransformationsEnabled(true);}
9.     public int getMaxRotationAngle() { return mMaxRotationAngle; }
10.    public void setMaxRotationAngle(int maxRotationAngle){mMaxRotationAng
       le = maxRotationAngle;}
11.    public int getMaxZoom() {return mMaxZoom;}
12.    public void setMaxZoom(int maxZoom) {mMaxZoom = maxZoom;}
13.    private int getCenterOfCoverflow() {
14.       return (getWidth() - getPaddingLeft() - getPaddingRight()) / 2+ get
          PaddingLeft();}
15.    private static int getCenterOfView(View view) {return view.getLeft() +
       view.getWidth() / 2;}
16.    protected boolean getChildStaticTransformation(View child, Transforma
       tion t) {
17.       final int childCenter = getCenterOfView(child); final int childWidth
          = child.getWidth();
18.       int rotationAngle = 0; t.clear();
19.       t.setTransformationType(Transformation.TYPE_MATRIX);
20.       if (childCenter == mCoveflowCenter) {transformImageBitmap((ImageView)
          child, t, 0);}
21.    else {rotationAngle = (int) (((float) (mCoveflowCenter - childCenter) /
       childWidth) * mMaxRotationAngle);
22.       if (Math.abs(rotationAngle) > mMaxRotationAngle) {
23.          rotationAngle = (rotationAngle < 0) ? -mMaxRotationAngle: mMaxRo
             tationAngle;  }
24.          transformImageBitmap((ImageView) child, t, rotationAngle);}
25.       return true;}
26.    protected void onSizeChanged(int w, int h, int oldw, int oldh) {
27.       mCoveflowCenter = getCenterOfCoverflow();super.onSizeChanged(w, h,
          oldw, oldh);}
```

```
28.    private void transformImageBitmap(ImageView child, Transformationt,
       int rotationAngle) {
29.       mCamera.save();    final Matrix imageMatrix = t.getMatrix();
30.       final int imageHeight = child.getLayoutParams().height;
31.       final int imageWidth = child.getLayoutParams().width;
32.       final int rotation = Math.abs(rotationAngle);
33.       //在Z轴上正向移动camera的视角,实际效果为放大图片。
34.       //如果在Y轴上移动,则图片上下移动; X轴上对应图片左右移动。
35.       mCamera.translate(0.0f, 0.0f, 100.0f);
36.       // As the angle of the view gets less, zoom in
37.       if (rotation < mMaxRotationAngle) {   float zoomAmount = (float)
          (mMaxZoom + (rotation * 1.5));
38.          mCamera.translate(0.0f, 0.0f, zoomAmount);}
39.       //在Y轴上旋转,对应图片竖向向里翻转。如果在X轴上旋转,则对应图片横向向里翻转
40.       mCamera.rotateY(rotationAngle); mCamera.getMatrix(imageMatrix);
41.       imageMatrix.preTranslate(-(imageWidth / 2), -(imageHeight / 2));
42.       imageMatrix.postTranslate((imageWidth / 2), (imageHeight / 2));
43.       mCamera.restore();}}
```

6.6.7 图文显示模块 Allshow

图文显示模块 Allshow 负责实现单击主界面图标后出现文字和图片浏览界面,不同的结点对应不同的浏览信息,主要包括:陶瓷器、青铜器、玉器和绘画等展区功能演示,其相应的布局文件为 tcshow.xml、qtshow.xml、cyshow.xml 和 hhshow.xml 等。如源代码清单 6.8～6.12 所示。

源代码清单 6.8 \Ch06_com.emdoor.smartBrowsing\src\com\emdoor\smartBrowsing\Allshow.java

```
1. public class Allshow extends Activity{
2.    private TextView TCTextView1 = null;   private TextView QTTextView1 =
      null;
3.    private TextView CYTextView1 = null;   private TextView HHTextView1 = null;
4.    private String Id = null;    boolean flag = false;    private int id = 0;
5.     protected void onCreate(Bundle savedInstanceState) {super.onCreate(
         savedInstanceState);
6.       Intent intent=getIntent();Id = intent.getStringExtra("Id"); id =
         ConvertHelper.getIntValue(Id);
7.       if(id==2){flag = true;setContentView(R.layout.tcshow);String value
         = "陶瓷器展示";
8.        TCTextView1 = (TextView)findViewById(R.id.TCTextView1);TCText
          View1.setText(value);
9.        image_button();
10.       ImageAdapter adapter = new ImageAdapter(this, getSD("/sdcard/
          Picture/TC_Picture"));
11.       adapter.createReflectedImages();
12.       GalleryFlow galleryFlow = (GalleryFlow) findViewById(R.id.TC_Gallery);
13.       galleryFlow.setAdapter(adapter);   }
14.    if(id==3){flag = true;setContentView(R.layout.qtshow);String value =
       "青铜器展示";
15.       QTTextView1 = (TextView)findViewById(R.id.QTTextView1);
16.       QTTextView1.setText(value);   image_button();
17.       ImageAdapter adapter = new ImageAdapter(this, getSD("/sdcard/
```

```
                Picture/QT_Picture"));
18.     adapter.createReflectedImages();
19.     GalleryFlow galleryFlow = (GalleryFlow) findViewById(R.id.QT_Gallery);
20.     galleryFlow.setAdapter(adapter);}
21.  if(id==4){flag = true;   setContentView(R.layout.cyshow);String value =
     "玉器展示";
22.     CYTextView1 = (TextView)findViewById(R.id.CYTextView1);
23.     CYTextView1.setText(value);  image_button();
24.     ImageAdapter adapter = new ImageAdapter(this, getSD("/sdcard/
        Picture/CY_Picture"));
25.     adapter.createReflectedImages();
26.     GalleryFlow galleryFlow = (GalleryFlow) findViewById(R.id.CY_Gallery);
27.     galleryFlow.setAdapter(adapter);}
28.  if(id==5){flag = true;setContentView(R.layout.hhshow);String value
     = "绘画展示";
29.     HHTextView1 = (TextView)findViewById(R.id.HHTextView1);HHText
        View1.setText(value);
30.     image_button();
31.     ImageAdapter adapter = new ImageAdapter(this, getSD("/sdcard/
        Picture/HH_Picture"));
32.     adapter.createReflectedImages();
33.     GalleryFlow galleryFlow = (GalleryFlow) findViewById(R.id.HH_Gallery);
34.     galleryFlow.setAdapter(adapter);}
35.     IntentFilter intentFilter= new IntentFilter(); intentFilter.addA
        ction("MY_ACTION");
36.     registerReceiver(idIntenal,intentFilter); /*广播接收动态注册*/ }
37.  private void image_button(){
38.     ImageButton media_player_button = (ImageButton) findViewById(R.
        id.video_memery);
39.     media_player_button.setOnClickListener(media_player_button_list
        ener);}
40.  private Button.OnClickListener media_player_button_listener = new
     Button.OnClickListener(){
41.    public void onClick(View v){ Intent intent = new Intent(); intent.put
       Extra("Id",Id);
42.         intent.setClass(Allshow.this,Allmovie.class);startActivity(
            intent); } };
43.  private BroadcastReceiver idIntenal = new BroadcastReceiver(){
44.    @Override
45.    public void onReceive(Context context,Intent intent)  { //从Intent
       中获取信息
46.       String msg = intent.getStringExtra("msg"); int Id_convert=Convert
          Helper.getIntValue(msg);
47.       if(Id_convert!=id&&flag){ flag=false;showDialogAlert();} } };
48.  protected List<String> getSD(String filefrom) { /* 设定目前所在路径 */
49.     List<String> it=new ArrayList<String>(); File f=new File(filefrom);
50.     File[] files = f.listFiles();   /* 将所有文件存入ArrayList中 */
51.     for(int i=0;i<files.length;i++)  { File file=files[i];
52.       if(getImageFile(file.getPath())) it.add(file.getPath()); }
53.     return it; }
```

```
54.      private boolean getImageFile(String fName) { boolean re;  /* 取得
         扩展名 */
55.         String end=fName.substring(fName.lastIndexOf(".")+1,fName.
         length()).toLowerCase();
56.      /* 按扩展名的类型决定 MimeType */
57.      if(end.equals("jpg")||end.equals("gif")||end.equals("png") |end.
         equals("jpeg")||end.equals("bmp"))
58.      { re=true; }
59.      else { re=false; }
60.      return re; }
61.   //广播接收如果在 onCreate 里面注册,那么在 onDestroy 里面取消注册
62.      protected void onDestroy() {super.onDestroy();Allshow.this.finish();
63.         android.os.Process.killProcess(android.os.Process.myPid());
64.         unregisterReceiver(idIntenal);}
65.      public void onBackPressed(){ Allshow.this.finish();
66.         android.os.Process.killProcess(android.os.Process.myPid());}
67.      private void showDialogAlert(){final Intent intent = new Intent();
68.         AlertDialog alertDialog = new AlertDialog.Builder(Allshow.this)
69.         .setMessage("您可能进入新的展区,点击选择相应的操作").setIcon(R.drawable
         .emdoor)
70.         .setPositiveButton("返回主页", new DialogInterface.OnClickListener() {
71.         public void onClick(DialogInterface dialog, int id) {Allshow.this.
         finish();
72.         intent.setClass(Allshow.this, IntelligentBrowsing.class);
73.         Allshow.this.startActivity(intent);
74.         android.os.Process.killProcess(android.os.Process.myPid());}})
75.            .setNegativeButton("继续浏览", new DialogInterface.OnClickLis
         tener() {
76.            public void onClick(DialogInterface dialog, int id) {dial
         og.cancel();}})
77.            .setCancelable(true).show();}}
```

源代码清单 6.9　\ch06_com.emdoor.smartBrowsing\res\layout\tcshow.xml

```xml
1. <?xml version="1.0" encoding="utf-8" ?>
2. <LinearLayout xmlns:android="http://schemas.android.com/apk/res/android"
3. xmlns:tools="http://schemas.android.com/tools" android:orientation="vertical"
4. android:layout_width="match_parent"  android:layout_height="match_parent"
5. android:background="#000000">
6. <RelativeLayout android:layout_width="match_parent"      android:layout_
   height="wrap_content"
7.    android:orientation="horizontal">
8.    <TextView android:id="@+id/TCTextView1" android:textSize="15pt"
9.       android:layout_width="fill_parent"      android:layout_height="50dp" />
10.   <ImageButton android:id="@+id/video_memery"    android:layout_width=
   "wrap_content"
11.      android:layout_height="wrap_content" android:src="@drawable/video_
      memery"
12.      android:layout_alignParentRight="true"/> </RelativeLayout>
13. <com.emdoor.smartBrowsing.show.GalleryFlow android:id="@+id/TC_Gallery"
14.      android:layout_width="fill_parent"     android:layout_height="wrap_
```

```
         content" />
15.  <!-- 为了实现文字跑马灯生成了一个 MarqueeText 类 -->
16.  <com.emdoor.smartBrowsing.text.MarqueeText android:id="@+id/TCTextView2"
17.     android:textSize="10pt"          android:layout_width="500dp"
18.     android:layout_height="50dip"   android:layout_marginLeft="80dip"
19.     android:singleLine="true"         android:focusable="true"
20.     android:focusableInTouchMode="true"  android:scrollHorizontally="true"
21.     android:marqueeRepeatLimit="marquee_forever"   android:ellipsize=
        "marquee"
22.     android:text="@string/tcshow" />   </LinearLayout>
```

源代码清单 6.10　\ch06_com.emdoor.smartBrowsing\res\layout\qtshow.xml

```
1.  <?xml version="1.0" encoding="utf-8" ?>
2.  <LinearLayout xmlns:android="http://schemas.android.com/apk/res/android"
3.   xmlns:tools="http://schemas.android.com/tools"  android:orientation=
    "vertical"
4.  android:layout_width="match_parent"  android:layout_height="match_parent"
5.  android:background="#000000">
6.  <RelativeLayout android:layout_width="match_parent"    android:layout_
    height="wrap_content"
7.   android:orientation="horizontal">
8.  <TextView android:id="@+id/QTTextView1" android:textSize="15pt"
9.     android:layout_width="fill_parent" android:layout_height="50dp" />
10. <ImageButton android:id="@+id/video_memery"    android:layout_width=
    "wrap_content"
11.    android:layout_height="wrap_content" android:src="@drawable/video
       _memery"
12.    android:layout_alignParentRight="true" />  </RelativeLayout>
13. <com.emdoor.smartBrowsing.GalleryFlow android:id="@+id/QT_Gallery"
14.   android:layout_width="fill_parent" android:layout_height="wrap_content" />
15. <!--为了实现文字跑马灯生成了一个 MarqueeText 类-->
16. <com.emdoor.smartBrowsing.MarqueeText         android:id="@+id/QTTextView2"
17.    android:textSize="10pt"              android:layout_width="500dp"
18.    android:layout_height="50dip"   android:layout_marginLeft="80dip"
19.    android:singleLine="true"            android:focusable="true"
20.    android:focusableInTouchMode="true" android:scrollHorizontally="true"
21.    android:marqueeRepeatLimit="marquee_forever"   android:ellipsize=
       "marquee"
22.    android:text="@string/qtshow" />     </LinearLayout>
```

源代码清单 6.11　\ch06_com.emdoor.smartBrowsing\res\layout\cyshow.xml

```
1.  <?xml version="1.0" encoding="utf-8" ?>
2.  <LinearLayout xmlns:android="http://schemas.android.com/apk/res/android"
3.   xmlns:tools="http://schemas.android.com/tools"  android:orientation=
    "vertical"
4.  android:layout_width="match_parent"  android:layout_height="match_parent"
5.  android:background="#000000">
6.  <RelativeLayout android:layout_width="match_parent" android:layout_
    height="wrap_content"
7.  android:orientation="horizontal">
```

```
8.  <TextView android:id="@+id/CYTextView1"        android:textSize="15pt"
9.      android:layout_width="fill_parent"      android:layout_height="50dp" />
10. <ImageButton android:id="@+id/video_memery"    android:layout_width="wrap_
    content"
11.     android:layout_height="wrap_content"      android:src="@drawable/
    video_memery"
12.         android:layout_alignParentRight="true" />  </RelativeLayout>
13. <com.emdoor.smartBrowsing.show.GalleryFlow android:id="@+id/CY_Gallery"
14.     android:layout_width="fill_parent" android:layout_height="wrap_content" />
15. <!--为了实现文字跑马灯生成了一个MarqueeText类-->
16. <com.emdoor.smartBrowsing.text.MarqueeText        android:id="@+id/CYTextView2"
17.     android:textSize="10pt"             android:layout_width="500dp"
18.     android:layout_height="50dip"       android:layout_marginLeft="80dip"
19.     android:singleLine="true"           android:focusable="true"
20.     android:focusableInTouchMode="true" android:scrollHorizontally="true"
21.     android:marqueeRepeatLimit="marquee_forever"     android:ellipsize=
    "marquee"
22.     android:text="@string/cyshow" />    </LinearLayout>
```

源代码清单6.12 \ch06_com.emdoor.smartBrowsing\res\layout\hhshow.xml

```
1.  <?xml version="1.0" encoding="utf-8" ?>
2.  <LinearLayout xmlns:android="http://schemas.android.com/apk/res/android"
3.  xmlns:tools="http://schemas.android.com/tools"  android:orientation=
    "vertical"
4.  android:layout_width="match_parent" android:layout_height="match_parent"
5.  android:background="#000000">
6.  <RelativeLayout android:layout_width="match_parent"       android:layout_
    height="wrap_content"
7.  android:orientation="horizontal">
8.  <TextView android:id="@+id/HHTextView1"        android:textSize="15pt"
9.      android:layout_width="fill_parent"      android:layout_height="50dp" />
10. <ImageButton android:id="@+id/video_memery"    android:layout_width=
    "wrap_content"
11.     android:layout_height="wrap_content"   android:src="@drawable/video_
    memery"
12.     android:layout_alignParentRight="true" />  </RelativeLayout>
13. <com.emdoor.smartBrowsing.show.GalleryFlow android:id="@+id/HH_Gallery"
14.     android:layout_width="fill_parent" android:layout_height="wrap_content" />
15. <!--为了实现文字跑马灯生成了一个MarqueeText类-->
16. <com.emdoor.smartBrowsing.text.MarqueeText        android:id="@+id/HHTextView2"
17.     android:textSize="10pt"             android:layout_width="500dp"
18.     android:layout_height="50dip"       android:layout_marginLeft="80dip"
19.     android:singleLine="true"           android:focusable="true"
20.     android:focusableInTouchMode="true" android:scrollHorizontally="true"
21.     android:marqueeRepeatLimit="marquee_forever"     android:ellipsize=
    "marquee"
22.     android:text="@string/hhshow" />    </LinearLayout>
```

6.6.8 视频显示模块Allmovie

视频显示模块Allmovie负责实现单击Allshow界面的纪录片图标后呈现视频显示界面，

不同结点对应不同视频信息，如源代码清单 6.13 所示。

源代码清单 6.13   \ch06_com.emdoor.smartBrowsing\src\com\emdoor\smartBrowsing\Allmovie.java

```java
1.  public class Allmovie extends Activity {
2.      private SurfaceView surfaceView;     private String path;
3.      private MediaPlayer mediaPlayer;     private boolean pause;
4.      private boolean flag = false;        private int id = 0;
5.      private String Id = null;            private String filename = null;
6.      public void onCreate(Bundle savedInstanceState) { super.onCreate(savedInstanceState);
7.          Intent intent = getIntent(); Id = intent.getStringExtra("Id");id = ConvertHelper.getIntValue(Id);
8.          if (id == 2) {flag = true;this.setContentView(R.layout.tcmovie);
9.              filename = "tcmovie.mp4";}//直接输入视频文件名称
10.         if (id == 3) {flag = true;this.setContentView(R.layout.qtmovie);
11.             filename = "qtmovie.mp4";}//直接输入视频文件名称
12.         if (id == 4) {flag = true;this.setContentView(R.layout.cymovie);
13.             filename = "cymovie.mp4";}//直接输入视频文件名称
14.         if (id == 5) {flag = true;this.setContentView(R.layout.hhmovie);
15.             filename = "hhmovie.mp4";}//直接输入视频文件名称
16.         mediaPlayer = new MediaPlayer();
17.         surfaceView = (SurfaceView) this.findViewById(R.id.surfaceView);
18.         //把输送给 surfaceView 的视频画面直接显示到屏幕上，不要维持它自身的缓冲区
19.         surfaceView.getHolder().setType(SurfaceHolder.SURFACE_TYPE_PUSH_BUFFERS);
20.         surfaceView.getHolder().setKeepScreenOn(true);//防止锁屏
21.         IntentFilter intentFilter = new IntentFilter();
22.         intentFilter.addAction("MY_ACTION");
23.         registerReceiver(idIntenal, intentFilter);}
24.     //Activity 销毁时停止播放，释放资源
25.     protected void onDestroy() {mediaPlayer.release(); mediaPlayer = null; super.onDestroy();
26.         Allmovie.this.finish();android.os.Process.killProcess(android.os.Process.myPid());
27.         unregisterReceiver(idIntenal);}
28.     public void onBackPressed() {Allmovie.this.finish();
29.         android.os.Process.killProcess(android.os.Process.myPid());}
30.     //定义一个 medisplay 方法
31.     @SuppressLint({ "ShowToast", "ShowToast" })
32.     public void medisplay(View v) {
33.         switch (v.getId()) {
34.         case R.id.playbutton:File file = new File(Environment.getExternalStorageDirectory()+ "/vedio", filename);
35.             if (file.exists()) {path = file.getAbsolutePath();play();
36.             } else { path = null;
37.                 Toast.makeText(this, R.string.filenoexist, 1).show();}
                                                //文件不存在的时候显示文件不存在
38.             break;
39.         case R.id.pausebutton:
40.             if (mediaPlayer.isPlaying()) {mediaPlayer.pause();pause = true;}
```

```
41.            else {if (pause) {mediaPlayer.start();pause = false;}
42.            } break;
43.        case R.id.resetbutton:
44.            if (mediaPlayer.isPlaying()) {mediaPlayer.seekTo(0);}
45.            else {if (path != null) {play();}}break;
46.        case R.id.stopbutton:if (mediaPlayer.isPlaying()) {mediaPlayer.stop
            ();}break;
47.        default:break;}}
48.    public void play() {
49.        try {mediaPlayer.reset();              //初始化播放器
50.            mediaPlayer.setDataSource(path);
51.            mediaPlayer.setDisplay(surfaceView.getHolder());
                                                   //把surfaceHolder传进来
52.            mediaPlayer.prepare();             //缓冲准备
53.            mediaPlayer.setOnPreparedListener(new PrepareListener());
54.        } catch (IOException e) {e.printStackTrace(); }}
55.    final class PrepareListener implements OnPreparedListener {
56.        // @SuppressLint({ "FloatMath", "FloatMath" })
57.        public void onPrepared(MediaPlayer player) {mediaPlayer.start();}}
                                                   //播放视频
58.    private BroadcastReceiver idIntenal = new BroadcastReceiver() {
59.        public void onReceive(Context context, Intent intent) {
                                                   //从Intent中获取信息
60.            String msg = intent.getStringExtra("msg");
61.            int Id_convert = ConvertHelper.getIntValue(msg);
62.            if (Id_convert != id && flag) {flag = false; abortBroadcast();
63.                showDialogAlert();} }};
64.    private void showDialogAlert() {final Intent intent = new Intent();
65.        AlertDialog alertDialog = new AlertDialog.Builder(Allmovie.this)
66.            .setMessage("您可能进入新的展区,点击选择相应的操作")
67.            .setIcon(R.drawable.emdoor).setPositiveButton("返回主页",
68.                new DialogInterface.OnClickListener() {
69.                    public void onClick(DialogInterface dialog, int id) {Allmovie.
                    this.finish();
70.                    intent.setClass(Allmovie.this,IntelligentBrowsing.class);
71.                    Allmovie.this.startActivity(intent);
72.                    android.os.Process.killProcess(android.os.Process.myPid());}})
73.            .setNegativeButton("继续浏览",new DialogInterface.OnClickListener() {
74.                public void onClick(DialogInterface dialog, int id) {dialog.
                cancel();}
75.            }).setCancelable(true).show();}}
```

本章小结

本章通过一个真实的企业级项目,一方面让学生了解 Java 本地接口(JNI)、警告对话框(AlertDialog)、定时器任务(TimerTask)等方面的基本知识;另一方面,让学生掌握意图、画廊、图片控件、网格视图控件和媒体播放器等的使用方法及技巧,从而提升学生综合利用

Android 的 UI 控件、API 框架及其活动、意图、BroadcastReceiver 等核心组件，解决问题的基本职业能力。

强化练习

1．填空题

（1）Android 中在 SharedPreferences 中采用（　　　）的方式来存储数据。

（2）文件（Files）数据存储主要是使用（　　　）配合 FileInputStream 或 FileOutputStream 对文件进行读写。

（3）在 Android 程序中可通过（　　　）方法来设置布局的背景颜色。

（4）对按钮事件进行事件监听的方法是（　　　）。

（5）Android 系统为用户提供了事件的处理机制。在处理单击事件时用户调用（　　　）方法，其回调方法为（　　　）。

（6）切换卡的实现有两种方式，分别是使用系统自带的（　　　）及继承自（　　　）和定义自己的（　　　）。

（7）线性布局的方向有两种不同的方向属性，分别为（　　　）和（　　　）。

（8）要从意图（　　　）中获取信息 String msg = intent.getStringExtra("msg")，必须调用（　　　）方法传递该信息。

（9）启动 Android 定时器（　　　）定时调度 TimerTask()的方法为（　　　）。

（10）设置文本视图显示文本颜色的方法为（　　　）。

（11）构造新线程 Thread readThead=new Thread()后，需要调用（　　　）方法启动该线程。

（12）Android 多媒体播放器在播放网上音乐前需要调用（　　　）方法设置 URL 数据来源。

2．单选题

（1）活动管理器（ActivityManager）属于 Android 系统架构中的层（　　　）。
    A．应用程序框架层　　　　　　　　B．用户接口层（UI）
    C．应用程序层　　　　　　　　　　D．Linux 操作系统内核层

（2）在 Android 的系统清单文件 AndroidManifest.xml 中，manifest 标签的 package 属性功能为：（　　　）。
    A．定义应用程序图标　　　　　　　B．声明命名空间
    C．设置应用程序名称　　　　　　　D．声明应用程序包

（3）在 Android 中，下列（　　　）方法不可以启动一个活动。
    A．startActivity (aIntent)　　　　　B．Activity.startActivityForResult()
    C．StartService()　　　　　　　　　D．Context.startActivity()

（4）在 Android 中，以下（　　　）组件可以作为应用程序间通讯的纽带。
    A．Intent　　　　　　　　　　　　　B．BroadcastReceiver
    C．ContentProvider　　　　　　　　D．Activity

（5）在 Android 的布局中，下列 TextView 的 id 属性赋值正确的是（　　　）。

A. Android:id="myTextView"　　　　　　B. Android:id="@+id/myTextView"

C. Android:id="+id/myTextView"　　　　D. Android:id="@id/myTextView"

（6）Button 对象方法名称 setHeight() 与布局文件中 XML 属性（　　）相对应。

A. android:layout_height　　　　　　　B. android:height

C. android:margin_height　　　　　　　D. android:button_height

（7）在 Android 的 XML 代码中，属性 android:layout_centerInParent 的功能为（　　）。

A. 设置水平居中　　　　　　　　　　　B. 定义垂直居中

C. 相对于子元素居中　　　　　　　　　D. 相对父元素完全居中

（8）在 Android 的 XML 布局文件中，控件属性 android:layout_alignParentRight 的含义为（　　）。

A. 紧贴父元素右边缘　　　　　　　　　B. 在某元素的右边缘

C. 在父元素左边缘　　　　　　　　　　D. 与某元素左边缘的距离

（9）Android 的存储数据方式不包括（　　）。

A. 多媒体　　　　　　　　　　　　　　B. 网路

C. 文件　　　　　　　　　　　　　　　D. SharePreferences

（10）在 Android 中，用于响应触摸事件的接口为（　　）。

A. onTouchEvent()　　　　　　　　　　B. onKeyUp()

C. onKeyDown()　　　　　　　　　　　D. onClickListener()

（11）在 Android 中，用于判断视图控件能否获取焦点的方法为（　　）。

A. requestFocus()　　　　　　　　　　　B. isFocusable()

C. setFocusable(boolean)　　　　　　　　D. isFocused()

（12）在 Android 中，设置活动布局的方法为（　　）。

A. startActivity()　　　　　　　　　　　B. setClass()

C. setContentView()　　　　　　　　　　D. startActivityForResult()

（13）在 Android 中，查找 XML 布局文件中 id 为 okButton 按钮控件对象的代码为（　　）。

A. Button button=(Button)this.findViewById(R.id. okButton);

B. Button button=view.findViewById(R.id. okButton);

C. Button button=(Button)findViewById(okButton);

D. Button button=this.view.findViewById(okButton);

（14）在 Android 的 Handler 案例中，用于接收 Handler 消息的方法为（　　）。

A. handleMessage(Message message);

B. dispatchMessage(Message message);

C. getMessageName(Message message);

D. sendMessageAtTime(Message message, long time);

（15）在 Android 中存储字符串资源的文件夹为（　　）。

A. values　　　　B. layout　　　　C. src　　　　D. drawable

（16）在 Android 中，可用来输入字符串的控件是（　　）。

A. Button　　　　B. CheckBox　　　　C. TextView　　　　D. EditText

（17）如果将 Button 对象的 android:layout_width 的属性设置为 match_parent，则该组件对

象的显示效果为（　　）。
    A．该按钮的高度仅占据该组件的实际高度
    B．该按钮的宽度仅占据该组件的实际宽度
    C．该按钮的高度将填充父容器高度
    D．该按钮的宽度将填充父容器宽度

（18）在 Android 中，以下（　　）不是从视图派生的。
    A．CheckBox   B．Adapter   C．Button   D．RadioGroup

3．问答题

（1）Android 系统的架构包括哪几层？
（2）Android 常用的是哪 5 种布局方式？
（3）Android 提供了哪 5 种方式存储数据？
（4）适配器（Adapter）起什么作用？活动（Activity）和意图（Intent）有什么区别？

4．编程题

（1）在布局界面中放置两个 TextView 控件，利用线程技术和定时器技术，编程实现图 6.18 所示的每隔 1 秒的文字闪烁效果（参见 Ch06_Coding01_Timer.rar）。

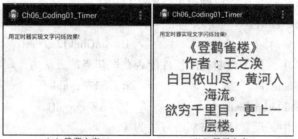

(a) 隐藏文字　　　　　　(b) 显示文字

图 6.18　每隔 1 秒的文字闪烁效果

（2）利用按钮单击事件处理机制和警告对话框，实现退出 Android 应用系统，跳转到桌面功能软件（见 Ch06_Coding02_AlertDialog.rar），如图 6.19 所示。

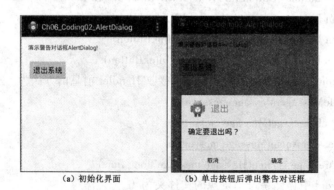

(a) 初始化界面　　　　　　(b) 单击按钮后弹出警告对话框

图 6.19　利用警告对话框退出应用系统

参 考 文 献

[1] 杨丰盛. Android 应用开发揭秘[M]. 北京：机械工业出版社，2010.
[2] 李宁. Android 开发权威指南[M]. 北京：人民邮电出版社，2011.
[3] 盖索林. Google Android 开发入门指南[M]. 2 版. 北京：人民邮电出版社，2009.
[4] 梅尔. Android 2 高级编程[M]. 2 版. 王超，译. 北京：清华大学出版社，2010.
[5] 余志龙，王世江. Google Android SDK 开发范例大全[M]. 2 版. 北京：人民邮电出版社，2010.
[6] 李宁. Android/OPhone 开发完全讲义[M]. 北京：水利水电出版社，2010.
[7] 李刚. 疯狂 Android 讲义[M]. 北京：电子工业出版社，2011.
[8] 汪永松. Android 平台开发之旅[M]. 北京：机械工业出版社，2010.
[9] E2EColud 工作室. 深入浅出 Google Android[M]. 北京：人民邮电出版社，2009.
[10] 梅尔. Android 高级编程[M]. 王鹏杰，霍建同，译. 北京：清华大学出版社，2010.
[11] 李华忠，梁永生，刘涛. Android 应用程序设计教程[M]. 北京：人民邮电出版社，2013.